NON–IONIZING RADIATION

Proceedings
2nd International Non–Ionizing Radiation Workshop

Vancouver, British Columbia, Canada
1992 May 10–14

Editor:

M. Wayne Greene
Occupational Health and Safety
University of British Columbia
Vancouver, British Columbia   W6T 1Z1

A joint project of the Canadian Radiation Protection
Association and the International Non–Ionizing
Radiation Committee of the International Radiation
Protection Association

ISBN 0-969-59580-8

Distributed by:

UBC Press
University of British Columbia
6344 Memorial Road
Vancouver, B.C. Canada V6T 1Z2
Tel:(604) 822-3259
Fax:(604) 822-6083

Printed in Canada

PREFACE

Workshop Objective

The 2nd International NIR Workshop was held in Vancouver, British Columbia, Canada, from 1992 May 10 to 14, at the University of British Columbia. The Workshop was a joint project of the Canadian Radiation Protection Association and the International Non-Ionizing Radiation Committee of the International Radiation Protection Association. The objective of the Workshop was to provide a comprehensive overview of non-ionizing radiation and ultrasound and to provide a forum for in depth analysis and discussion of current research and international standards.

The Workshop preceded the Eight Congress of the International Radiation Protection Association held in Montreal, Quebec, Canada, 1992 May 17-24.

Panel of Speakers

The following were speakers, panel members, and moderators during the Workshop.

David Agnew	Gordon Hester
L. Anderson	H. Jammet
Ulf Bergqvist	Bengt Knave
Jürgen H. Bernhardt	John A. Leonowich
Stephen H.P. Bly	Peter A. Lewin
Harvey Checkoway	Rüdiger Matthes
B. Ralph Chou	M.L. McBride
D. Courant	Alastair McKinlay
L. Court	William D. O'Brien, Jr.
Anthony P. Cullen	Brian Phillips
Yvon Deslauriers	Michael H. Repacholi
K.E. Donnelly	Mark E. Schafer
A.S. Duchene	David H. Sliney
Richard P. Gallagher	Maria A. Stuchly
Kelly Gibney	Mays Swicord
Martino Grandolfo	L.D. Szabo
M. Wayne Greene	T.S. Tenforde
Gerald R. Harris	M.L. Walsh
Stuart M. Harvey	

International Non-Ionizing Radiation Committee

The International Non-Ionizing Radiation Committee of IRPA reviews the scientific literature on non-ionizing radiation and makes assessments of the health risks of human exposure. The Committee recommends guidelines on exposure limits, drafts codes of safe practice, and works in conjunction with other international organizations to promote safety and standardization. Many of the papers presented in this monograph and lectures given at the workshop were contributed by members of the INIRC. Committee members are:

M.H. Repacholi, Chair, Australia
H. Jammet (Chair Emeritus), France
A.S. Duchene, Scientific Secretary, France
J.H. Bernhardt, Germany
B.F.M. Bosnjakovic, Hungary
L.A. Court, France
M. Grandolfo, Italy
B.G. Knave, Sweden
A.F. McKinlay, Great Britain
M.G. Shandala, C.I.S.
D.H. Sliney, U.S.A.
J.A.J. Stolwijk, U.S.A.
M.A. Stuchly, Canada
L.D. Szabo, Hungary

Members of the Organizing Committee

The 2nd International NIR Workshop was a joint project of the Canadian Radiation Protection Association (CRPA) and the International Non-Ionizing Radiation Committee of the International Radiation Protection Association (IRPA). The CRPA Organizing Committee members worked with the INIRC members to put together a program covering all areas of non-ionizing radiation and ultrasound. The Canadian Organizing Committee members are:

M. Wayne Greene, Chair
C. Cardinale
Kelly Gibney
Brian Phillips
Maria Stuchly
Murray Walsh

Acknowledgements

The success of this workshop depended upon the efforts of many people. The Canadian Organizing Committee members put a lot of consideration and effort into the program to ensure that the whole area of non-ionizing radiation was covered. In this regard, I would like to specifically thank Maria Stuchly for her many suggestions and assistance. The Local Organizing Committee worked tirelessly to ensure that the Workshop went smoothly and that the needs of the speakers and delegates were taken care of. Members of the Local Organizing Committee are: Leigh Baker, Wayne Greene, Kelly Gibney, Lars Larsson, Lutz Moritz, Brian Phillips, Richard Piskor and Craig Smith. Their efforts were greatly appreciated. Members of the Department of Occupational Health and Safety at the University of British Columbia are thanked for their assistance in making the Workshop a success. Special thanks are extended to Sue Bryant and Noni Brown. The financial support of many generous sponsors is appreciated.

FOREWORD

We live in a sea of radiation varying from cosmic radiation to microwaves used in communications to sunlight and to the induced fields produced by transmission lines. Humans have evolved in this background radiation and more recently they have been exposed to additional fields generated by devices. The effects of ionizing radiation have been the subject of studies since their earliest discovery. On the other hand, the interaction of non-ionizing radiation with biological materials is not generally well understood. In recent years this lack of understanding has led to speculation on interaction mechanisms and effects and has resulted in many research activities.

The biological effects of non-ionizing radiation and ultrasound are of considerable scientific and public interest. Members of the public and the work force are being exposed to radiation fields from many sources. What are the effects of this exposure and what can be done to minimize the exposures are questions that public health officials are being continuously asked. The answers to such questions are often difficult to interpret.

This monograph has brought together a series of papers that cover the subject of non-ionizing radiation and ultrasound. These papers review the most recent scientific developments and changes in standards. People involved in public health, occupational hygiene, safety, radiation protection, regulation development, communications and administration will find this volume of interest.

These papers formed the basis for the 2nd International NIR Workshop held in Vancouver, British Columbia, Canada, 1992 May 10-14. In addition to the papers there were several panel discussions and forums for discussion of major health and safety issues. Unfortunately these discussions could not be included in the monograph.

The reader is encouraged to review the papers in this volume and to make use of the references given by the authors.

M. Wayne Greene
University of British Columbia
February 1992

CONTENTS

Contents

VI POPULATION STUDIES and STANDARDS   415

PART I

INTRODUCTION

NON-IONIZING RADIATION

M H Repacholi

Visiting Scientist
Australian Radiation Laboratory
Yallambie, Victoria

Introduction

Non-ionizing radiation (NIR) exists throughout our entire environment and, except for the narrow spectrum of visible radiation, it is unperceived by any of the human senses unless its intensity becomes so great that it is felt as heat.   Differences in wavelengths, even within a single wave band, are particularly important when evaluating hazards from exposure to NIR. The ability of the radiation to penetrate into the human body and the sites of absorption differ significantly from one type of radiation to another.

The electromagnetic spectrum is nominally divided into frequency or wavelength bands (see Table 1) and includes: static electric and magnetic fields, extremely low frequency (ELF) fields (normally having frequencies greater than zero up to about 300 Hz), radiofrequencies (RF) (although various ranges are defined, they can include frequencies of 300 Hz and up to 300 GHz), and optical radiations: infrared (IR wavelengths range from 0.78 to 100 μm), visible (400-780 nm), ultraviolet (100-400nm), and ionizing radiations (wavelengths less than 100 nm).

Figure 1 gives a representation of all the NIRs with respect to wavelength, frequency and major application.

Table 1 gives details of the subdivisions within each band of radiation. For example, ultraviolet has been divided into UV-A, UV-B and UV-C since each of these radiations has varying penetrating abilities into tissues and interacts with biological matter with differing effects.   Similarly, the infrared region has been subdivided into IR-A, IR-B, IR-C.   The International Commission on Illumination (Commission International de l'Eclairage-CIE) subdivided the UV, visible and IR into these radiation bands since they have particular biological responses or significance. Another subdivision of the UV is "near UV", "far UV" and "extreme UV". The term "near" is used to identify the UV subdivision nearest the visible

TABLE 1: Ranges of frequency, wavelength and energy for non-ionizing electromegnetic radiations and fields (From WHO, 1982; EPA, 1984)

Type of Radiation	Frequency range[a]	Wavelenth range	Energy range per photon[a]
Ionizing	>3000 THz	<100nm	>12.40 eV
Ultraviolet(UV) (nonionizing part)	3000-750 THz	100-400 nm	12.40-3.10 eV
extreme (vacuum)	3×10^5 to 30000-1580	1 to 10-190	1240 to 124-6.53
far	1580-1000	190-300	6.53-4.13
near	1000-750	300-400	4.13-3.10
UV-C[b]	3000-1070	100-280	12.40-4.43
UV-B[b]	1070-952	280-315	4.43-3.94
UV-A[b] ("black light")	952-750	315-400	3.94-3.10
Visible light[c]	750-385 THz	400-1000μm	3.10-1.59 eV
Infrared (IR)	385-0.3 THz	0.78-1000μm	1590-1.24 eV
IR-A[b]	385-214	0.78-1.4	1590-886
IR-B[b]	214-100	1.4-3.0	886-413
IR-C[b]	100-0.3	3-1000	413-1.24
near	385-100	0.78-3.0	1590-413
middle	100-10	3-30	413-41.33
far	10-0.3	30-1000	41-1.24
Lasers	1500-15 THz	0.2-20 μm	6200-62 meV
Class 1 non-risk laser devices Class 2 low risk, low-power laser devices Class 3a low risk, medium-power laser devices Class 3b moderate risk, high-power laser devices Class 4 high risk, high-power laser devices			
Radiofrequency(RF)	300 GHz-300 Hz	1 mm-1000 m	1240 μeV-124 feV
Microwave(MW)	300-0.3 GHz	1-1000 mm	1240-1.24 μeV
EHF (extremely high frequency)	300-30 GHz	1-10	1240-124 μeV
SHF (super-high frequency)	30-3	10-100	124-12.4
UHF (ultra-high frequency)	3-0.3	100-1000	12.4-1.4
VHF (very high frequency)	300-30 MHz	1-10 m	1240-124 neV
HF (high frequency)	30-3	10-100	124-12.4
MF (medium frequency)	3-0.3	100-1000	12.4-1.24
LF (low frequency)	300-30 kHz	1-10 km	1240-124 peV
VLF (very low frequency)	30-3	10-100	124-12.4
VF (voice frequency)	3-0.3	100-1000	12.4-1.24
ELF (extremely low frequency)	300-30 Hz	10^2-10^4	1240-124 feV
Sub ELF	30-0	10^4-∞ km	124-0
Static electric & magnetic fields	0	∞	0

Note: The ranges given are only approximations, since no precise limits can be defined. Lower limits are exclusive, upper limits are inclusive.

a. The figures given have generaly been rounded up or down to the third significant digit.

b. Radiation bands of biological significance designed by the International Commission on Illumination (CIE).

c. The visibility limits of the human eye very amoung individuals between about 380-400nm and 750-780nm.

Values of prefixes used on units.

$T = 10^{12}$ $G = 10^{9}$ $M = 10^{6}$ $k = 10^{3}$ $m = 10^{-3}$ $\mu = 10^{-6}$ $n = 10^{-9}$ $p = 10^{-12}$ $f = 10^{-15}$

Figure 1.   **Electromagnetic spectrum showing the frequency and photon energy distinction between ionizing and non-ionizing radiations.**

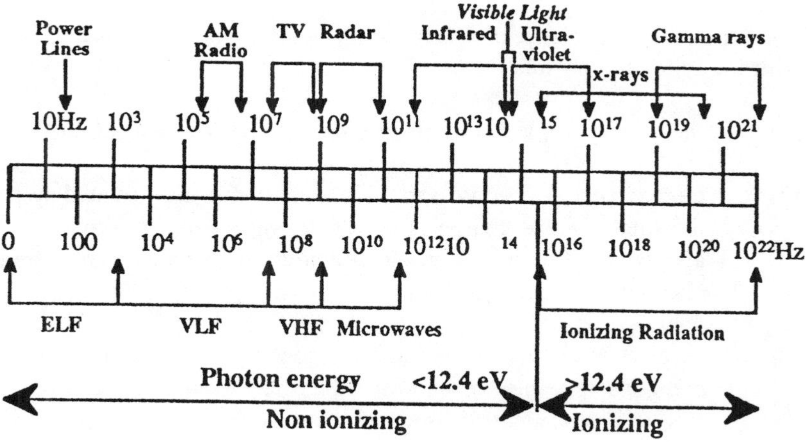

region.   The extreme (or vacuum) UV band is furthest from the visible region.

Non-ionizing electromagnetic radiations incorporate all radiations and fields of the electromagnetic spectrum that do not normally have enough energy to produce ionization in matter. In other words, non-ionizing radiations are incapable of imparting enough energy to a molecule or atom to disrupt its structure (breaking chemical bonds) by removing one or more electrons, leaving positive and negative charges or ions (hence ionization). NIRs have an energy per photon less than 12.4eV, wavelengths longer than 100nm and frequencies lower than 3000THz.

Maxwell's Equations for Electricity and Magnetism

Faraday first conceived that the space surrounding an electrical charge was filled with "lines of force" that indicate, at any point in space, the direction and magnitude of the force acting on a charge. Maxwell used Faraday's ideas, and mathematical descriptions of electric and magnetic fields to formulate the famous field equations and to theoretically predict the existence of electromagnetism. A full description of electromagnetism can be obtained by simply stating a few basic postulates and definitions of physical quantities and the other laws which relate these quantities to each other.

The experimental evidence on electric and magnetic phenomena is consistent with the following postulates:

(1)   there exist two kinds of electric charge: positive and negative;

(2)   electric charge is conserved. Meaning that in any isolated system the total electric charge is constant: whenever a positive charge appears or disappears, an equal amount of negative charge must also appear or disappear;

(3)   an electric charge in motion undergoes a force consisting of two components, either or both of which may eventually be zero, the first is independent of its speed whereas the other is proportional to its velocity and perpendicular to it. Indicating the electric charge by q and its velocity, $\mathbf{v}$ the force $\mathbf{F}$ can be expressed by the relation.

$$\mathbf{F} = q\,(\mathbf{E} + \mathbf{v} \times \mu_o\mathbf{H}) \tag{1}$$

μ_o being a constant. The two vector quantities $\mathbf{E}$ and $\mathbf{H}$, which are defined by this relation, are termed the electric and magnetic field respectively.

The characteristics of these fields are completely described by the four relations known as Maxwell's equations. These are usually expressed in terms of $\mathbf{E}$ and $\mathbf{H}$ and other quantities are directly related to them.

The electric current I flowing in a conductor is the electric charge Q flowing through any section in a unit time (t), i.e.:

$$I = dQ/dt \qquad (2)$$

I is a scalar quantity, whereas the current density

$$\mathbf{J} = \mathbf{n}\, dI/ds \qquad (3)$$

is a vector whose direction is in the same as the motion of positive charges and whose magnitude gives the current following through a unit area. $\mathbf{n}$ is a positive normal vector of the surface element.

Charge density (ρ) is defined as the electric charge per unit volume, i.e.:

$$\rho = dQ/dv \qquad (4)$$

The quantities ρ and $\mathbf{J}$ are point functions, since they account for the electric charge and its motion in any point of space; the more meaningful quantities Q and I over finite regions or through finite areas are obtained by volume or surface integration, respectively.

Maxwell combined the existing knowledge concerning electromagnetism into a coherent, unified theory and was the first to realize that Gauss, Ampere and Faraday laws, when written in a general mathematical form, yield a set of four equations from which all electric and magnetic phenomena can be derived. Maxwell's equations can now be written in terms of the above quantities. They are most frequently expressed in differential form:

$$\nabla \cdot \boldsymbol{E} = \frac{\rho}{\varepsilon_0} \qquad (5a)$$

$$\nabla \cdot \boldsymbol{H} = 0 \qquad (5b)$$

$$\nabla \times \boldsymbol{E} = \mu_0 \frac{\partial \boldsymbol{H}}{\partial t} \qquad\qquad (5c)$$

$$\nabla \times \boldsymbol{H} = \boldsymbol{J} + \epsilon_0 \frac{\partial \boldsymbol{E}}{\partial t} \qquad\qquad (5d)$$

By using vector analysis, these laws can be converted to integral form to express the laws of electromagnetism in free space, i.e.

Equation 5(a) is Gauss' law; stating the conservation of free charge. 5(b) states that isolated magnetic poles do not exist. Equation 5(c) is Faraday's induction law; time varying magnetic fields induce electric fields with resulting potential differences and currents through conducting paths. 5(d) is Ampere's law where the well known formula V=IR can be derived with V= voltage, I-current, R-resistance.

To complete the relationship between field variables, the electric flux density **D** and the magnetic flux density **B** (also called dielectric and magnetic induction, respectively) may be introduced and related to the electric and magnetic fields by:

$$\mathbf{D} = \epsilon_0 \mathbf{E} \qquad\qquad (6)$$

$$\mathbf{B} = \mu_0 \mathbf{H} \qquad\qquad (7)$$

The constants ϵ_o and μ_o are termed dielectric constant and magnetic permeability. The final relation is:

$$\mathbf{J} = \sigma \mathbf{E} \qquad\qquad (8)$$

expressing the familiar Ohm's law, in which σ is the conductivity of the medium.

A detailed discussion of Maxwell's equations is outside the scope of this paper. Further information can be obtained from NCRP (1981), Grandolfo and Vecchia (1985) or Polk and Postow (1986).

Electromagnetic Waves:

Oscillating electric charges induce an electromagnetic field within the region surrounding the charge source. In turn, this oscillating "induction field" (often called the near field of the source) generates an electromagnetic wave that radiates energy from the region surrounding the charges. The radiated wave consists of coupled electric and magnetic fields that oscillate at the same frequency as the source and the wave propagates outward from the source at the velocity of light in the medium (see Figure 2). In free space, the velocity of light is 3×10^8 m/s, whereas in a medium with low electromagnetic energy dissipation, the velocity is this value divided by the square root of the material's dielectric constant **relative** to that of free space. In a low-loss material such as fatty tissue for example, with a relative dielectric constant of 4, the velocity of light is 1.5×10^8 m/s.

For an electromagnetic wave travelling at velocity v, the wavelength in the medium is the distance between the same points on successive waves. If f is the oscillation frequency of the wave in cycles per second, or hertz (Hz), the wavelength λ is expressed as

$$\lambda = v/f \tag{9}$$

Electromagnetic waves of all frequencies (except static fields) carry energy. According to quantum mechanics, they can also be thought of as packets of energy called photons. The energy E of a photon equals hf, where $h = 6.63 \times 10^{-34}$ joule seconds (Planck's constant) and f = frequency.

The energy of a photon is thus directly proportional to the frequency of the radiation. Thus, the higher the frequency (and conversely, the shorter the wavelength), the higher is the photon energy. When the frequency approaches or exceeds 3×10^{15} Hz, the photon energies equal or exceed 2×10^{-18} J, or 12.4 eV, and become comparable to the binding energy of electrons to atoms. Thus, high-frequency (high energy photon) radiations (X-rays, gamma rays, etc.) are able to break electrons away from atoms to form ions and are thus referred to as ionizing radiations.

Figure 2.   An   electromagnetic   monochromatic   wave. Electromagnetic waves consist of electric and magnetic forces that move, in the far-field, in a consistent wave-like pattern at 90° to one another.

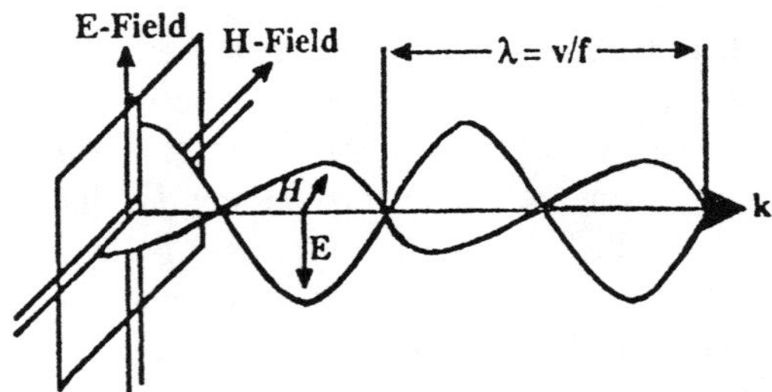

Wave Propagation

In free space, electromagnetic waves spread uniformly in all directions from a theoretical point source. The wavefront, or the wave surface joining all points of identical phase, is spherical in this case. As the distance from the point source increases, the area of the wavefront surface increases as a square of the distance, so that the source power is spread over a larger area. This is the origin of the well-known inverse square law of intensity with distance from the radiating source.

If one considers the example of radiofrequency fields, the term power density is sued and is defined as the ratio of the total radiated power to the spherical surface area enclosing the source. The power density W is inversely proportional to the square of the distance from the source, and can be expressed:

$$W = P/4\pi r^2 \tag{10}$$

where P = transmitted power and r = distance from the source.

This equation states the inverse square law for wave propagation from an isotropic source. The source radiates uniformly in all directions in space. However, there is no source that has this property, although there are close approximations. The inverse-square law can apply when the source is anisotropic (directional), provided the medium is homogeneous and isotropic (e.g. free space or a medium in which the velocity of wave propagation does not change with direction or distance).

Ultrasound

Ultrasound is included amongst the NIR with the electromagnetic radiations and fields described above. However, unlike these NIRs which can be propagated in a vacuum, ultrasound has a mechanical nature and needs a material medium in which to propagate. Sound or acoustic energy is a form of vibrating energy propagated in the form of waves through a medium by motion of the particles within the medium. The characteristics of these waves are functions of both the wave and the medium through which it travels. Ultrasound refers simply to that sound with frequencies higher than the audible range. The frequency of 20 kHz is usually chosen as an arbitrary lower limit for the ultrasound frequency range.

In many ultrasound applications, the waves employed are periodic. This is, the force or disturbance is repeated as a function of time and is usually sinusoidal. Under these conditions, particle displacement from equilibrium and particle positions will be as shown in Fig. 3. In addition, at any one point in the medium, particle displacement will vary with time. The wavelength (λ) is the distance between consecutive points having identical displacement in the medium. The time required for the wave to move a distance λ is the period T. The frequency is defined as the number of cycles of the wave that pass a point per unit time (usually one second). Thus, frequency and period are related.

$$f = 1/T \tag{11}$$

The frequency and wavelength are related by the propagation velocity c

$$c = f\lambda \tag{12}$$

The constant c contains information about the medium through which the wave passes, in particular about the coupling between the particles of the medium. In fact, the propagation velocity can be related to the properties of the medium.

$$c = (K_a/\rho)^{1/2} \tag{12}$$

where K_a depends on the elasticity of the medium and ρ is the density. For water with $\rho = 1.0$ x 10^3 kg/m^3, the velocity (speed of sound) is 1.43 x 10^3 m/s. Note that the propagating medium, but is independent of frequency.

The direction of ultrasound energy propagation may be parallel or perpendicular to the direction of oscillation of the particles. The corresponding waves are termed longitudinal and transverse, respectively. Transverse waves are especially important in splids, but other waves such as shear waves, torsion waves, flexural (or Lamb) waves, surface Rayleigh waves, and Love waves also exist. However, only shear waves are of interest in ultrasonics. Transverse waves travel only through solids, because liquids and gases do not support shear stresses under normal conditions.

Longitudinal waves on the other hand can pass through all types of media and are more important with regard to interactions with biological systems. In longitudinal waves, the collective motion of particles creates alternate regions of compression and rarefaction, i.e. a periodical pressure variation as shown in Figure 3. This variation has the same propagation speed and frequency of oscillations of particles.

Figure 3.   A longitudinal wave as a function of distance: (top) particle position; (bottom) particle displacement in acoustic wave propagation.

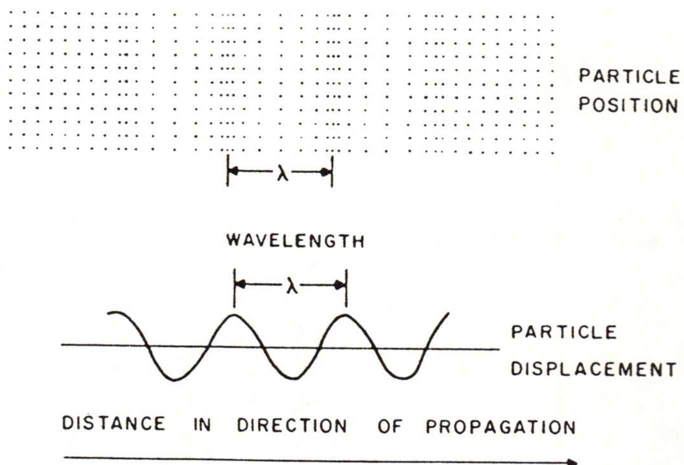

References

EPA, (1984), Environmental Protecton Agency, Biological effects of radiofrequency radiation, J.A. Elder and D.F. Cahill (Eds.), US EPA, Pub. EPA-600/8-83-026f. Research Triangle N.C. 27711.

GRANDOLFO, M. and VECCHIA, P. (1985), Physical description of exposure to static and ELF electromagnetic fields. In Biological effects and dosimetry of static and ELF electromagnetic fields. Grandolfo, M., Michaelson, S.M. and Rindi, A. editors. Plenum Press (New York and London) pp 49-70.

POLK, C. and POSTOW, E., (1986), (Editors), Handbook of biological effects of electromagnetic fileds. CRC Press Boca Raton, Florida.

WHO (1982), World Health Organisation. Non-ionizing radiation protection. M.J. Suess, editor. WHO Regional Publications European Series No. 10, WHO Copenhagen.

PART II

ELECTROMAGNETIC FIELDS

INTRODUCTION TO BIOELECTROMAGNETICS

Maria A. Stuchly
Department of Electrical and
Computer Engineering
University of Victoria,
Victoria, B.C., V8W 3P6

INTRODUCTION

The field of bioelectromagnetics encompasses practically all aspects of scientific and practical endeavour involving electric, magnetic and electromagnetic fields in the non-ionizing range. More customarily higher frequencies, above 300 GHz, are treated separately. This approach is adapted in this chapter, as an introduction to infrared, visible and ultraviolet radiations is provided in another chapter. The frequencies and wavelengths associated with the non-ionizing part of the electromagnetic spectrum are illustrated in Figure 1, and the band designation is given in Table 1. Often frequencies from about 10 kHz to 300 MHz are referred to as radiofrequencies (RF) and above 300 MHz as microwaves.

The areas of bioelectromagnetics can be divided into four groups as shown in Figure 2. It is well recognized that these groups are linked with each other. Exposure assessment provides the starting point for dosimetric evaluations. Studies of interaction mechanisms heavily rely on dosimetry and are aimed at explanation of observed in vitro and in vivo biological effects. The latter are then considered from two aspects. One of the important aspects is to assess whether and under what circumstances these biological effects pose a risk to human health and if they do how it can be prevented or limited. The other aspect involves beneficial uses of the interactions, both physical and biological, in medical diagnosis and therapy.

Historically, the early studies in bioelectromagnetics were mostly of exploratory nature, frequently only qualitative in assessment and often subject to artifacts. Current studies are in many cases conducted by inter-disciplinary teams. These studies include a careful character-ization of exposure conditions, dosimetric evaluation where possible, and a meticulous control of artifacts. Many of the studies are aimed at testing and extending specific hypotheses. Since the Second World War till the seventies, most of the investigations were conducted with radiofrequency and microwave fields. As a lot has been learnt, and exposure standards have been

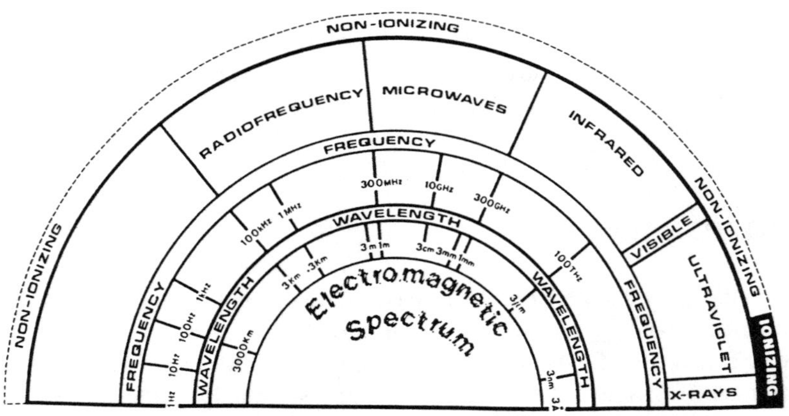

Figure 1. Electromagnetic Spectrum

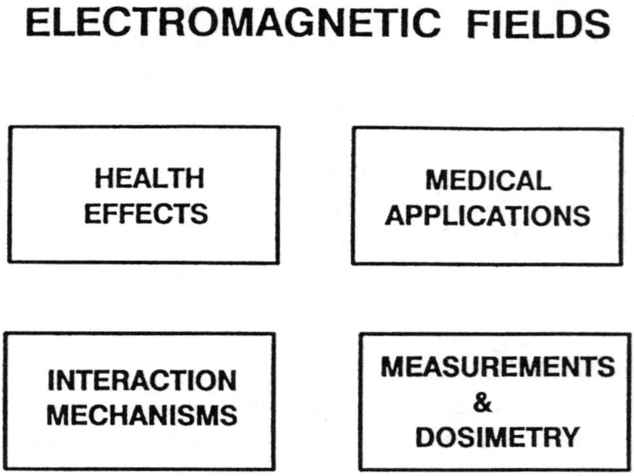

ELECTROMAGNETIC FIELDS

HEALTH EFFECTS	MEDICAL APPLICATIONS
INTERACTION MECHANISMS	**MEASUREMENTS & DOSIMETRY**

Figure 2. Areas of Bioelectromagnetics

Table 1
Nomenclature of Frequency Bands
(Reference Data, 1974)

Designation	Frequency Range	Metric Subdivision
ELF Extremely low frequency	30-300 Hz	Megametric waves
VF Voice frequency	300-3000 Hz	-
VLF Very-low frequency	3-30 kHz	Myriametric waves
LF Low frequency	30-300 kHz	Kilometric waves
MF Medium frequency	300-3000 kHz	Hectrometric waves
HF High frequency	3-30 MHz	Decametric waves
VHF Very-high frequency	30-300 MHz	Metric waves
UHF Ultra-high frequency	300-3000 MHz	Decimetric waves
SHF Super-high frequency	3-30 GHz	Centimetric waves
EHF Extremely high frequency	30-300 GHz	Millimeter waves

Prefixes: k = kilo = 10^3, M = Mega = 10^6, G = Giga = 10^9

established in many countries (Stuchly, 1987). Some of the remaining activities at RF address many unresolved questions related to transient and pulsed fields, and fields amplitude modulated at ELF. Starting in the eighties the main research effort has been at ELF. Research activities and knowledge in bioelectromagnetics are vast and diverse.

In this chapter fundamental principles of bioelectromagnetics will be briefly reviewed and a selected glimpse at some of the research areas will be given. More detailed reviews are contained in other chapters dealing with specific areas of bioelectromagnetics.

BASIC PRINCIPLES

Electric Field

An electric field is a vector field in which an infinitely small charge is subject to a force, according to the formula (Cheng, 1989):

$$E = \lim_{q \to o} (F/q) \tag{1}$$

where q is the charge, F is the force, and E is the electric field. Both E and F are vector quantities, i.e. they are characterized by the magnitude and direction (Note: all vectors in this text are denoted by bold print). In the International System of Units (SI), the unit of charge is coulomb (C), the unit of the electric field is volt per meter (V/m), and the unit of force is Newton (N). The smallest quantity of electric charge observed in nature is that of an electron or proton, and is equal to 1.6×10^{-19}C. In practice, as loHx as the charge is small enough in order not to perturb the charge distribution of the source, eq.(1) simplifies to

$$E = F/q \tag{2}$$

and the force acting on a charged particle q in the field E is

$$F = qE \tag{3}$$

Electric fields can be produced therefore be electric charges located in space. At any point in such space the electric field depends on the spatial distribution of the charge. Specifically the two postulates have to be satisfied. In free space:

$$\nabla \cdot E = \rho/\epsilon_o \tag{4}$$

where ρ is the volume charge density, or in the integral form (Gauss law):

$$\oint_s E \cdot ds = Q/\epsilon_o \tag{5}$$

and

$$\nabla \times E = 0 \tag{6}$$

or equivalently (Kirchnoff's voltage law):

$$\oint_c \mathbf{E} \cdot \mathbf{dl} = 0 \tag{7}$$

where ∇ is the del operator, Q is the total charge contained in volume V bounded by surface S, ϵ_0 is the permittivity of free space ($\epsilon_0 = 1/36\pi \times 10^{-9}$ F/m), C is an arbitrary enclosed contour, ds and dl are infinitely small surface and length, respectively, within S and along C.

For a point charge q, the electric field is:

$$\mathbf{E} = \mathbf{a}_R \frac{q}{4\pi\epsilon_o R^2} \tag{8}$$

where a_R is a unit vector in the direction from the charge to the point of observation, and R is the distance between the charge and the point of observation where the electric field is given by eq(8).

The electric field can also be expressed in terms of a scalar electric potential, V, as:

$$\mathbf{E} = -\nabla V \tag{9}$$

The SI unit of electric potential is volt. Eq(9) is used to calculate the electric field produced by charged surfaces. For instance, for two parallel plates separated by the distance d, with one plate grounded (potential zero) and the other having potential V, the electric field magnitude in the space between the plates is the same and equal to:

$$E = V/d \tag{10}$$

and the direction is perpendicular to the plates surfaces pointing towards the positively charged plate.

The behaviour in electric fields in other dielectrics than free space is characterized by their permittivity, ϵ_r where:

$$\epsilon_r = \epsilon/\epsilon_0 \tag{11}$$

In addition to the electric field vector, the field in a dielectric is characterized by the electric flux density, or electric displacement, D:

$$D = \epsilon E \tag{12}$$

Magnetic Field

A magnetic field is a vector field in which moving electric charges are subject to a force, according to Lorentz force equation (Cheng, 1989):

$$F = qv \times B \tag{13}$$

where: F is the force (N), q is the charge (C), v is the charged particles velocity (vector) and B is the magnetic flux density. The SI unit of the magnetic flux density is tesla (T) or weber per square meter (W/m^2) (1 tesla $= 10^4$ gauss, gauss is a unit in CGS). In free space:

$$\nabla \cdot B = 0 \tag{14}$$

or in the integral form

$$\oint_S B \cdot ds = O \tag{15}$$

and

$$\nabla \times B = \mu_0 J \tag{16}$$

or

$$\oint_C Bdl = \mu_0 I \tag{17}$$

where μ_0 is the permeability of free space ($\mu_0 = 4\pi \times 10^{-7}$ H/m), J is the current density and I is the total current through the surface S encompassed by contour C.

The magnetic flux density produced by an infinitely long straight wire carrying current I is equal to

$$B_\phi = a_\phi \frac{\mu_0 I}{2\pi r} \tag{18}$$

where: a_ϕ is the unit vector in the circumferential direction ϕ, and r is the distance from the wire.

For a circular loop of a radius b carrying current I, the magnetic field on its axis is equal to (Cheng, 1989):

$$\boldsymbol{B_z} = a_z \frac{\mu_o I b^2}{2(z^2 + b^2)^{3/2}} \qquad (19)$$

where a_z is the unit vector in the direction perpendicular to the plane of the loop, and z is the distance to the plane of the loop.

The behaviour of a material in the magnetic field is characterized by its permeability μ or the relative permeability, μ_r, where

$$\mu_r = \mu/\mu_o. \qquad (20)$$

The magnetic flux density, B, is equal to:

$$\mathbf{B} = \mu\mathbf{H} \qquad (21)$$

where **H** is the magnetic field strength. The SI unit of the magnetic field is ampere per meter (A/m).

Electromagnetic Field

Static electric fields are fully defined by eqs. 4, 6 and 12, while static magnetic fields are defined by eqs. 14, 16, and 21. For static fields vectors **E** and **D** are not related to **B** and **H**. For alternating fields these quantities are inter-related by Faraday's law of induction

$$\nabla \times \mathbf{E} = -(\partial \mathbf{B}/\partial t) \qquad (22)$$

indicating that a time-varying magnetic field produces an electric field.

A complete description of time-varying electromagnetic fields is provided by Maxwell's equations:

$$\nabla \times \mathbf{E} = -(\partial \mathbf{B}/\partial t)$$
$$\nabla \times \mathbf{H} = \mathbf{J} + (\partial \mathbf{D}/\partial t)$$
$$\nabla \cdot \mathbf{D} = \rho$$
$$\nabla \cdot \mathbf{B} = 0$$

(23)

or in the integral form:

$$\oint_c \mathbf{E}dl = -(d\phi/dt)$$

$$\oint_c \mathbf{H}dl = I + \int_s \int (\partial \mathbf{D}/\partial t)\, ds$$

(24)

$$\oint_s \mathbf{D}ds = Q$$

$$\oint_s \mathbf{B}ds = O$$

PHYSICAL INTERACTION OF FIELDS WITH BIOLOGICAL MATTER

Electrical Properties

As indicated earlier the permittivity and permeability describe on the macroscopic level interactions of electric and magnetic fields with matter. For biological materials with a few exceptions the magnetic permeability is the same as that of free space. The permittivity, on the other hand, is significantly different for various tissues, and also varies with frequency.

The permittivity contains two components, the dielectric constant, ϵ', and the loss factor, ϵ'':

$$\hat{\epsilon} = \epsilon' - j\epsilon''$$

(25)

The dielectric constant, ϵ', is a measure of the ability to store electric field energy. The loss factor, ϵ'', describes the fraction of the energy dissipated in the material. The relative permittivity, i.e. the permittivity normalized to that of free space (vacuum) is:

$$\hat{\epsilon}_r = \epsilon'_r - j\epsilon'' = \hat{\epsilon}/\epsilon_o \tag{26}$$

The loss factor, ϵ'', is related to the conductivity of the material, σ, in the following way

$$\epsilon_r'' = \sigma/\omega\epsilon_o \tag{27}$$

where $\omega = 2\pi f$, f is the frequency. The unit of conductivity σ is siemens per metre (S/m).

Typical changes of the permittivity of tissue with frequency are shown in Figure 3. Extensive studies have been conducted on dielectric properties of various tissues and molecular and cellular mechanisms contributing to those properties (Foster and Schwan, 1986, Stuchly and Stuchly, 1990).

ELF Fields

Because of the slow rate of time change and very long wavelength of ELF fields compared to dimensions of biological bodies, the interactions can be considered as quasi-static. The solution of field problems are simplified and involve Laplace's equation rather than Maxwell's equations.

At these frequencies for biological tissues the following relationship is satisfied leading to simplifications:

$$\omega\epsilon_o\epsilon' << \sigma \tag{28}$$

Approximate calculations of the electric fields induced in human and animal bodies exposed to ELF electric field indicate that these fields are between 10^{-7} to 10^{-5} times the value of the external exposure field (Tenforde and Kaune, 1987). Correspondingly the induced current densities are very low. For instance, for a vertical electric field of 10 kV/m at 60 Hz, the induced current densities are of the order of a fraction of $\mu A/cm^2$, with the highest values in the ankles (2 $\mu A/cm^2$) and the neck 0.55 $\mu A/cm^2$). Internal patterns of current flow are highly complex due to the heterogeneous structure and electrical properties of tissues.

Electric ELF fields are considerably perturbed by human and animal bodies. The perturbation results in a local enhancement of the external field on the

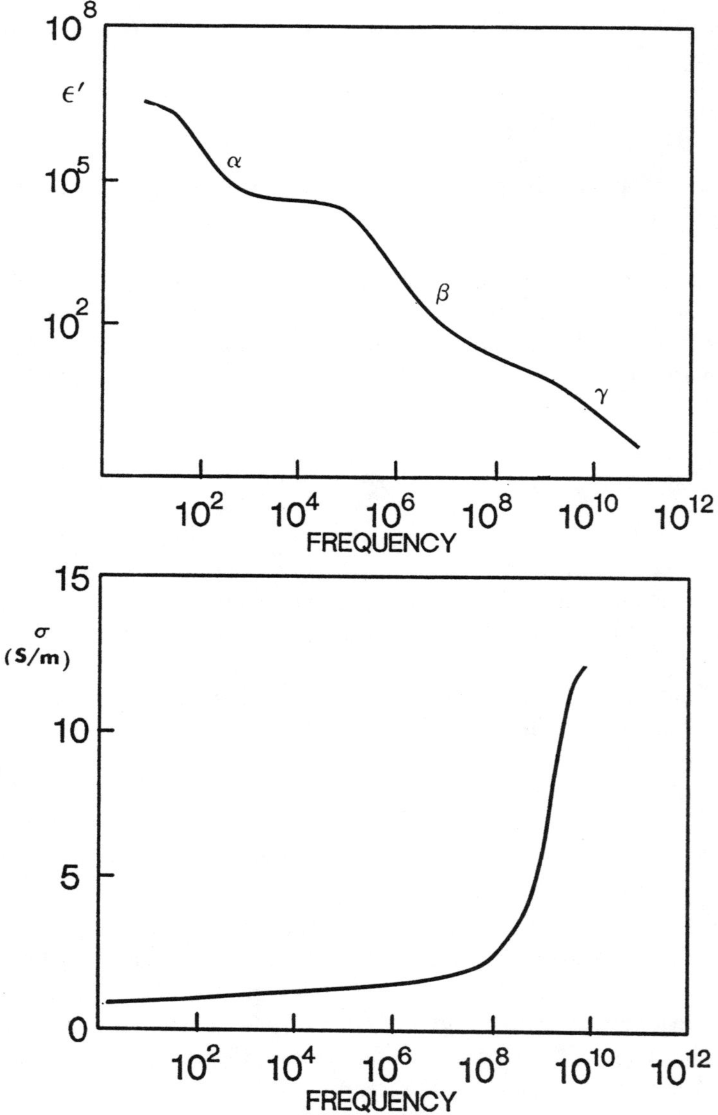

Figure 3. Electrical Properties of Tissue

exposed body surface.

Magnetic ELF fields are not significantly perturbed by biological bodies, as tissue permeabilities are equal to that of free space. A minimal field modification is due to the fact that the fields are not static and therefore the body induced currents in turn produce magnetic fields. However, their magnitude is very small, usually less than 1% of the exposure field.

ELF magnetic fields induce electric fields and currents. The currents form closed loops. The specific patterns of these loops are rather complex because of variations in the tissue conductivity.

RF Fields

To determine induced electric fields, Maxwell's equations have to be solved at higher frequencies. Solutions are found only for approximate models. In general, induced electric fields are highly non-uniform and vary with frequency, as described in another chapter. The induced fields result in energy deposition. Interactions are described in terms of the specific absorption rate (SAR) defined as

$$SAR = (E_i^2 \, \sigma/\rho) \tag{29}$$

where E_i is the strength of the electric field and ρ is the tissue density. The initial rate of tissue heating is proportional to SAR, while the resultant temperature depends on heat conduction by physical and thermoregulatory mechanisms.

OVERVIEW OF CURRENT STATUS

Exposure Assessment

Exposure assessment at ELF and in particular at power line frequencies (50 or 60 Hz) is an area of a currently active pursuit. This is particularly important in support of epidemiological studies. A number of sophisticated personal dosimeters have been developed for this purpose (Renew et al, 1990). One of the problems in this area is that the existing biological data do not clearly indicate what dosimetric parameters are of importance in eliciting interactions. Some studies suggest that the waveform, or perhaps transients or changes are important, which would indicate that measurements extending beyond the amplitudes of the electric and magnetic fields might be required.

Various field probes and instruments have been developed for RF measurements (Tell, 1981). There is also a relatively large data base on the electromagnetic field levels in various occupational and environmental settings (Stuchly and Mild, 1987). One of the challenges is posed by the electromagnetic environment consisting of multiple sources of radiation operating at various frequencies in a space containing reflectors and scatterers (e.g. buildings, trees, large metallic objects).

Measurements of transient fields and pulsed electric and magnetic fields have recently become of growing interest, as there is a need to assess exposures associated with the electromagnetic pulse (EMP), as well as many other exposure situations such as large switching current at switchyards and transformer stations, arc-welding, induction heating, RF heating, and radars. Several novel active and electro-optical sensors and measurement techniques have been recently developed (Thansandote et al, 1991, Stuchly et al, 1991).

Dosimetry

For ELF electric and magnetic fields, dosimetric measures have not yet been established. Although induced current densities from exposure to these fields have been calculated and can be measured (Tenforde and Kaune, 1987), at present there is no compelling evidence that the current density in tissue is the critical dosimetric measure. Perhaps, calculations of current densities and induced potentials on the microscopic (subcellular) level will prove useful when used in conjunction with biophysical models of interactions.

At RF, the whole-body average SAR was the first and remains the most universally used dosimetric measure. Both theoretical and experimental methods for evaluation of average SARs are well established (Durney, 1980, Stuchly and Stuchly, 1986). More recent work has been aimed at calculation and measurements of spatial SAR distributions in the human body (Spiegel et al 1989, Dimbylow and Gandhi, 1991). The SAR distribution plays an important role in some biological effects and has recently been taken into account in some protection standards. Another area of current interest is dosimetry of the electromagnetic pulse (EMP).

Health Effects

Currently most of the studies are being conducted with ELF fields, with the predominance of the power line frequencies (50 and 60 Hz), but other extremely low frequencies are often also used in the same studies. A number of studies are motivated by beneficial effects of exposure rather than a potential hazard. These studies often utilize complex waveforms with energy contained below 1000 Hz. Magnetic fields are more often used rather than electric fields. The on-going investigations can be divided into three groups: (1) human studies, (2) animal studies, and (3) in vitro studies. Recently, at least a large part of the effort within North America and Europe has been coordinated through information exchange among the research sponsors. There also is an increasing effort to maintain links between the three types of studies in search of answers to a specific question, e.g. do ELF fields affect cancer development?, in what way?, and by what mechanism?

On-going human studies are mostly epidemiological and concerned with cancer. Environmental exposures of children to electric and magnetic fields at 50 or 60 Hz are being scrutinized for a possible association with childhood cancers, particularly leukaemia and brain cancer. Epidemiological assessment is also underway for various occupational groups that are more exposed to electromagnetic fields than workers in other occupational groups. What distinguishes these studies from a number of already published studies is that actual measurements of exposure are performed, if only on the sample basis. In these studies, personal dosimeters are used for a certain period of time (e.g. 48 h) to determine an average exposure of the studied cohorts to electric and magnetic fields. Usually, current studies also have more statistical power than the previous studies, and some are prospective rather than retrospective. A very interesting human study with volunteers looks at effects on cardio-vascular and nervous system responses (Graham et al 1990). This study is characterized by careful dosimetry, as well as elimination of artifacts.

To elucidate the question of a possible association between exposure to ELF fields and cancer, a number of animal studies have been undertaken. Long-term chronic studies which encompass an animal lifespan are presently underway and include lymphoma development in mice, brain cancer in rats, liver cancer in mice and skin cancer in mice. In some of these studies, a liver cancer in mice and skin cancer in mice. In some of these studies, a specific cancer is initiated (with a chemical agent or ionizing radiation) and the invest-igations are aimed at determining whether the field acts as a promoter. Also short-term studies looking at specific interaction mechanisms (e.g. melatonin suppression, or co-promotion) have been undertaken.

The number of cellular in-vitro studies in progress is relatively large, many of them focused on a biological interaction mechanism. There appears to be a better convergence among the studies undertaken by researchers concerned mostly with potential hazards from environmental, occupational or medical exposures and those who are primarily interested in beneficial uses of ELF fields.

Medical Applications

Electromagnetic fields in a broad frequency range from ELF to high RF have found numerous applications in medical diagnosis and therapy. At low frequencies, magnetic fields are often replacing previously used electric fields produced directly in tissue by surface electrodes (Stuchly, 1990). This is the case in such diverse applications as brain cortex stimulation by single strong pulses for diagnostic purposes and in bone repair therapy by relatively weak repetitive pulses.

At RF, progress in dosimetry to a large degree benefiting from very similar dosimetric techniques developed for the hazard assessment has resulted in significant improvements in RF hyperthermia used in cancer treatment, and in magnetic resonance imaging (MRI) used for diagnosis. Other improvements in the medical field are anticipated from development of ablation and angioplasty catheters using RF energy at hundreds of megahertz and few gigahertz rather than electrosurgical catheters operating at hundreds of kilohertz.

REFERENCES

Cheng, D.K. (1989) Field and wave electromagnetics, Second edition
        Reading, MA, Addison - Wesley Publ. Co.
Dimbylone, P.J and Gandhi, O.P. (1991) Finite-difference time domain
        calculations of SAR in a realistic heterogeneous model of the head

for phone-wave exposure from 600 MHz to 3 GHz. Phys. Med. Biol., 8:1075-1089.

Durney, C.H. (1980) Electromagnetic Dosimetry for Models of Humans and Animals. A Review of Theoretical and Numerical Techniques, Proc. IEEE, 68:33-40.

Foster, K.R. and Schwan, H.P. (1989) Dielectric Properties of Tissues and Biological Materials: A Critical Review, CRC Crit. Rev. Biomed. Eng., 17:25-104.

Graham, C., Cohen, H.D. and Cook, M.R. (1990) Immunological and Biochemical Effects of 60-Hz Electric and Magnetic Fields in Humans, Midwest Research Institute Report No. RA-338-C.

Renew, D.C., Male, J.C. and Maddock, B.G. (1990) Power-Frequency Magnetic Fields: Measurement and Exposure Assessment, CIGRE Congress Record, Publ. No. 36-105.

Spiegel, R.J., Fatini, M.B., Stuchly, S.S. and Stuchly, M.A. (1989) Comparison of finite-difference time dimension SAR calculations with measurements in a heterogeneous model of man, IEEE Trans Biomed, 3G:849-855.

Stuchly, M.A. (1990) Application of Time-Varying Magnetic Fields in Medicine, CRC Crit. Rev. Biomed. Eng., 18, Sept.

Stuchly, M.A. and Mild, K.H. (1987) Environmental and Occupational Exposure to Electromagnetic Fields, IEEE Eng. Medicine Biol. Magazine, 6:15-17.

Stuchly, M.A. and Stuchly, S.S. (1986) Experimental Radio and Microwave Dosimetry, In: CRC Handbook of Biological Effects of Electromagnetic Fields, Polk, C. and Postow, E., Eds., CRC Press Inc., Boca Raton, Fl., 229-272.

Stuchly, M.A. and Stuchly, S.S. (1990) Electrical Properties of Biological Substances, In: "Biological Effects and Medical Applications of Electromagnetic Energy", O.P. Gandhi, Ed., Prentice Hall, New Jersey, 75-112.

Tell, R.A. (1981) Instrumentation for Measurement of Electromagnetic Fields: Equipment, Calibration and Selected Applications, In: "Biological Effects and Dosimetry of Nonionizing Radiation", Grandolfo, M. et al., Eds., Plenum Press, New York, 95-161.

Tenforde, T.S. and Kaune, W.T. (1987) Interaction of Extremely Low Frequency Electric and Magnetic Fields with Humans, Health Physics, 53:585-606.

BIOEFFECTS OF RADIOFREQUENCY FIELDS

Jürgen H. Bernhardt

Institute of Radiation Hygiene
Federal Office for Radiation Protection
D-8042 Munich-Neuherberg

1.   Introduction

The radiofrequency part of the electromagnetic spectrum is used in various areas of human activity. Equipment utilizing this energy today is to be found in industry, medicine, research, commerce, and private homes.

During the past century, due to the rapid increase of electric power in industrial society, the level of exposure of biological systems to electromagnetic fields has grown by orders of magnitude over a frequency range extending from zero to hundreds of GHz.

It is both appropriate and important to evaluate possible interactions between the man-made electromagnetic environment and living organisms, including man, and wether such interactions are beneficial or detrimental, transient or permanent. In the past two decades, research programmes throughout the world have made significant progress in defining the physical interactions of electric, magnetic and electromagnetic fields with living organisms and in describing biological effects resulting from the interactions.

At sufficiently high intensities, exposure to electromagnetic fields can produce a variety of adverse health effects. Such effects include cataracts of the eye, overloading of the thermoregulatory response, thermal injury, altered behavioral patterns, convulsions, decreased endurance, stimulation of nerve and muscle cells, electric shocks and radiofrequency burns from touching ungrounded metal objects in strong electric fields. Exposure limits are needed to protect against these adverse health effects of electromagnetic field exposure.

Numerous weak and subtle biological responses to electromagnetic field exposure have been reported. In many cases, the data available are not sufficient to permit a comprehensive explanation of the observed responses. Additional investigations are necessary to improve our understanding of how electromagnetic energy interacts with living matter and what are the consequences of "weak" effects for human health. Therefore, this review has been restricted mainly to well established mechanisms.

There are comprehensive reviews providing detailed information on the biological effects of radiofrequency electromagnetic fields on animals and man (UNEP 1992, Elder, et al. 1989: Guy, 1987, NCRP, 1986, Polk and Postow, 1986, Saunders et al., 1991). It is the purpose of this paper to give only a short survey of the biological effects and to concentrate on a discussion of the principal experimental and theoretical bases which allow the assessment of data on potential hazards to human health resulting from exposure to electromagnetic fields.

The physical properties of radiofrequency radiation are treated in detail in NCRP (1981, 1986), Stuchly and Stuchly (1983). Radiofrequency Radiation (RFR) is defined by the International Telecommunication Union as being part of the electromagnetic spectrum used for wireless communication, including frequencies that range between 3 Hz and 300 GHz. The 300 MHz to 300 GHz frequency range is also known as microwave range. This chapter deals with the frequency range between 300 Hz and 300 GHz.

## 2.	Mechanisms of interaction

### 2.1	General

Electromagnetic fields in the frequency range 300 Hz to 300 GHz interact with biological systems through direct and indirect pathways. A direct ineraction produces effects in the exposed organisms directly from exposure to the electromagnetic field. An indirect interaction is mediated through the presence of other bodies in the electromagnetic field, and occurs as a result of an interaction (usually physical contact with) between the biological system and another body, such as an automobile, fence or even other biological systems. Indirect interactions are important at frequencies below 100 MHz.

Direct interactions can be considered as resulting from induced currents and internal electric fields. The macroscopic spatial distribution of these currents and fields within an exposed biological system is of importance and is determined by theoretical and experimental dosimetry. The spatial distributions of the currents and fields at the cellular level are also important. Patterns of induced currents and fields within biological systems usually are highly non-uniform. They depend on the geometry and electrical properties of the exposed system, as well as the field frequency, and for lower frequencies on the type of field, whether electric or magnetic (where spatial separation of the electric and the magnetic field is realistic).

For frequencies below about 100 kHz an established interaction mechanism is the stimulation of excitable tissues by induced currents. For higher frequencies thermal interactions predominate. At frequencies below approximately 1 kHz and at higher frequencies amplitude modulated at extremely low frequencies (1-300 Hz) there is experimental evidence that interactions occur through mechanisms other than thermal or cell excitation. These mechanisms are not understood.

In the context of direct and indirect interaction mechanisms electrical properties of tissues have to be considered. Macroscopic electrical properties of tissues play a major role in defining induced currents and fields and their patterns inside the body. Microscopic electrical properties provide an insight into events at the molecular and cellular level that result from exposure of the biological system to an electromagnetic field.

2.2   Mechanisms on molecular and cellular levels

Many effective mechanisms can be comprehended on a molecular and cellular level by examining the macroscopic electric properties of the material (dielectric constant and electric conductivity). The highfrequency energy is transmitted to biological tissue by several basic mechanisms within the range of molecular structure (see reviews Schwan, 1957, 1963, Pethig and Kell, 1987, Foster and Schwan, 1986).

The main basic mechanisms are polarisation of bound charges, orientation of permanent dipoles and displacement of free charge carriers (electronic as well as ionic). For example, the energy dissipation may occur from friction losses along with the orientation of molecular dipoles (e.g. water) and the induced oscillation- and rotation movements within the molecules.

Orientation polarisation is the main cause of energy absorption in biological aqueous substances for frequencies between 1 and approx. 100 GHz. The maximum energy dissipation of free water molecules lies by near 20 GHz. Bound water, peptides, proteins or side-schains of larger molecules have their maximum absorption at frequencies between 1 and 20 GHz. The orientation movements do not lead to molecular changes; the absorbed energy is completely transformed into heat.

Another mechanism occurs mainly in the frequency range between 100 kHz and 100 MHz and can be described as a bounding layer polarisation frequently occurring in structures with different electric properties. Due to the poor electric conductivity of the cell membrane, in comparison to the interior and exterior cell space, the electric highfrequency current produced by the electromagnetic field in tissue essentially flows around the cell for lower frequencies (e.g. 10 kHz). Due to charge displacements, the cell acts here like a large dipole. At higher frequencies, on account of the capacity of the cell membrane, a large part of the highfrequency current is then flowing through the cell membrane and cytoplasm.

Charge displacements around and within a biological cell result in differences of the electric potential between the internal and external space of the cell which superimposes the existing resting potential of the membrane and shows the same time-dependence as the external field. Since the cells behave similarly to dipoles, these field-induced dipoles transfer forces to other dipoles in their environment, as was verified by appropriate experiments (Schwan, 1977).

The maximum voltage across the membrane for spherical cells is related to the electric field strength by

$$V_m = 1.5 \, RE$$

where R is the cell radius, and E is the electric field strength in extracellular fluid (Foster & Schwan, 1986). For ellipsoidal cells similar equations have been derived by Bernhardt and Pauly (1973). Their results show that electric fields axial to a cell induce a voltage across the membrane which is proportional to the length of the cell and to the extracellular electric field strength. Thus asymmetrical muscle cells exhibit dimension-dependent induced voltages when exposed to electric fields. At high field strengths (voltages across the membrane) pores are formed in the membrane, and eventually at a few hundred mV across the membrane, breakdown occurs (Foster & Schwan, 1986).

Gradients in the induced surface charge can also affect molecules and cells in solution. Polar molecules (e.g. water, proteins) align themselves with the field at high electric field strengths of the order of 10^6 V/m. Also, non-spherical cells align themselves with the field and form "pearl chains". The larger the cell, the lower the field strength required for orientation and formation of pearl chains. For instance, for a cell of a radius of $1\mu m$, an electric field strength of 10 kV/m is required (Foster & Schwan, 1986).

The interaction mechanisms on an atomic and molecular level and the subsequent effects on the microscopic level - heat effects, the generation of electrical potential differences and field-induced force effects - are not independent from each other and are capable of occurring simultaneously according to frequency and exterior field strength, as shown by in vitro studies. Under given physiological conditions, the heat effect is predominant in most cases. For molecular structures, further quantum-mechanical or cooperative absorption effects are postulated (see reviews in Polk and Postow 1986) but bear no significance for risk assessments. They are, however, of certain merit where a better understanding of effect mechanisms is concerned.

2.3   Direct interactions with excitable tissues

Well established interaction mechanisms for the direct effects of electric and magnetic fields can be divided into two types, each dependent on the field frequency. For frequencies below approximately 100 kHz, the interactions (stimulation) with excitable tissue are of primary interest. Above about 100 kHz, the current density thresholds for stimulation and other effects due to interactions with excitable tissue are higher than those required to produce energy deposition rates of about 1 W/kg. At such rates of energy deposition in tissue, thermal interactions become important. In both frequency ranges, other forms of interactions are also observed for induced currents and fields below those associated with stimulation or heating.

In tissues the induced electric currents and fields are amplified across the cell membranes. At sufficiently high field strengths, these affect the electrical excitability of nerve an muscle cells. This interaction occurs up to hundreds of kilohertz (Lacourse et al., 1985). Changes in the membrane potential cause changes in the permeability to ions, conformational changes in the embedded proteins, a number of ion gates open, and eventually membrane depolarization results in an action potential. Threshold current densities for subtle modulations of excitable cells and their biological significance are not well understood. There is a substantial amount of data on tissue stimulation, extra-systole elicitation and ventricular fibrillation.

Data are summarized by Bernhardt (1985, 1988). The ventricular fibrillation thresholds are above those needed for stimulation. Threshold for stimulation of excitable tissue depend not only on the current density and frequency, but also on the waveform. In the case of pulsed fields they depend on pulse duration and other parameters (Reilly, 1988).

2.4   Mechanisms of field interactions on a macroscopic level

For frequencies below several MHz, a spatial separation of the electric and magnetic field is realistic considering the coupling mechanisms with biological objects.

The exposure of living organisms from *electric* fields is normally characterized by an undisturbed electric field strength (measured or calculated field strength *without* an object). Because biological bodies are good electrical conductors in comparison to air, the introduction of the object to the field results in a change of the field. Two interaction mechanisms are predominant. Firstly, a strong increase in effective field strength on the surface of the biological object, depending on its position and size, together with a charge redistribution on the surface of the object. Secondly, a subsequent production of body currents within the object.

The electric field strength on the body surface causes a charging of the relatively high ohmic body hair, so that active forces between hair and the body surface cause a vibration of the hair shaft. The strong field increase on the body surface may produce a prickly sensation between clothing and skin as well as a direct stimulation of peripheral receptors in the skin. The sensation of electric fields due to surface irritations, as hair vibration and itch of the skin, depends on such factors as body field orientation, position of the extremities, clothing and grounding conditions, etc. This effect has some significance mainly at 50/60 Hz. In this case avoidance of annoyances due to electric field effects may be considered.

Internal electric body currents can be calculated or measured in models. Studies on the distribution of the electric field in different parts of the human body have shown that, with frequencies below 10 MHz, the internal field strength increases proportional with the frequency with a predetermined external electric field strength. This means that a simple relationship is possible between the internal and the external electric field strengths, depending only on the body part or organ considered and, furthermore, on the exposure conditions.

The electrically induced current densities within the body may be described by J=k f E, using suitable k-values for different body parts (Bernhardt, 1985). The longitudinal axis of the body parallel to the external E vector must be considered as "worst case" with external electric fields. The k-values can be determined by different methods (discussed in detail in Bernhardt, 1985). The current densities for the head and the thorax may be calculated by using approximately the same value $k=3 \times 10^9$ S/(Hz m). For other parts of the body the values of k may be larger, depending also on the exposure conditions, e.g. a. threefold value for the neck and a tenfold value for the ankles for a human standing on and in electrical contact with ground (Kaune and Phillips, 1980; Guy et al., 1982; Kaune and Forsythe, 1985). Detailed data may be obtained from the calculations of Dimbylow (1987, 1988).

For the time-varying *magnetic* fields inductive effects are predominant compared to effects caused by other mechanisms. Therefore the electric field strength or current density that is induced in the body may be regarded as the basic physical quantity for biological effects that occur on a cellular basis. This selection allows an extrapolation of animal experimental results on conditions in man by comparing the effective current densities (Bernhardt, 1979, 1985). For frequencies below 10 MHz, the current density induced in the biological object by the time-varying magnetic field can approximately be calculated by the equation

$$J = \sigma E = \frac{1}{2} R \cdot \sigma \cdot \frac{dB}{dt} \quad \text{(for sinusoidal fields } J = \pi R \sigma f B_0 \text{)}$$

where

J = induced electric current density (A/m^2);
σ = conductivity (S/m)
E = induced electric field strength (V/m);
R = radius of the current loop;
f = frequency;
B = magnetic flux density (T).

There is only very little experimental or theoretical work dealing with the coupling of magnetic fields to models of living organisms (e.g., Spiegel, 1976, Ghandi et al., 1984). However, using "worst case" assumptions, an estimate of the order of magnitude for "safe" and "dangerous" magnetic field strengths and their frequency dependence is possible (Bernhardt, 1979, 1985, 1988).

For *radiofrequency radiation*, the penetration depth can be derived from the frequency-dependent dielectric tissue data. For muscle tissue, the penetration depth (distance over which about 2/3 of the radiation energy is absorbed) measures for instance about 10 cm at 30 MHz and 3 cm at 1 GHz (Schwan and Foster, 1980). Concerning penetration into tissue, microwaves practically behave like infrared radiation above 20 GHz (Justesen et al., 1982).

The absorbed energy can be calculated from the field strength in tissue, taking dielectric tissue data and conductivity into account. Variously structured and combined models are available from calculations and experimental studies (Dosimetry handbooks of Durney et al., 1978, 1980, 1986, NCRP, 1986). The dosimetry handbooks show the essential principles of the dependence of energy absorption on exposure conditions and frequency. Almost all absorption curves apply to the far field, and thus are inappropriate for dosimetric purposes in the near field range. The absorbed energy is averaged over the whole body and divided by the body mass. This quantity is described as mean specific absorbed rate (SAR) in Watt per kg. SAR can also be averaged over smaller masses (e.g. organ, body portion, 10 g).

The frequency dependent SAR-curves can be devided into three sections. In the lower frequency range (subresonance range below 30 MHz), the absorption of energy decreases rapidly at decreasing frequency. In order to produce heat in the body, the power flux densities must be high. The distribution of absorbed energy or induced body field strength is particularly inhomogeneous.

In the resonance range (about 30 to 400 MHz), object (whole body or body parts) and wavelength are of similar order. High absorption cross sections are possible, and exposure limits must be set at lower values. The resonance frequency is ranging higher for children than for adults, due to the smaller body configuration of the former. Shifts in resonance frequency are not only due to body dimensions, but also to the influence of conducting and bodyconnected surfaces. The resonance frequency is hereby moved toward lower values.

In the upper frequency range, the wavelength is shorter in comparison to the object. The penetration depth is low. An increased frequency will only heat up layers that are close to the surface. In the frequency range between 200 and 3000 MHz refraction may furthermore result in focussing effects producing spatially limited "hot spots" in the body. For example, a considerable rise in temperature may occur in certain areas of the head (Kritikos and Schwan, 1975).

The SAR-curves for animals (e.g. mice, rats) differ by 1-2 orders from human SAR-curves where resonance frequency and SAR-values at identical frequencies are concerned. This must be taken into consideration when extrapolating results from animal experiments to man.

2.5 Thermal interactions

Exposure to an electromagnetic field can result in a spatially nonuniform SAR in the body. The initial rate of temperature increase, when heat losses are neglected, is directly proportional to the SAR:

$$dT/dt = SAR/C$$

where T is the temperature, t is time and C is the specific heat capacity of tissue.

The rapid rate at which heating can occur, and a uniquely non-uniform spatial pattern of energy deposition are important and unique to thermal interactions of electromagnetic energy. The rate of initial heating appears to be very important for pulsed fields. These two features make biological responses due to electromagnetic thermal loading unlike those due to other thermal agents. Thermal interactions are not necessarily accompanied by significant local or whole-body temperature increases.

One of the most prominent thermally induced effects where the temperature increases are very small is the microwave hearing effect (Guy et al., 1975; Lin, 1978). Exposure to one pulse of electromagnetic energy results in the perception of a click, and exposure to repeated pulses in a buzzing or hissing sound. The energy threshold for human beings is very low, 16 mJ/kg, and the resulting temperature increase is estimated to be only about 5×10^{-6} ^{0}C (Guy et al., 1975). The simplified mechanism of interaction is as follows: absorption of electromagnetic energy causes a rapid temperature increase, which in turn produces thermal expansion pressure initiating an acoustic wave which is detected by cochlea (Guy et al., 1975; Lin, 1978).

For determining a temperature increase it is important to not only observe the heating process due to highfrequency radiation but also the cooling of tissue from heat conduction, heat radiation, transport via blood and perspiration (e.g. Adair, 1983, Michaelson, 1974, Stolwijk, 1980). Corresponding to homogenous spheric models or models from various layers, a power flux density of 100 W/m² increases the average temperature from 0.5⁰C to 2⁰C (depending on frequency), taking heat transport and normal blood flow into consideration (see e.g. Adair, 1986). Ambient temperature and humidity play an important role in heat emission, so that unfavorable ambient conditions are lowering the maximum tolerable thermal stress. Conditions that impair thermoregulatory functions include, e.g., fever, diabetes, cardiovascular desease, obesity, old age, special medications. At a temperature increase of about 0.5⁰C, the normal thermoregulation in the human hypothalamus is beginning to respond.

For other parts of the body it is also likely that the absorbed energy deviates considerably from the average within the body. This is true particularly for frequencies below the resonance frequency where SAR-values may differ by two orders of magnitude for individual parts of the body like the ankles (Ghandi et al., 1985, Dimbylow, 1988, see also review Elder et al., 1989).

In poorly circulated organs, as for instance the eye lens, the temperature increase is greater; power flux densities of more than 1 kW/m² may lead to the formation of a cataract (Carpenter, 1979, Cleary, 1980, Guy et al., 1975). Accordingly, the eyes, for example, must be protected during HF-therapy in the head region or the applied HF-power must be reduced. No cataracts occurred in rabbits that were chronically exposed with 100 W/m² (maximum SAR in head of 17 W/kg) for 180 days (Guy et al., 1980).

A completely different energy distribution may be taking place if metallic implants are present in the body. For large implants, on the contact surfaces of metal-tissue, it is possible that locally increased highfrequency currents in the tissue may produce burns. This effect, for instance, must be taken into consideration for HF-therapy.

Special circumstances apply for the absorption of highfrequency energy during NMR-examination due to the fact that the patient is here practically part of the transmitter coil at nearfield conditions. Only few results are so far available from measurements of the absorbed HF-energy (Shellock and Crues, 1987); theoretical assessments supply contradictory data.

2.6   Indirect interactions

Secondary short term effects must be considered when evaluating health risks resulting from electric or magnetic field exposure at frequencies below about 100 MHz. Hazardous thresholds for some indirect effects are lower than the thresholds for biological effects due to the direct influence of electromagnetic fields. The main points are

- Contact currents enter a person through electrical conductors contacting the skin,
- spark discharges introduce transient currents into the body through an arc gap when the electrical break-down potential of air is exceeded,
- electric or magnetic fields interfere with the performance of unipolar cardiac pacemakers.

Electric Fields at frequencies below a few MHz can induce an RF charge on ungrounded or poorly grounded metallic objects such as cars, trucks, cranes, wires, fences, etc. When a person comes in contact with such objects, current to ground flows through the body: the person practically shortcircuits the object. Short-circuit currents from different objects to ground were calculated by Guy and confirmed by measurements in the kHz range (Guy and Chou, 1982, Guy, 1985).

Threshold levels for electric field strength, producing different indirect effects are published by Guy (1985), Chatterjee et al. (1986) and Bernhardt (1988).

There are several reactions and sensations that are of interest in assessing the coupling effects of low frequency electric fields. The sensations and reactions in the general order of increasing stimulus are: Perception, Annoyance, Startle, Aversion and (the following apply only to direct contact currents for frequencies below 100 kHz) Let-go, Respiratory tetanus, Fibrillation.

The stimulation threshold for nerve and muscle tissue as a function of frequency for sinusoidal currents was shown to be a monotonically increasing value. Between 100 kHz and about 10 MHz an overlapping of both ot the effects of heat sensation and stimulation is possible. Lacourse et al. (1985) have shown that excitable tissue can be stimulated up to frequencies of several MHz, if the current density is great enough. In this frequency range, when the current was adjusted to a value equal to the perception threshold, pain was reported typically within 10 to 20 s (Ghandi and Chatterjee, 1982, Chatterjee et al., 1986).

Another point of concern are reactions to spark discharges. In these cases, the person and/or the object act as capacitors on which a charge is induced by the action of the electric field. When the person and the conducting object come close together (a fraction of a centimeter), the electric breakdown potential of air may be exceeded, and an electric arc results. This topic is important mainly at 50/60 Hz (see review by Bernhardt, 1988).

3.   Biological effects

3.1   General

Many aspects of bioeffects of electromagnetic fields are already treated in the sections 2.3, 2.5 and 2.6 of this chapter. This section is limited to a summary of laboratory and human studies especially for frequencies above 10 MHz and is based on different comprehensive reviews (Elder and Cahill, 1984, Elder et al., 1989, NCRP, 1986, Saunders et al., 1991, UNEP, 1992).

The current density and the specific absorption rate (SAR) can frequently be applied as a dosimetric quantity for a quantitative description of biological effects induced by electromagnetic radiation. Next to the many factors determining these quantities in a biological object, it must also be considered that, i.e., the SAR-distribution is very inhomogenous in the body and may lead to temperature gradients in the body.

Biological effects were observed also under conditions where a highfrequency-induced temperature rise is not likely to occur. This applies mainly to Extremely-Low-Frequency (ELF)-modulated highfrequency fields or microwave in the millimeter range (Adey and Bawin, 1982, Blackman et al., Myers and Ross, 1981). Numerous hypotheses exist, but there are no adequately reliable data that would be of use for a risk assessment of such effects or derivation of individual radiation protection values.

3.2   Laboratory studies

Many of the biological effects of acute exposure to electromagnetic fields are consistent with responses to induced heating, resulting in rises in tissue or body temperature of about 1⁰C or more. Most responses have been reported at SARs above about 1-2 W/kg in different animal species exposed under various environmental conditions (Adair, 1983, 1986). These animal (particularly primate) data indicate the types of responses that are likely to occur in humans subjected to a sufficient heat load. However, direct quantitative extrapolation to humans is difficult given species differences in responses in general, and in thermoregulatory ability in particular.

The most sensitive animal responses to heat loads are thermoregulatory adjustments, such as reduced metabolic heat production and vasodilation, with thresholds ranging between about 0.5-5 W/kg, depending on environmental conditions. However, these reactions form part of the natural repertoire of thermoregulatory responses that serve to maintain normal body temperatures. Transient effects seen in exposed animals which are consistent with responses to increases in body temperature of 1⁰ C or more (and/or SAR's in excess of about 2 W/kg in primates and rats) include reduced performance of learnt tasks and increased plasma corticosteroid levels. Effects on the blood building system and the immune system were reported to occur with SAR-values of more than 0.4 W/kg (Czerski, 1975, Liburdy, 1979, Shandala et al., 1983, Smialowicz et al., 1982). Other heat-related effects include temporary haemopoietic and immune responses, possibly due to elevated corticosteroid levels. The most consistent effects observed are reduced levels of circulating lymphocytes and increased levels of neutrophils, and altered natural killer cell and macrophage function.

There are very few reports about chromosomal- or DNA changes in highfrequency irradiated experimental animals without reference to a significant increase in temperature (Edwards et al., 1984, Swicord et al., 1983). Individual reports on DNA- and chromosomal changes occurring with non-thermic intensities could not be verified (Foster et al., 1987). Other effects that were reported to have occurred at insignificant temperature rises include changes in the ion exchange over cell membranes in electrophysiological properties of nerve cells (Wachtel et al., 1975).

According to animal experiments, nerve cells of the central nervous systems as well as operant behaviour were acutely influenced by high power flux densities, but also by chronic exposure to low energy with

SAR-values of more than 2 W/kg (d'Andrea et al., 1977, De Lorge, 1984, Mitchell et al., 1988). Modulated HF-fields may produce changes in the EEG (Takashima et al., 1979). Pulsed HF-radiation may release an increased effect of drugs and pharmaceuticals on the function of the central nervous system (Baranski and Edelwejn, 1968).

Most animal data indicate that implantation and the development of the embryo and fetus unlikely to be affected by exposures which increase maternal body temperature by less than 1^0 C. Above these temperatures, adverse effects, such as growth retardation and post-natal changes in behaviour may occur, with more severe effects occuring at higher maternal temperatures.

Most animal data suggest that RF exposure low enough to keep body temperatures within the normal physiological range is not mutagenic: Such exposure will not result in somatic mutation nor in hereditary effects.

There is much less information describing the effects of chronic low level exposure. So far, however, it is not apparent that there are any long-term effects which can result from exposures below thermally significant levels. A study of 100 rats exposed for most of their lifetime at about 0.4 W/kg did not show an increased incidence of non-neoplastic lesions or total neoplasias compared to control animals (Guy et al., 1985).

The possibility that exposure to RF fields might influence the process of carcinogenesis is of particular concern. So far, there is no definite evidence that irradiation does have an effect, but there is clearly a need for further studies to be carried out. Many experimental data indicate that RF fields are not mutagenic, and so they are unlikely to act as initiators of carcinogenesis; the few studies carried out have looked mostly for evidence of an enhancement of the effect of a known carcinogen. Chronic exposure of mice at 2-8 W/kg resulted in an increase in the progression of spontaneous mammary turmours and of skin tumours in mice whose skin was treated with a chemical carcinogen (Szmigielski et al., 1982). Repeated RF exposure followed by a "sub-carcinogenic" dose of carcinogen resulted in an increased number of skin tumours; however this study has been reported only briefly, and the authors note the need for experimental confirmation.

A substantial body of data exists describing biological responses to amplitude-modulated RF or microwave fields at SAR's too low to involve any response to heating (athermal effects). Some studies have reported effects after exposure at SAR's of less than 0.01 W/kg, occurring within modulation frequency "windows" (usually between 1-100 Hz) and sometimes within power density "windows"; similar results have been reported at frequencies within the VF range. It is important that these studies be confirmed and that the health implications are determined. Of particular importance would be studies which link extremely low frequency, amplitude-modulated RF or microwave interactions at the cell surface with changes in DNA synthesis or transcription. It is worth noting that this interaction implies a "demodulation" of the RF signal at the cell membrane.

3.3   Human studies

There are some studies of controlled exposures to volunteers that have provided valuable information on responses to RF exposure. These studies include warming and pain thresholds for RF heating of the skin, RF hearing and RF shocks and burns (see 2.6). Clinical studies of accidental overexposures do provide information on acute exposure responses.

Exposure of the human body to RF fields can cause heating that is detectable by the temperature-sensitive receptors in the skin. Several investigators have experimentally determined the threshold intensities that cause sensations of perceptible warmth, pain and delay in response to the stimulus in human subjects. Adair (1983) noted that RF exposures to frequencies of 30 GHz and above would probably be similar to infrared in their perception threshold values. Cutaneous perception depends on the frequency of the incident RF field. In the resonance region particularly, internal organs may suffer thermal damage (burns) without any sensation of warmth during the exposure.

Studies in the laboratory have described cutaneous perception of fields in the 2 - 10 GHz range. Threshold for just noticable warming have been reported at power densities of 270 W/m^2 to 2000 W/m^2, depending on the area irradiated (13 - 100 cm^2) and the duration of exposure (1 - 180 s).

When human volunteers are exposed to SAR's of 4 W/kg for 20 minutes their average body temperature rises by 0.2 - 0.5^0 C, which is quite acceptable to healthy people. It is unknown what impact this added thermal load would have on thermoregulatory impaired individual in environments that minimize the perspiration based cooling mechanisms.

Recently, Meister et al. (1989) reported effects on perception, performance and well-being in eight volunteers exposed to a 2.45 GHz field with power densities of up to 10 W/m². Changes in visual perception thresholds were reported at 5 and 10 W/m², acute symptomes were also found at 10 W/m². Although the health implication of these results seems to be questionable, replication studies should be done to validate the findings. As was already mentioned, some people can perceive individual pulses of RF as audible clicks, chirping or buzzing sounds, depending on the pulsing regime and intensity of the field. This phenomenon was first investigated by Frey (1961). Since that time, there have been many studies of auditory responses of volunteers. Frequencies between 6.5 and 8.9 GHz seem most effective in producing these responses.

Induced current effects, aspects of thermoregulation and contact current effects are described in the sections 2.3, 2.5 and 2.6, respectively.

The few epidemiological studies that have been carried out in populations exposed to RF fields have failed to produce significant associations between such exposures and outcomes of shortened life span, or excesses in particular causes of death, except for increased incidence of death by cancer where there may have been chemical exposure as a confounder (UNEP, 1992). Such studies tend to suffer from poor exposure assessment and poor ascertainment and determination of other risk factors.

There are a number of problems with epidemiological studies on man as well as clinical reports about RFR-effects that concern quantitative data as much as uncertainties about "dose", health effects, latencies, dose-effect-relationships and interactions with other physical or chemical agents (Robert and Michaelson, 1985, Roberts et al., 1986, Silverman, 1980). In summary, the epidemiological and clinical comparative studies do not provide clear evidence of detrimental health effects in humans from exposure to RF fields (UNEP, 1992). Some occupational groups, eg exposed physiotherapists and industry workers, should be studied further. The question whether RF might act as a carcinogen should be further evaluated in epidemiological studies.

Occupational exposure to RF will be at higher levels than encountered by the general public, and so there is less likelihood of health effects in the general population as a whole.

4.   **Risk assessment and requirements for health protection**

For a risk assessment of the effects from radiofrequency electromagnetic fields it has so far been common use to apply acute (nonstochastic) effects that occur not earlier than after certain minimum thresholds. While numerous data are at hand about acute effects, the data situation concerning "late effects" likely to occur with chronic exposure is insufficient and inappropriate for setting limits. For risk analysis, different possible effects that vary in severity can be useful as a point of orientation. The "dose"-or "dose rate" areas can be differentiated as follows:

- Area in which no effects are clearly verifiable or expected to occur,
- area in which subtle biological effects are verifiable; including perceivable effects,
- area in which biological effects can be detrimental to health, including distress, nervousness or irritation,
- area in which the effects represent a hazard but are of reversible nature,
- area in which injury can be medically demonstrated.

The following categories of effects must be considered for risk assessment.

a.   Absorption of RF energy causes tissue heating. This is recognized and has been well studied. This effect occurs from absorption of RF fields, especially at the higher end of the frequency range (above about 1 MHz). RF heating is not directly equivalent to heating by other forms of energy, because of very non-uniform, energy deposition that occurs in biological systems.

b.   At frequencies from 300 Hz to approximately 100 kHz, currents induced or caused by charging of metallic objects in the RF field, may result in stimulation of electrically excitable tissues. At frequencies between approximately 100 kHz and 100 MHz, contact currents of sufficiently high density may cause burns.

c.   For frequencies below several hundred kHz, the predominant effect is stimulation of excitable tissue resulting from currents induced by the RF fields. At these lower frequencies, thermal interactions occurs only at energy levels much higher than interactions with excitable tissue.

The first two of these are sufficiently well understood to be used in risk assessment and the development of recommended limits of exposure. The third category is reasonably well understood but there are only few quantitative data.

The effects noted in three other categories are elaborately described and, especially the last two of these, poorly understood. In view of their importance they must be considered. However, the lack of understanding and the total absence of quantitave relations of these effects to either exposures or the outcomes in question makes it impossible to derive recommended limits of exposure (UNEP, 1992).

- *Pulsed fields.*
  It has been shown under a number of conditions that the thresholds for biological effects at frequencies above several hundred MHz are decreased when the energy is delivered in short (1 - 10 μs) pulses. As an example auditory effects occur when pulses of less than 30 μs duration deliver more than 400 mJ/m^2 per pulse. With such pulses the available evidence does not provide a precise identification of a safe limit.

- *Amplitude modulated RF fields.*
  The effect described for this type of fields at the cellular, tissue and organ level cannot be related to adverse health effects. No dose-effect relationships can be formulated that demonstrate threshold levels; therefore, the available information cannot lead to specific recommendations.

- *RF field exposures on tumour induction and promotion.*
  The reports of effects of RF exposure in certain cell lines on cell transformation, enzyme activity, and tumour incidence and progression in animals do not permit one to conclude that RF exposure has any effect on the incidence of cancer in humans or the necessity to have specific recommendations to limit such fields to reduce cancer risks.

  For establishing exposure limits, it is necessary to introduce safety factors due to a number of uncertainties (e.g. different absorption according to body size, allowing for different exposure conditions as well as for possible reflexion, focussing and scattering; adaptation of animal studies to man). Such safety factors may differ for different population groups (IRPA, 1988).

For frequencies of more than about 10 MHz it is today common practice to take the specific absorption rate (SAR) averaged over the whole body mass as a basic quantity for the purpose of assessing biological effects. For frequencies of less than a few MHz, the body current density may be used as the basic quantity that determines the biological effect. Of predominance for frequencies below 100 kHz are non-thermic effects that depend on the tissue field strength or current density on a microscopic level (e.g. effects on cell membranes, stimulation effects on nerve- and muscle cells). SAR, body field strength E_i (effective value) and body current density J_i (effective value) are interlinked in the following equations:

$$SAR = \frac{\sigma \cdot |E_i|^2}{\rho} - \frac{|J_i|^2}{\rho \cdot \sigma}$$

(σ: conductivity, ρ: density)

For frequencies above several MHz, the exposure limits are primarily derived from the basic quantity SAR (averaged over the entire body mass). They are being modified by additional considerations as, for instance, the different SAR- and J-distribution in various parts of the body. A different aspect of modifying limits for the electric field strength is the limitation of hazards from HF-burns and electroshocks, especially for frequencies below 100 MHz.

Today, it must be assumed that RFR-exposures resulting in SAR-values of 1 - 4 W/kg (averaged over the entire body) cause effects that are detrimental to health already after one hour of exposure at unfavorable conditions (e.g. overload of thermoregulation of impaired risk persons, change in behavioral patterns, reduced resistance) and result in injury after prolonged exposure (days, weeks) (NCRP, 1986, Elder et al., 1989).

An occupational RF exposure guideline of 0.4 W/kg (averaged over 6 min. intervals) based on thermal consideration leaves a considerable margin of safety for other limiting conditions such as high ambient temperature, humidity or physical activity. Higher energy absorption rates in extremities and limited body regions, do not appear to cause adverse effects, for SAR values below thresholds dependent on the body part and the volume.

The time factor of 6 min may be taken as a "thermic time constant". For exposure times under 6 min, the resulting temperature rise in a volume element is given by the energy deposited in that field. Only after a longer exposure time than 6 min will such mechanisms as heat conduction, blood circulation and thermoregulation influence the temperature change. Exposure times of under 6 min permit higher limits for the power flux density in the presence of a simultaneously constant deposit of energy.

In infants, the frail elderly and in individuals taking certain drugs, the thermoregulatory capacity may be much reduced and as a result their tolerance for the combined effect of RF exposure, exercise, solar radiation and high ambient temperature is much lower. Recognition that this tolerance is lower dictates that guidelines for population exposure to RF fields be reduced. A whole body average SAR of 0.08 W/kg offers an additional safety factor.

Particularly at frequencies below 10 MHz and above 10 GHz and in the near-field range of antennas or other equipment, the SAR-distributions are very inhomogeneous. Additional precautions must be exercised for situations which might cause large peak values of the SAR in order to eliminate rapid elevations of local temperature by more than 1°C. This requires that the peak (or local) SAR's not exceed about 2 W/100g in the extremities and 1 W/100g in any other part of the body. The eye may need special consideration (averaging mass 10 g).

The derivation of exposure field strengths from the frequency-dependent SAR-value is not successful in the near field range. Examples are low power devices with antennas operating near the body (mobile phones), where high local SAR values were calculated or measured. The maximum permissible power of such equipment should hereby be derived from the maximum permissible local SAR-values (e.g., 0.1 W per 10 g). Accordingly, emission limits for other equipment and facilities are likewise to be derived from the exposure limits (e.g. for microwave ovens).

Hazardous indirect effects are avoidable, so that limiting the current for frequencies below about 100 MHz may exclude HF-burns or electroshocks, respectively. The threshold value for HF-burns from finger contact with conductive surfaces is about 200 mA (Rogers, 1981, Guy, 1985), the threshold values for the stimulation of nerve- and muscle cells below 100 kHz depend very much on the frequency (Bernhardt, 1988).

The most important aspects in determining exposure limits can be summarized as follows:
Human RF exposure limits should be derived from "basis limits". Basic limits have been developed using different scientific bases for frequencies above and below about 10 MHz.

The *basic limit* value is the permissible value specified on the basis of biological effects for

- the specific absorption rate (SAR),
- the electric current density in the body,
- current flowing through the human body.

These quantities are often referred to as basic "dosimetric" quantities which cannot be obtained directly by means of a measuring instrument.

In practice, values of "field" quantities are required: The *derived limit* value is the permissible value for the electric or magnetic field strength, the power flux density and the touch voltage which is derived from basic limits. Derived limit values are given to provide a method of assessment for the effect of electromagnetic fields and to demonstrate compliance with the basic restrictions of exposure. Said limit values have been specified in such a manner that the basic limit values are not exceeded even if the most unfavourable conditions of field exposure are taken as a basis.

The *general requirement* is that the basic limits specified shall be observed. If the derived limit values are observed, the basic limits are complied with as well. The derived limit values can be exceeded in the individual case, if it is ensured that the basic limit values will be observed under all conditions which may occur.

For electromagnetic fields the main points are the following:
- above about 10 MHz, the specific absorbed rate SAR (in some cases also the absorbed energy) serves as a basic quantity; the field strengths or the power flux density are the derived limits,
- below about 10 MHz, the exposure limits should be derived from local SAR values or from current densities for field strengths induced in the body; the external electric and magnetic field strengths are the derived limits that are measurable,

-       the avoidane of RFR-skin burns and electroshocks is a further
        criterion in determining limits for the electric field strength below
        100 MHz.

5.   References

Adair, E. R., ed. (1983) Microwaves and thermoregulation. New York,
        Academic Press.

Adair, E. R. (1986) Thermoregulation in the presence of microwave fields.
        In: CRC Handbook of Biological Effect of Electromagnetic Fields
        (Polk, C. and Postow, E., eds). Boca Raton, Florida, CRC Press:
        315-337.

Adey, W. R., Bawin, S. M. (1982) Binding and release of brain calcium
        by low electromagnetic fields: a review, Radio Sci., 17:149-157.

Baranski, S., Edelwejn, Z. (1968) Studies on the combined effect of micro-
        waves and some drugs on bioelectric activity of rabbit central ner-
        vous system, Acta physiologica polonica 19:31-41.

Bernhardt, J.H., Pauly, H (1973) On the generation of potential differences
        across the membranes of ellipsoidal cells in an alternating
        electrical field, Radiat. Environ. Biophys. 10: 89-98.

Bernhardt, J.H. (1979) The direct influence of electromagnetic fields on
        nerve- and muscle cells of man within the frequency range of 1
        Hz to 30 MHz, Radiat. Environ. Biophys. 16:309-323.

Bernhardt, J.H. (1985) Evaluation of human exposure to low frequency
        fields. In: The impact of proposed radio frequency radiation stan-
        dards on military operations, proceedings of a NATO-workshop.
        92200 Neuilly-surseine, France: AGARD, 7 rue Ancelle; AGARD
        Lecture Series No. 138: Proceedings 8.1-8.18.

Bernhardt, J.H. (1988) The establishment of frequency dependent limits for
        electric and magnetic fields and evaluation of indirect effects,
        Radiat. Environ. Biophys. 27:1-27.

Blackman, C.F. et al. (1980) Induction of calcium ion efflux from brain
        tissue by radiofrequency radiation: effect of sample number and
        modulation frequency on the power-density window, Bioelectroma-
        gnetics 1:35-43.

Blackman, C.F., Benane, S.G., House, D.E., Joines, W.T. (1985) Effects
        of ELF (1-120 Hz) and modulated (50 Hz) Rf fields on the efflux
        of calcium ions from brain tissue in vivo, Bioelectromagnetics 6:1-
        11.

Carpenter, R.L. (1979) Ocular effects of microwave radiation, Bull. N.Y. Acad. Med 55:1048-1057.

Chatterjee, I., Wu, D., Gandhi, O.P. (1986) Human body impedance and threshold currents for perception and pain for contact hazard analysis in the VLF-MF band, IEEE Trans. Biomed. Eng. BME-33:486 --494.

Cleary, S.F. (1980) Microwave cataractogenesis, Proc. IEEE 68:49-55.

Czerski, P. (1975) Microwave effects on the blood-forming system with particular reference to the lymphocyte, Ann. N.Y. Acad. Sci. 247: 232-242.

D'Andrea, J.A., Gandhi, O.P., and Lords, J.L. (1977) Behavioral and thermal effects of microwave radiation at resonant and nonresonant wavelengths, Radio Sci. 12 (S):251-256.

De Lorge, J.O. (1984) Operant behaviour and colonic temperature of Macaca Mulatta exposed to radio frequency fields at above resonant frequencies, Bioelectromagnetics 5:232-246.

Dimbylow, P.J. (1987) Finite difference calculations of current densities in a homogenous model of a man exposed to extremely low frequency electric fields, Bioelectromagnetics 8:355-375.

Dimbylow, P.J. (1988) The calculation of induced currents and absorbed power in a realistic, heterogeneous model of the lower leg for applied electric fields from 60 Hz to 30 MHz, Phys.Med. Biol 33: 1453-1468.

Durney, C.H. et al. (1978) Radiofrequency radiation dosimetry handbook, 2nd ed. USAF School of Aerospace Medicine (Report SAM-TR-78-22) Brooks Air Force Base, Texas.

Durney, C.H. (1980) Electromagnetic dosimetry for models of humans and animals: a review of theoretical and numerial techniques, Proc. IEEE 68:33-40.

Durney, C.H. et al. (1980) Radiofrequency radiation dosimetry handbook, 3rd ed. USAF School of Aerospace Medicine (Report SAM-TR-80-32) Brooks Air Force Base, Texas.

Edwards, G.S. et al. (1984) Resonant microwave absorption of selected DNA molecules, Phys. Rev. Letters 53:1284-1287.

Elder, J.A., Cahil, D.F. (1984) Biological effects of radiofrequency radiation. United States Environmental Protection Agency Publication EPA-600/8-83-026 F, Research Triangle Park, NC.

Elder, J.A. et al. (1989) Radiofrequency radiation
In: Nonionizing radiation protection, M.J. Suess and D.A. Benwell-Morison (eds.) 2.ed. WHO Regional Public. European Series No. 25, Copenhagen: 117-173.

EPA (1986) Environmental Protection Agency. Federal Radiation

Protection Guidance; Proposed alternatives for controlling public exposure to radiofrequency radiation; Notice of proposed recommendations. Federal Register Part II Vol 51 No 146, July 30, 1986:27318-27339.

Foster, K.R., Schwan, H.F. (1986) Dielectric properties of tissue. In: Polk C. and Postow E., ed. CRC Handbook of Biological Effects of Electromagnetic Fields. CRC Press, Boca Raton, Fl.: 28-96.

Foster, K.R. et al. (1987) "Resonances" in the dielectric absorption of DNA? Biophys. J. 52:421-425.

Gandhi, O.P., Chatterjee, I. (1982) Radiofrequency hazards in the VLF to MF bands, Proc. IEEE 70:1462-1469.

Gandhi, O.P. et al. (1984) Impedance method for calculation of power deposition patterns in magnetically induced hyperthermia, IEEE Trans. Biomed. Eng. BME-31:644-651.

Gandhi, O.P. et al. (1985) Likelihood of high rates of energy deposition in the human legs at the ANSI recommended 3-30 MHz RF safety levels, Proc. IEEE 73:1145-1147.

Guy, A.W. et al. (1975) Microwave-induced acoustic effects in mammalian auditory systems and physical materials, Ann. N.Y. Acad. Sci. 247: 194-215.

Guy, A.W., Chou, C.K. (1982) Hazard analysis: very low frequency through medium frequency range. Report USAFSAM 33615-78D-0617. USAF School of Aerospace Medicine, Aerospace Medical Division, Brooks Air Force Base, TX 78235.

Guy, A.W. et al. (1982) Determination of electric current distributions in animals and humans exposed to an uniform 60-Hz high-intensity electric field, Bioelectromagnetics 3:47-71.

Guy, A.W., Lin, J.C., Kramar, P.O., Emery, A.F. (1975) Effect of 2450 MHz radiation on the rabbit eye, IEEE Trans. Microwave Theory Tech. MTT-23:492-498.

Guy, A.W., Kramar, P.O., Harris, C.A., Chou, C.K. (1980) Long-term 2450-MHz CW microwave irradiation of rabbits: Methodology and evaluation of ocular and physiologic effects, J. Microwave Power 15:37-44.

Guy, A.W. (1985) Hazards of VLF electromagnetic fields. AGARD Lecture Series No 138: The impact of proposed radiofrequency radiation standards on military operations, pp 9.1-9.20.

Guy, A.W., Chou, C.K., Kunz, L.L., Crowley, J. and Krupp, J. (1985) "Effects of Long-Term Low-Level Radiofrequency Radiation Exposure of Rats," Vol.9, Summary, Report No. USAFSAM-TR-

85-64, USAF School Aerospace Medicine, Brooks AFB, TX 78235-5301.

Guy, A.W. (1987) Dosimetry associated with exposure to nonionizing radiation: very low frequency to microwaves, Health Physics 53:569-584.

IRPA (1988) Guidelines on limits of exposure to radiofrequency electromagnetic fields in the frequency range from 100 kHz to 300 GHz, Health Physics 54:115-123.

Justesen, D.R. et al. (1982) A comparative study of human sensory thresholds: 2450-MHz microwaves VS far-infrared radiation, Bioelectromagnetics 3:117-125.

Kaune, W.T., Phillips, R.D. (1980) Comparison of the coupling of grounded humans, swine and rats to vertical, 60-Hz electric fields, Bioelectromagnetics 1:117-129.

Kaune, W.T., Forsythe, W.. (1985) Current densities measured in human models exposed to 60 Hz electric fields, Bioelectromagnetics 6:13-23.

Kritikos, H.N., Schwan, H.P. (1975) The distribution of heating potential inside lossy spheres, IEEE Trans. Biomed. Eng. BME-22:457-463.

Lacourse, J.R., Miller, W.T., Vogt, M., Selikowitz, S.M. (1985) Effect of high frequency current on nerve and muscle tissue, IEEE Trans. Biomed. Eng. BME-32:82-86.

Lidurdy, R.P. (1979) Radiofrequency radiation alters the immune system: modulation of T- and B-lymphocyte levels and cell mediated immunocompetence by hyperthermic radiation, Radiation Research 77:34-46.

Lin, J.C. (1978) Microwave auditory effects and applications. Charles C. Thomas Springfield, IL.

Meister, A., Eggert, S., Richter, J., Ruppe I., (1989) Die Wirkung eines höchstfrequenten elektromagnetischen Feldes (2,45 GHz) auf Wahrnehmungsprozesse, psychische Leistung und Empfindung, Z. gesamt Hyg. 35:203-205.

Michaelson, S.M (1974) Thermal effects of single and repeated exposures to microwaves - a review. In: Biological Effects and Health Hazards of Microwave Radiation (Czerski, P., Ostrowski, K., Shore, M.L., Silverman, Ch., Suess, M.J., Waldeskog, B. (eds) Warsaw Polish Medical Publishers:1-14.

Mitchell, C.L., McRee, D.I., Peterson, J., Tilson, H.A. (1988) Some behavioral effects of short-term exposure of rats to 2.45 GHz microwave radiation, Bioelectromagnetics 9:259-268.

Myers, R.D., Ross, D.H. (1981) Radiation and brain calcium: A review and critique, Neurosci. Behav. Rev. 5:503-543.

NCRP (1981) Radiofrequency electromagnetic fields: properties, quantities and units, biophysical interaction, and measurements. NCRP Report No. 67 National Council on Radiation Protection and Measurements, Washington, D.C.

NCRP (1986) Biological effects and exposure criteria for radiofrequency electromagnetic fields. NCRP Report No. 86. National Council on Radiation Protection and Measurement, Bethesda,, MD.

Pethig, R., Kell, D.B. (1987) The passive electrical properties of biological systems: their significance in physiology, biophysics and biotechnology, Phys. Med. Biol. 32:933-970.

Polk, C., Postow, E. (eds) (1986) CRC Handbook of biological effects of electromagnetic fields. CRC Press inc., Boca Raton, Florida.

Reilly, J.P.)1988) Electrical models for neural excitation studies. John Hopkins APL Techn. Digest. 9:44-59.

Roberts, N.J. Jr., Michaelson, S.M. (1985) Epidemiological studies of human exposures to radiofrequency radiation, Internat. Arch. Occupat. Environm. Health 56:169-178.

Roberts, N.J. Jr., Michaelson, S.M., Lu, S-T. (1986) The biological effects of radiofrequency radiation: A critical review and recommendations, Int. J. Radiat. Biol. 50:379-420.

Rogers, S.J. (1981) Radiofrequency burn hazards in the MF/HF band (Aeromedical Review 3-81) 76-89. In: Proc. of a Workshop on the Protection of Personnel Against Radiofrequency Electromagnetic Radiation, USAF/SAM Aerospace Medical Division, Brooks Air Force Base, TX.

Saunders, R.D., Kowalczuk C.I. and Sienkiewicz Z.J. (1991) The biological effects of non-ionizing electromagnetic fields and radiation: III Radiofrequency and microwave radiation. National Radiological Protection Board Publ. Oxon, U.K. (in press).

Schwan, H.P. (1957) Electrical properties of tissue and cell suspensions. In: Advances in Biological and Medical Physics, vol. V: 147-209.

Schwan, H.P. (1963) Electric characteristics of tissue. Biophysik 1:198-208.

Schwan, H.P. (1977) Field interaction with biological matter. Annals N.Y., Acad. Sci. 103:198-213.

Schwan, H.P., Foster, K.R. (1980) RF-field interactions with biological systems, Proc. IEEE 68:104-113.

Silverman, C. (1980) Epidemiologic studies of microwave effects, Proc. IEEE 68:78-84.

Smialowicz, R.J., Weil, C.M., Kinn, J.B., Elder, J.A. (1982) Exposure of rats to 425-MHz (CW) radiofrequency radiation: Effects on lymphocytes, J. Microwave Power 17:211-221.

Spiegel, J.R. (1976) ELF coupling to spherical models of man and animals, IEEE Trans. Biomed. Eng. BME-23:387-391.

Stolwijk, J.A.J. (1980) Mathematical model of thermal regulation, Annals, N.Y. Acad. Sci. 33:309-325.

Stuchly, M.A., Stuchly, S.S. (1983) Industrial,, scientific, medical and domestic applications of microwaves, IEEE Proc. A 130:467-503.

Swicord, M.L. et al. (1983) Chain-length-dependent microwave absorption of DNA., Biopolymers 22:2513-2516.

Szmigielski, S., Szudzinski, A., Pietraszek, A., Bielec, M. Wrembel, J.K. (1982) Accelerated development of spontaneous and benzopyrene-induced skin cancer in mice exposed to 2450 MHz microwave radiation, Bioelectromagnetics 3:179-191.

Takashima, S. et al. (1979) Effects of modulated RF energy on the EEG of mammalian brains: effects of acute and chronic irradiations, Radiat. Environ. Biophys. 16:15-27.

UNEP United Nations Environment Programme (1992) Environmental Health Criteria Nr. XX, Electromagnetic fields (300 Hz to 300 GHz). WHO Geneva (in press).

Wachtel, H., Seaman, R., Joines, W. (1975) Effects of low-intensity micro waves on isolated neurons, Ann. N.Y. Acad. Sci. 247:46-62.

ELF AND RF ELECTROMAGNETIC SOURCES

Jürgen H. Bernhardt and Rüdiger Matthes

Institute of Radiation Hygiene
Federal Office for Radiation Protection
D 8042 Munich-Neuherberg

1.   Introduction

In the last few decades the use of devices that emit electromagnetic fields has increased considerably . This proliferation has been accompanied by an increased concern about possible health effects of exposure to these fields. This chapter surveys sources of electromagnetic fields, both natural and human-made, in the 50 Hz to 300 GHz frequency range. The human-made electromagnetic environment consists of electromagnetic fields that are produced either intentionally or as by-products of the use of other devices. Human-made sources in the spectrum considered here, however, produce local field levels many orders of magnitude above the natural background. For all practical purposes of hazards assessment, therefore, the electromagnetic fields on the earth's surface arise from human-made sources

2.   Aspects of exposure assessment

Similar to other forms of energy, electromagnetic energy has the potential to interact with biological systems. The outcome may be of no significance, may cause different degrees of harm, or may be beneficial.
The electromagnetic fields are quantified in terms of electric field strength (E), expressed as volts per metre (V/m) and magnetic field strengths (H), expressed as amperes per metre (A/m). Often the magnetic flux density (B) is used to describe the magnetic field generated by currents in conductors. B is expressed in units of tesla (T); 1 A/m is corresponding 1.256 μT. E and H are vector fields, i.e. they are characterized by magnitude and direction at each points. In the far field of an antenna, the high frequency electromagnetic field is often quantified in terms of power flux density (S), expressed in units of watt per metre squared (W/m^2). Reviews of dosimetric concepts, quantities and units are

available (INIRC, 1985; NCRP, 1981), and can be found in other chapters of this book.

The basic dosimetric quantities which are of primary significance when considering the biological effects of exposure are
- for frequencies up to several MHz,the electric field strengths within the body (or, as a substitute, the current density, which is connected to the field strength via the conductivity),
- the electric current, to describe the effects of secondary effects such as shocks and burns (for frequencies below 100 MHz), and
- for frequencies above 10 MHz, the specific absorption rate or SAR (in units of W/kg); SAR may be specified as a value normalized over the whole body or over a small volume of tissue ("localized SAR").

For practical purposes of exposure assessment the measurement of the external electric and magnetic field strength and power flux density is necessary. The relationship between the internal basic dosimetric quantities and the derived practical dosimetric quantities is complicated by the interactions of electromagnetic fields with objects. In general, when electromagnetic fields interact with biological objects, a portion of the incident energy is reflected, external electric fields are distorted, a part of the energy is absorbed, and some is transmitted. The proportion of energy transmitted, absorbed, distorted or reflected depends on the frequency and polarization of the field, and the electrical properties and shape of the object. A superimposition of the incident and reflected waves results in standing waves, and spatially non-uniform field distribution. Since waves are totally reflected from metallic objects, standing waves are formed close to such objects. Since interactions of electromagnetic fields with biological systems depend on many different field characteristics, and because of the complexities of exposure fields encountered in practice, the following factors should be considered in describing exposure to electromagnetic fields:

- both, the electric and magnetic field strengths for the near-field, or one of these or the power flux density for the far-field;

- spatial variations of the magnitude of the field(s);

- field polarization, i.e. the direction of the electric field with respect to the direction of wave propagation or with respect to the orientation of the object.

3. Natural Sources of Exposure

The natural electromagnetic environment originates from the properties of the Earth, from processes such as discharges in the earth's atmosphere (terrestrial sources) or in the sun and deep space (extra-terrestrial sources). Atmospheric fields of frequencies less than 30 MHz originate predominantly from thunderstorms. Their strengths and range of frequencies vary widely with geographical location, time of day and season. Overall, atmospheric fields have an emission spectrum with the largest amplitude components having frequencies of between 2 and 30 kHz. Generally, the atmospheric field level decreases with increasing frequency. The geographical dependence is such that the highest levels are observed in equatorial areas and the lowest in polar areas.

The electric fields of the Earth consist of a static component, which is dominant, and a time-varying component, which is smaller than the static component by several orders of magnitude. The natural electric field strength at the power frequencies of 50 Hz or 60 Hz is about 10^{-4} V/m, which means that fields in the close vicinity of high-voltage (HV) transmission lines are 10^8 times stronger, and the fields introduced into homes by wiring or appliances are still about 10^3 - 10^6 times stronger than the natural background (UNEP, 1984).

The natural electric field near the Earth's surface is a static field of about 130 V/m. This is due to a separation of electric charges between the atmosphere and the ground. Daily changes in the natural electric field are attributed to factors, such as thunderstorms, that affect the rate of charge transfer between the ground and the upper atmosphere. With thunderstorms, electric fields of 3 - 20 kV/m have been observed.

The alternating fields at low frequency are related to thunderstorm activity and magnetic pulsations that produce currents within the Earth. The strength of the Earth's electric field varies in time and over the frequency range 0.001 - 5 Hz. In the frequency range of 5 - 1 000 Hz field strengths of 10^{-4} - 0.5 V/m can be measured. These fields are related to atmospheric changes (atmospherics), and their amplitudes decrease with increasing frequency.

The natural magnetic field consists of one component due to the earth acting as a permanent magnet as well as several other small components which differ in characteristics and are related to such influences as solar activity and atmospheric events. The earth's magnetic field originates from electric current flow in the upper layer of the earth's core. There are significant local differences in the strength of this field. At the surface of the earth, the vertical component is maximal at the magnetic poles, amounting to about 6.7 x 10^{-5} T (67 μT) and is zero at the magnetic equator. The horizontal component is maximal at the magnetic equator, about 3.3 x 10^{-5} T (33 μT), and is zero at the magnetic pole.

The naturally occurring time-varying fields in the atmospheric have several origins, including diurnally varying fields of the order of 3 x 10^{8} T associated with solar and lunar influences on ionospheric currents (UNEP, 1987). The largest time-varying atmospheric magnetic fields arise intermittently from intense solar activity and thunderstorms, and reach intensities of the order of 5 x 10^{7} T during large magnetic storms.

The natural RF radiation result from terrestrial emissions and from extra terrestrial fields.

The earth emits electromagnetic radiation (black body radiation) as do all media at a temperature T which is different from absolute zero. In the RF range the black body radiation follows the Rayleigh-Jeans law and the thermal noise from the earth (T about 300 K) is 0.003 W m^{-2} when integrated up to 300 GHz (UNEP, 1992).

The human body also emits electromagnetic fields at frequencies of up to 300 GHz at a power density of approximately 0.003 W m^{-2}. For a total body surface area of about 1.8 m^2 the total radiated power is approximately 0.0054 W.

The atmosphere, ionosphere and magnetosphere of the earth shield it from extra-terrestrial sources of electromagnetic energy. Electromagnetic waves which are able to penetrate this shield are limited to frequency windows, one optical and the other encompassing radiowaves of frequencies from about 10 MHz to 37.5 GHz. The short-wave boundary of the RF-window is due to energy absorption by molecules contained in the atmosphere (primarily O_2 and H_2O), whereas the long-wave boundary is related to the shielding action of the ionosphere.

RF radiation of cosmic origin observed with earth satellites ranges in magnitude from 1.8 x 10^{-20} W m^{-2} Hz^{-1} at 200 kHz to 8 x 10^{-20} W m^{-2} Hz^{-1} at 10 MHz.

4. Environmental and home related exposures

The time-varying electromagnetic fields originating from man made sources generally have much higher intensities than the naturally occuring fields. Although general public exposure to electromagnetic fields must be considered at every part of the spectrum, the most dominant exposure is due to the distribution and the domestic use of electric energy with frequencies of 50 or 60 Hz. Typical exposure levels are shown in the Tables 1 and 2.

Beside this dominant exposure in the ELF region, electromagnetic exposure may result from high frequency sources such as microwave ovens, telecommunication equipment, TV and broadcasting stations, security and surveillance systems and traffic radars. Typical exposure levels are shown in Table 3.

In contrast to high voltage overhead transmission lines which represent an extremely extended source where the decrease of field strength down to 10% of the peak value needs distances of more than 50 m depending on the technical design, most of the other common sources of electromagnetic fields are relatively small and low powered. Therefore a relevant exposure is only expected in the extreme vicinity of these devices. The radiation power density from a microwave oven, for example, has declined to about 1/10 th at a distance of 30 cm. For other appliances that produce magnetic flux densities up to a few mT at a distance of 3 cm, the decline of the field strength at 30 cm may be more than 1/100 th depending on the structure of the source. Relevant exposure levels (e.g. in mobile telecommunication) may occur within distances of a few cm up to about half a meter depending on technical details such as power, frequency and antenna impedance.

The average ELF exposure in homes with a 115 V electric installation, ranges from 0.05 up to over 10 μT in the case of electric heating with high values of some mT in the close vicinity of some appliances. The electric field may range from ambient 1 - 10 V/m up to more than 500 V/m near appliances. In countries with higher nominal voltage of the domestic installation electric fields are higher but magnetic fields from comparable appliances may be lower.

High frequency exposure at home is caused by microwave cooking with average power densities of approximately 0.5 W/m² or less or from mobile telefone in the frequency range from 20 MHz to 1 GHz. In this case close to the antenna power densities of more than 1000 W/m² are possible (measured with electric field probes). The environmental exposure caused by TV or broadcasting services is in general low, less than 0.05 mW/m²; only in some extreme rare cases this exposure may reach some hundred mW/m² (Tell and Mantiply, 1980, Table 3).

5. Exposures at the workplace

Levels of occupational exposure vary considerably, and are strongly dependent upon the particular application.

5.1 50/60 Hz electric and magnetic field sources

The principal man-made sources of 50/60 Hz electric and magnetic fields are high voltage transmission lines, and all devices containing current carrying wires, operating at power frequencies of 50 Hz in most countries and of 60 Hz in North America.

The electric fields associated with high voltage transmission lines, substations, and certain industrial equipment are the strongest electric fields to which persons, animals, and plants may be routinely exposed.

Magnetic fields associated with overhead transmission lines are usually relatively low and have so far attracted less attention. Stronger magnetic fields occur near large magnets, electrical furnaces or other devices using high currents. The potential use of cryogenic or superconducting transmission facilities would introduce another localized source of strong magnetic fields.

Utility employees working in substations or for maintenance of transmission lines form a special group continually exposed to large fields.

Small selected groups of workers in the electrochemical industry and in research laboratories may be exposed to high electric and magnetic field strengths. For instance, near induction furnaces and industrial electrolytic cells magnetic flux densities can be measured as high as 50 mT.

Occupational exposure to magnetic fields stems predominantly from working near industrial equipment using high currents. Such devices include various types of welding machines, electroslag refining, various furnaces, induction heaters, and stirrers. Details of surveys of magnetic field strengths in industrial settings are given in Table 1. Information on average and maximum values are given by taking notice of the fact that magnetic fields often are very inhomogeneous. Surveys on induction heaters used in industry show magnetic flux densities at operator locations ranging from 0.0007 to 6 mT, depending on the frequency used and the distance from the machine. In their study of magnetic fields from industrial electrosteel and welding equipment, Lövsund et al. (1982) found that spot welding machines and ladle furnaces produced fields up to 10 mT at distances up to 1 m.

Conductor height, geometric configuration, lateral distance from the line, and the voltage of the transmission line are by far the most significant factors in considering the maximum electric field strength at ground level. At lateral distances of about twice the line height, electric field strength decreases with distance in an approximately linear fashion (Zaffanella and Deno, 1978).

Tell (1983) reported results of extensive calculations and measurements of electric field strengths near ground level under extra-high-voltage-electric power lines. Both calculated and measured data are given showing an overall good agreement. Nominal maximum values of electric field of about 10 kV/m can exist beneath 765 kV lines at 1 m above ground.

Inside buildings near HV transmission lines, the field strengths are typically lower than the unperturbed field by a factor of about 10-100, depending on the structure of the building and the type of materials.

Occupational exposures that occur near high voltage transmission lines depend on the worker's location either on the ground, or at the conductor during live-line work at high potential. Typical exposure levels in HV-generating stations are shown in Table 2. When working under live-line conditions, protective clothing may be used to reduce the electric field strength and current density in the body to values similar to those that would occur for work on the ground. Protective clothing does not weaken the influence of the magnetic field.

In the absence of ferromagnetic materials, the magnetic field lines are solenoidal; that is, they form concentric circles around the conductor. Apart from the geometry of the conductor, the maximum magnetic flux density is determined only by the magnitude of the current. The magnetic field beneath HV transmission lines is directed mainly transverse to the line axis. The maximum flux density at ground level may be under the

Table 1: Typical exposure levels from 50/60 Hz magnetic field sources (from Allen, 1991, Bernhardt, 1988, Krause, 1988, Lövsund et al., 1982 and Stuchly, 1986)

Magnetic Field Source	Magnetic Flux Density	
	Average Level mT	Peak Level mT
Earth	10^{-9}	
High Voltage AC power transmission		
- Overhead transmission lines (380 & 765 kV)	0.01 - 0.04	0.4 (during failure)
- 380 kV, 25 m from midspan	0.008	
- Generating stations	0.02 - 0.04	0.27
Office and Household		
- Standard dwellings	10^{-5} - 10^{-3}	10^{-3} - $4\ 10^{-2}$
- Residence with electric heating	$1.2\ \ 10^{-2}$	-
- 30 cm from appliances		10^{-2} - 1
- 30 cm from VDU's	$8\ \ 10^{-4}$	0.005-0.018
Industry and Medicine		
- Electrolytyc processes	1 - 10	50
- Welding machines	0.1 - 5.8	130
- Induction heating	1 - 6	25
- Ladle furnace	0.2 - 8.0	-
- Arc furnace	up to 1	-
- Electroslag welding	0.5 - 1.7	-
- Therapeutic equipment	1 - 16	-

centre line or under the outer conductors, depending on the phase relationship between the conductors. The maximum magnetic flux density at ground level for a double circuit 500 kV overhead transmission lines system is approximately 35 μT per kiloampere.

Typical values for the magnetic flux density at working places near overhead lines, in substations and in power stations (16 2/3, 50, 60 Hz) range up to 50 μT (Krause, 1986).

5.2 Induction Heating

Magnetic fields below a few MHz are used in industry for induction heating of metals and semiconductors. Such heating is used mainly for forging, annealing, tempering, brazing and soldering. The equipments usually operate from 50 or 60 Hz up to about 10 kHz. Surveys of the magnetic field strength to which the operators are exposed have shown that these exposures are, in many instances, high compared with recommended exposure limits (Stuchly & Lecuyer, 1985; Stuchly, 1986; Stuchly & Lecuyer, 1989; Conover et al., 1986)

Since the dimensions of coils producing the magnetic fields are often small there is seldom high-level whole-body exposure, but rather local exposure of, for instance, the hands. Magnetic flux density to the hands of the operator may reach 25 mT (Lövsund & Mild, 1978; Stuchly and Lecuyer, 1985). In most cases the flux density is less than 1 mT. The electric field strength near the induction heater is usually low.

5.3 Dielectric Heating

Dielectric or RF heaters are widespread in many industries. RF energy produces heat directly within the processed material. The technique is commonly used in sealing plastics, drying glue for joining wood, etc. According to several surveys sealing or welding of PVC is the most common use for RF dielectric heaters. All RF heaters have a higher efficiency in comparison with conventional ovens and are operating at one of the ISM-frequencies, usually at 13.56, 27.12 or 40.68 MHz.

However, during the operating cycle this value may vary by several megahertz. The RF output power ranges from fractions of a kilowatt to about 100 kW. The relatively high output power, and the use of unshielded electrodes can produce relatively high fields around RF heaters. Surveys in the United States have indicated that 60% of the devices measured exposed the operator to E fields greater than 200 V/m, and 29% to H fields greater than 0.5 A/m (Conover et al. 1980). Similar results were obtained by Stuchly et al. (1980). The stray fields are localized in the immediate vicinity of the sealers, so that exposure of the body is highly inhomogeneous.

Since exposure of heater sealer operators and other personnel working in the same area takes place in the near-field, both E and H field strengths must be measured in the absence of the operator to evaluate exposure .

Table 2: Typical exposure levels from 50/60 Hz electric field sources (from Allen 1991, Bernhardt 1988, Grandolfo et al. 1985, UNEP/WHO/IRPA 1984)

Electric field source	Electric field strength kV/m
Earth	10^{-7}
High voltage AC power transmission - Overhead transmission lines	
110 kV	1-2 kV/m
245 kV	2-3 kV/m
380 kV	5-6 kV/m
	1 kV/m 25 m from midspan
800 kV	10-12 kV/m
- Generating stations	
110 kV	5-6 kV/m
245 kV	9-10 kV/m
380 kV	14-16 kV/m
800 kV	14-16 kV
Office and household levels - Standard dwellings	
30 cm from appliances	10 - 500 V/m
ambient, distant appliances	1-10 V/m
- 30 cm from VDU's, DC	0.1-30 kV/m
ELF	0.030-1.5 kV/m

levels. However, to demonstrate compliance with basic restrictions on exposure the development of body current measurement techniques should prove to be useful (Allen et al, 1986). For a complete assessment of the potential risk of industrial applications it is necessary to take also into account: significant harmonic components of the fundamental frequency, RF field leakages due to electrical connection of RF generator and cabinet to earth, and, furthermore, low frequency amplitude modulation of the RF carrier field. There is major concern for operators working with RF heaters, particularly since such heaters operate at frequencies in or near the region of human resonant absorption for a person in electrical contact with the ground. Hence, workers near some of these devices absorb RF energy at rates above the recommended limits. To reduce stray fields to acceptable levels it is necessary to shield the electrode system, divert the RF energy, or reduce exposure by a administrative measures (Eriksson and Mild, 1985).

5.4 Communication purposes and radar

While most communication and radar workers are exposed to only relatively low-intensity fields, some can be exposed to high levels of RF. Workers climbing FM radio or TV broadcast towers may be exposed to E fields up to 1 kV m^{-1} and H fields up to 5 A m^{-1} (Mild & Lovstrand, 1990)

High exposures can arise transiently in close proximity to lower power mobile sources such as CB or mobile radio antennas. E-field strengths of 200 to 1350 V/m have been measured 2 cm from low power mobile antennas (27-450 MHz, Allen, 1991).

Radar Systems produce strong EM fields on the axis of the antenna. However, in most systems, average field strengths are reduced typically by a factor of 100 to 1000 because of antenna rotation. With stationary antennas, which represent the worst case, peak power flux densities of 10 MW m^{-2} may occur on the antenna axis up to a few meters from the source. In areas surrounding air traffic control radars (ATCRs), workers can be exposed to power flux densities of up to tens of W m^{-2}, but normally those encountered are in the range 0.03 - 0.8 W m^{-2}. In an exposure survey of civilian airport radar workers in Australia it was found that, unless working on open waveguide slots, or within transmitter cabinets when high voltage arcing was occurring, personnel were, in general, not exposed to levels of radiation exceeding the specified limits (Joyner & Bangay, 1986).

6. Medical practice

Patients suffering from bone fractures that do not heal well or unite have been treated with pulsed magnetic fields (Bassett et al., 1982). Various devices generating magnetic field pulses are used for bone growth stimulation. A typical example is the device that generates an average magnetic flux density of about 0.3 mT, a peak strength of about 2.5 mT

Table 3: Typical exposure levels from electromagnetic field sources for the general public.

Source	Frequency	Distance	Exposure levels	Remarks
Microwave-oven	2.45 GHz	0.3 m 0.5 m 1 m	< 5 W/m^2 < 2 W/m^2 < 1 W/m^2	Worst case assumed, i.e., 50W/m^2 in 5 cm distance is approached
Traffic radar	9 - 35 GHz	3 m 10 m	< 250 mW/m^2 < 10 mW/m^2	Power 0,5 - 100 mW
Security systems	0,9 - 10 GHz		< 2 mW/m^2	Within system
Walkie-Talkies	27 MHz	5 cm 12 cm	up to 1000 V/m up to 0.2 A/m up to 200 V/m up to 0.1 A/m	Power:several watts
Powerful RF-Stations FM VHF-TV	87.5 - 108 MHz 47 - 68 MHz and 174 - 230 MHz	~ 1.5 km ~ 1.5 km	< 50 mW/m^2 < 20 mW/m^2	Power up to 100 kW 100 - 300 kW
UHF-TV	470 - 890 MHz	~ 1.5 km	< 5 mW/m^2	up to 5 MW
Short wave	3.95 - 26.1 MHz	220 m 50 m	27,5 V/m or 2 W/m^2 121 V/m or 40 W/m^2	Power 750 kW
AM and Longwave	130 - 285 kHz 415 - 1606.5 kHz	300 m 50 m	90 V/m 450 V/m	Power 1,8 MW
HF-Exposure in cities RF and TV-Stations	1 - 1000 MHz		> 200 mW/m^2 > 10 mW/m^2 > 0.05 mW/m^2 > 0.02 mW/m^2	US-Population (1980) 0.02 % 1 % 50 % 90 %
Radar-Stations	1 - 10 GHz	0.1 - 1 km > 1 km	0,1 - 10 W/m^2 < 0.5 W/m^2	Power 0.2 - 20 kW

Applications of magnetic field devices in medicine are rapidly expanding. Own measurements yielded average levels of 1-16 mT within treatment coils at frequencies of 12, 15, 20, 50 or 75 Hz.

The earliest therapeutic application of radiofrequency electromagnetic fields was in diathermy. Two types of diathermy are commonly used, short-wave (usually at about 27 MHz) and microwave. Only a part of the patient's body is exposed to RF energy and exposure duration is limited (typically 15-30 minutes). However, exposure intensity is high and sufficient to cause a sustained increase in tissue temperature. Exposures to operators of short-wave diathermy devices may exceed 60 V/m and/or 0.16 A/m for operators standing in their normal positions (in front of the diathermy console) for some treatment regimes. Stronger fields are encountered close to the electrodes and cables (Stuchly et al., 1982). In the "worst case" (i.e., high power settings), the exposure limits for the staff may be exceeded at distances lower than 1.5 - 2 m (27.12 MHz) or 1 m (433 MHz and 2.45 GHz, Veit and Bernhardt, 1984).

Electromagnetic fields have been also successfully used in inducing local hyperthermia for cancer therapy. As in diathermy, the patient is exposed to intense fields for a short time. There is relatively little information on operator exposure. One of the devices around which exposure fields have been measured operates at 13.56 MHz and employs coils for heating the torso, neck or thigh. Depending on the power to the coil and the coil type, exposure may exceed the recommended exposure limits, when the operator is 25 cm-1 m from the coil.

In medical diagnosis, RF fields are used in conjunction with static and slowly varying magnetic fields in magnetic resonance imaging (MRI). Systems used at present operate at frequencies ranging from about 6 to 100 MHz. In Canada, the Federal Republic of Germany, the United Kingdom and the United States guidelines have been issued that limit patient exposure. The RF field in MRI devices is almost fully contained within the MR-system. The exposure level of the patient is up to 1 W/kg (averaged SAR-value for the whole body). The exposure of the operator is negligible. Typical exposure levels from electromagnetic fields for medical applications are summarized in Table 4.

Table 4:   Typical exposure levels from electromagnetic fields for medical applications

Source	Frequency (MHz)	Distance (m)	Exposure level	Remarks
Shortwave Diathermia	27,12	0,2 0,5 1	up to 1000 V/m or 2500 W/m^2 up to 500 W/m^2 up to 140 V/m or 50 W/m^2 up to 0.40 A/m or 50 W/m^2	Staff
	27,12		100-1000 V/m or 25-2500 W/m^2 up to 1.6 A/m or 1000 W/m^2	Patient, untreated body parts
Microwave Treatment	433 2450	0,5 1 0.3 - 3	25 W/m^2 10 W/m^2 50-200 V/m or 6-100 W/m^2	Staff
	433 2450		20 - 140 W/m^2	Patient, untreated body parts
Magnetic Resonance (MR)	60 - 100	within MR-System	up to 1 W/kg	Averaged value for the whole body
Therapeutic equipment, ELF magnetic fields	12; 15; 20; 50; 75 Hz	within coil	1 - 16 mT	Patient, treated body parts

References

Allen, S.G., Blackwell, R.P. and Unsworth, C. (1986). Field intensity measurements, body currents, specific absorption rates and their relevance to operators of dielectric PVC welding machines, Proc of BNCE Conf. on Heating and Processing 1-3000 MHz, St. John's College Cambridge (London: BNCE).

Allen, S.G. (1991). Radiofrequency field measurements and hazard assessment. J. Radiol. Prot. 11: 49-62.

Bassett, C.A.L., Mitchell, S.N. and Gaston, S.R. (1982). Pulsing electro magnetic field treatment in ununited fractures and failed artrodeses. J.Am. Med. Assoc., 247: 623-628.

Bernhardt, J.H. (1988). Extremely Low Frequency (ELF) Electric fields. Bernhardt, J.H. Extremely Low Frequency (ELF) Magnetic fields. In: Non-Ionizing Radiation: Physical Characteristics, Biological Effects and Health Hazard Assessment. M.H. Repacholi (ed.). Proceedings of the International Non-Ionizing Radiation Workshop, Melbourne, 5.-9. April 1988. Available from: Austrialian Radiation Laboratory, Lower Plenty Road, Yallambie, Victoria, Australia: 235-253 and 273-289.

Conover, D.L. et al. (1980). Measurement of electric and magnetic-field strengths from industrial radiofrequency (6-38 MHz) plastic sealers. Proceedings of the Institute of Electrical and Electronics Engineers, 68: 17-20.

Conover, D.L., Murray, W.E., Lary, J.M. and Johnson, P.H. (1989). Magnetic field measurements near RF induction heaters. Bioelectromagnetics 7: 83-90.

Eriksson, A. and Mild, K.H. (1985). Radiofrequency electromagnetic leakage fields from plastic welding machines. Measurements and reducing measures. Journal of Microwave Power, 20: 95-107.

Grandolfo, M., Michaelson, S.M. and Rindi, A. (1985). Biological effects and dosimetry of static and ELF electromagnetic fields. Plenum Press, New York and London.

INIRC (1985). International Radiation Protection Association/ International Non-Ionizing Radiation Committee. Review of concepts, quantities, units and terminology for non-ionizing radiation protection. Health Phys. 49: 1329-1362.

Joyner, K.H. and Bangay, M.J. (1986). Exposure Survey of Civilian Airport Radar Workers in Australia, J. Microwave Power, 21: 209-219.

Krause, N. (1986). Exposure of people to static and time variable magnetic fields in technology, medicine, research and public life; dosimetric aspects. In: "Biological Effects of Static and ELF-Magnetic Fields", BGA-Schriftenreihe 3/86, (ed. J.H. Bernhardt), MMV Medizin Verlag München: 57-71.

Lövsund, P. and Mild, K.H. (1978). Low frequency electromagnetic field near some induction heaters. Stockholm Board of Occupational Health and Safety (in Swedish).

Lövsund, P., Oberg, P.A. and Nilsson, S.E.G. (1982). ELF magnetic fields in electrosteel and welding industries, Radio-Science, 17 (5S): 355-385.

Mild, K.H. and Lovstrand, K.G. (1990). Environmental and Professionally Encountered Electromagnetic Fields. In: Gandhi O.P., ed. Biological effects and medical applications of electromagnetic fields, Engelwood Cliffs, New Jersey, Prentice Hall Inc.

NCRP (1981). Radiofrequency Electromagnetic Fields. Properties, Quantities and Units, Biophysical Interaction, and Measurement NCRP Report 67 (Washington, DC: National Council on Radiation Protection and Measurements).

Stuchly, M.A. et al. (1980). Radiation survey of dielectric heaters in Canada. Journal of Microwave Power, 15: 113-121.

Stuchly, M.A. Repacholi, M.H., Lecuyer, D.W. and Mann, R.D. (1982). Exposure to the operator and patient during short wave diathermy treatments. Health Physics, 42: 341-366.

Stuchly, M.A. and Lecuyer, D.W. (1985). Induction heating and operator exposure to electromagnetic fields. Health Physics, 49: 693-700.

Stuchly, M.A. (1986). Exposure to static and time-varying magnetic fields in industry, medicine, research and public lilfe; dosimetric aspects. In: "Biological effects of static and ELF-magnetic fields", BGA-Schriftenreihe 3/86, (ed. J.H. Bernhardt), MMV Medizin Verlag München: 39-56.

Stuchly, M.A. and Lecuyer, D.W. (1989). Exposure to electromagnetic fields in arc welding. Health Physics, 56: 297-302.

Tell, R.A. and Mantiply, E.D. (1980). Population exposure to VHF and UHF broadcast radiation in the United States. Proc. IEEE, 68: 6-12.

Tell, R.A. (1983). Instrumentation for measurement of electromagnetic fields: equipment, calibrations, and selected applications. In: Biological Effects and Dosimetry of Nonionizing Radiation, Radiofrequency and Microwave Energies, eds M. Grandolfo, S.M. Michaelson and A. Rindi (New York: Plenum).

UNEP (1984). United Nations Environmental Programme/World Organization/International Radiation Protection Association. Environmental Health Criteria 35. Extremely low frequency (ELF) fields. Geneva: World Health Organization.

UNEP (1987). United Nations Environmental Programme/World Organization/International Radiation Protection Association. Environmental Health Criteria 69. Magnetic fields. Geneva: World Health Organization.

Veit, R. and Bernhardt, J.H. (1984). Strayfield measurements around microwaves and shortwave-diathermy devices in the Federal Republic of Germany. In: Radiation-Risk-Protection, Proceedings of the 6th International Congress of IRPA, Eds.: Kaul, A.; Neider, R.; Pensko, J.; Stieve, F.-E.; Brunner, H., Verlag TÜV Rheinland GmbH, Köln, Vol. III, 1323-1326.

Zaffanella, L.E., Deno, D.W. (1978). Electrostatic and electromagnetic effects of ultra-high-voltage transmission lines. Palo Alto, CA: Electric Power Research Institute Final report, EPRI EL-802.

ELECTROMAGNETIC FIELD DOSIMETRY

Martino Grandolfo

National Institute of Health
Department of Physics
Rome, Italy

In a broad sense the term dosimetry is used to quantify an exposure to radiation. Quantitative descriptions of an exposure to radiation, for the purpose of formulating protection standards and exposure limits (Duchêne et al., 1991), require the use of adequate quantities; that means that the quantities should represent, as well as possible, those physical processes that are closely linked to the biological effects of radiation.

Whether such quantities can simply be related to the amount of energy imparted to tissue such as in the case of ionizing radiation, or whether other quantities such as field strengths are more appropriate, is one of the most important questions in non ionizing radiation (NIR) dosimetry.

In the case of NIR, different characteristics of physical interaction mechanisms, measurement conditions and techniques, as well as the limited knowledge of biological response mechanisms have led to a diversity of quantities used for the specification of exposure.

In general, it can be stated that, across the NIR spectrum, the temporal characteristics of exposure are of critical importance, and the contributions of ambient factors such as temperature must be taken into account. For example, if the emphasis is on the limitation of thermal effects, many data in the radiofrequency (RF) region at present seem to support the introduction of exposure limits that are based on the rate of energy deposition, as opposed to absorbed dose in the case of ionizing radiation (IRPA/INIRC, 1988).

A clear understanding of the biological responses of RF and microwave (MW) radiation (Franceschetti et al., 1989) must be based on a quantitative understanding of the relationship between the field impinging on the subject, the energy absorbed in the tissues, and the observed effect. However, only the fields inside the tissues and biological bodies can interact with them, and therefore it is necessary to determine these fields for any meaningful and general quantification of biological data obtained experimentally. Internal fields and absorbed energy, in principle, can be directly measured *in vivo* by implanted probes transparent to electromagnetic radiation. However, accurate, reliable implanted probes still remain to be developed. Currently, there are two distinctive approaches toward quantification of induced fields and absorbed energy. These are the whole-body absorption and the spatial distribution of absorbed energy. While each has its own merit, clarification of the biological responses and hazards of RF and MW radiation demands knowledge of both in most situations.

Dosimetry in bioelectromagnetic research has been developing in two interacting and complementary lines: experimental and theoretical. Experimental dosimetry concerns the development of methods and instruments suitable for: (i) measurements of the total specific absorption rate (SAR) and its distribution in live or dead experimental animals, (ii) measurements of the average SAR and SAR distribution in models of animals and humans, being SAR the time rate of energy absorption per gram of tissue, and (iii) development of exposure devices for animals and *in vitro* preparations in which quantification of the exposure conditions and internal fields is possible. Experimental dosimetry enables comparison of experiments conducted under different exposure conditions, and for different species, and plays an important role in verifying the predictions of theoretical dosimetry. Employing the physical scaling principle, dosimetric data can be used to extrapolate the results of animal experiments to human exposures. For instance, on the basis of the scaling principle, the equivalent human exposure at a frequency f_p for a person of a height L_p and an animal of a height L_a exposed at a frequency f_a is given by the relationship $f_p = f_a L_a / L_p$, provided that both have the same orientation in respect to the exposure field (M. Stuchly and S. Stuchly, 1986; Michaelson and Lin, 1987). Theoretical dosimetry deals with the analysis of simplified models, and several methods have been described in the literature for numerical calculations of SAR and temperature distributions in the human body in response to electromagnetic exposures. Although experiments must be done to verify the accuracy of any theoretical procedure, the numerical methods allow a detailed modeling of the anatomically relevant human inhomogeneities that are not easy to model experimentally. Maxwell's equations have been solved in both the integral and differential equation forms for the distribution of electric and magnetic fields and mass-normalized rates of energy deposition. To simplify computational requirements, several authors have used two-dimensional formulations for individual cross sections of the body, but several techniques have been also developed that lend themselves to high-resolution 3-D modeling of the body for either far-field or near-field electromagnetic exposures (Gandhi, 1990).

Quantities and units that have been selected for the specification of electromagnetic field dosimetry are described in the following section.

Quantities and units

Physical quantities are used to describe and characterize physical phenomena in a quantitative way. For the purpose of radiation protection, physical quantities are needed (i) to describe sources and fields of radiation as well as (ii) the interaction of radiation with matter. Some quantities have a special significance, because they may be needed to describe the exposure of the human body to non-ionizing radiation and to estimate the absorbed energy and its distribution inside the body (dosimetric quantities).

Since the early 70's, many different concepts and quantities have been proposed and used to characterize propagation and absorption of electromagnetic energy in biological systems (Justesen, 1975).

Power density is the time rate of energy flow per unit area across a surface and is given by the value of the Poynting vector in watts per square metre (SI unit), milliwatts per square centimetre, or microwatts per square centimetre. It expresses exposure in terms of incident power per unit area.

A dosimetric measure that has been widely adopted is the specific absorption rate, defined as the time derivative of the incremental energy, dW, absorbed by or dissipated in an incremental mass, dm, contained in a volume element, dV, of a given density ρ:

$$SAR = dW'/dt; \qquad W' = dW/dm = dW/(\rho dV) \qquad (1)$$

The SAR is expressed in watts per kilogram (W/kg) or derived units (e.g., mW/g). The average SAR is defined as a ratio of the total power absorbed in the exposed body to its mass. The local SAR refers to the value within a defined unit volume or unit mass, which can be arbitrarily small. The term whole-body SAR is used to specify whole-body absorbed-power density to recognize the fact that the same SAR for whole-body and local absorption does not necessarily produce the same degree of biological response.

The term SAR distribution is used to indicate the SAR pattern inside the body. Incident radiation, body geometry, and orientation as well as dielectric properties affect the SAR distribution. A highly nonuniform SAR distribution may result from a plane wave impinging on a biological object.

At lower frequencies, and under near-field irradiation conditions, the electric field strength and the magnetic field strength are used instead of power density to express the exposure. The field strengths may either be calculated or measured in the absence of the experimental object.

Furthermore, field strengths are useful measures for quantifying biological interactions that are field specific. In this case, the terms internal electric and internal magnetic field strengths may be used. They are expressed in volts per metre and amperes per metre, respectively. They are appropriate units for specifying incident as well as internal fields.

This compilation should not be considered exhaustive nor definitive, and it can not replace existing more specialized documents pertaining to non ionizing radiation, particularly radiofrequency and microwave radiation from a few kilohertz to 300 GHz (IRPA/INIRC, 1985). The fundamental quantities and SI units used in experimental and theoretical dosimetry are summarized in Table 1.

Energy absorption in humans and animals

Electromagnetic absorption cross section is defined as the power absorbed divided by the incident power density. Unlike the field of ionizing radiation, where the absorption cross section of a biological target is directly related to its geometrical cross section, the whole-body energy absorption has been shown to be strongly dependent on frequency, polarization, and physical environment, such as conductivity ground at other reflecting surfaces (Gandhi, 1974; Durney et al., 1975).

The knowledge about whole-body absorbed dose is well established for the following: (i) far-field irradiation conditions typically used for animal experiments (Durney et al., 1978, 1980, 1986); (ii) different frequencies and polarizations; (iii) isolated and regularly spaced multiple animals (Gandhi, 1980); (iv) some near-field exposure conditions (Chatterjee et al., 1980, 1982, 1985; Durney et al., 1980; M. Stuchly et al., 1986, 1987); and (v) parallel-plate radiations representative of RF sealers (Chen and Gandhi, 1989).

Table 1 - Quantities and units (SI) used in experimental dosimetry

Quantity	Symbol	Unit	Symbol
Conductivity	σ	siemens per metre	S/m
Current density	**J**	ampere per square metre	A/m^2
Electric field strength	**E**	volt per metre	V/m
Electric flux density, Displacement	**D**	coulomb per square metre	C/m^2
Energy	W	joule	J
Energy density	w	joule per cubic metre	J/m^3
Frequency	f	hertz	Hz
Magnetic field strength	**H**	ampere per metre	A/m
Magnetic flux density, Magnetic induction	**B**	tesla	T
Permittivity	ε	farad per meter	F/m
Power density, Energy flux density	S	watt per square metre	W/m^2
Poynting vector	**S**	watt per square metre	W/m^2
Propagation vector	**k**	Reciprocal metre	m^{-1}
Relative permittivity	ε_r	This quantity is dimensionless	-
Specific absorption	SA	joule per kilogram	J/kg
Specific absorption rate	SAR	watt per kilogram	W/kg
Wave impedance	Z	ohm	Ω

The distribution of absorbed dose, or *distributive dosimetry*, is an area where the knowledge is still in a fairly elementary stage, although the picture is improving dramatically with the development of new high resolution numerical procedures (see next Section) and a computer-controlled scanning system that can be used to determine internal electric field distributions in a full-scale human model (M. Stuchly et al., 1986, 1987; S. Stuchly et al., 1986, 1987).

The whole-body absorption of electromagnetic waves by biological bodies is strongly dependent on frequency and the orientation of the electric field **E** relative to the longest dimension L of the body. The highest rate of energy deposition occurs for **E**//**L**   (E-orientation) for frequencies such that the major length is approximately 0.36 to 0.40 times the free-space wavelength, λ, of radiation (Gandhi, 1974). Peaks of whole-body SAR for the other two polarizations (major length oriented along the direction **k** of propagation, **k**//**L** or k-orientation, or along the vector **H** of the magnetic field, **H**//**L** or H-orientation) have also been reported. The results obtained by theoretical analysis of the prolate spheroidal models of man, monkey and mouse (confirmed experimentally) are shown in Fig. 1 for an incident power density of 1 W/m^2 (M. Stuchly, 1983).

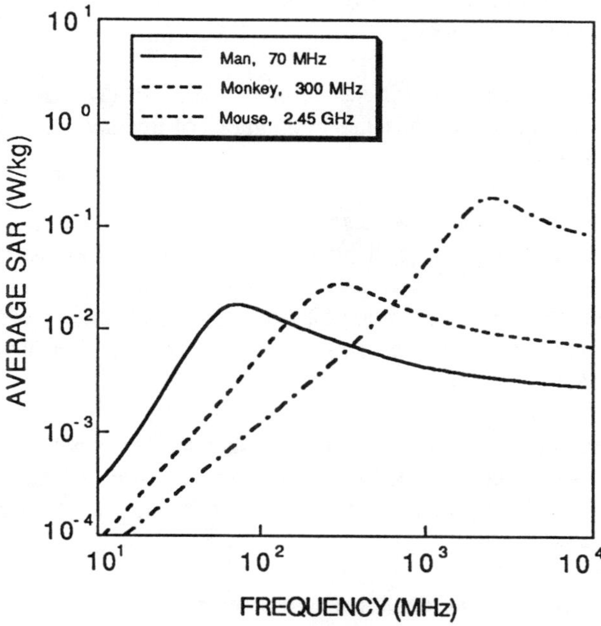

Fig. 1 - Average SAR values for some prolate spheroidal models (for 1 W/m^2 incident plane-wave field).

For the E-orientation the whole-body absorption curves may be discussed in terms of five regions (Gandhi, 1990):

(i) Frequencies well below resonance ($L/\lambda < 0.1$ to 0.2). An f^2-type dependence has been derived theoretically and checked experimentally by Durney et al. (1978).

(ii) Subresonant region ($0.2 < L/\lambda < 0.36$). An $f^{2.75}$ to f^3 dependence of total power deposition has been experimentally observed for this region (Gandhi et al., 1977; Hill, 1984).

(iii) Resonant region ($L/\lambda = 0.36\text{-}0.40$). The resonance frequency f_r (in MHz) is approximately $114/L$, where L is the height of the individual in meters.

(iv) Above-resonance region ($f_r < f < 7f_r$), where a whole-body absorption reducing as f_r/f from the resonance values has been experimentally observed.

(v) Frequencies well above resonance. The absorption cross section should asymptotically approach the *quasioptical* value, i.e. about one half the physical cross section.

Empirical equations for whole-body-averaged (WBA) specific absorption rates of human models under conditions of free-space irradiation are given in Table 2.

Minor peaks in the above-resonance region at 150 and at 350 MHz are ascribed to maxima of energy deposition in the various body parts such as the arm and the head, respectively. It can be seen that the highest SARs are calculated for the sections through the neck, knee, and ankle, the former because of a smaller cross section and the latter two because they also have a high bone content.

For a standing person with feet in conductive contact with a perfect ground, there is a drastic alteration in the whole-body-average SAR and its distribution. For E-polarization the new resonant frequency is lower than that given in Table 2.

Near-field exposure conditions are perhaps of greater interest than far-field irradiation conditions because they are encountered in industrial and occupational situations. A point to note is that, for near-field exposures corresponding to leakage-type fields such as those from industrial RF sealers, SARs considerably lower than those for far-field exposure conditions are observed.

Table 2 - Empirical equations for WBA-SARs of human models (E-polarization, free-space irradiation)

Resonant frequency, f_r (MHz) = 114/L(m)

For **subresonant region** $(0.5\ f_r < f < f_r)$:

SAR (in W/kg for 1 W/m^2 incident plane-wave field) =

$$= 0.52\ L^2\ (m)\ M^{-1}\ (kg)\ (f/f_r)^{2.75}$$

For **above-resonant region** $(f_r < f < 1.6\ S_{rcs}\ f_r)$:

SAR(in W/kg for 1 W/m^2 incident plane-wave field) =

$$= 59.5\ L\ (m)\ f^{-1}\ (MHz)\ M^{-1}\ (kg)$$

where L (m) is the long dimension of the body in meters, M (kg) is the mass in kilogram, and the relative absorption cross section is:

$$S_{rcs} = 4.8\ L\ (m)\ (10\ L\ (m)\ /M\ (kg))^{0.5}$$

Numerical methods

Several methods have been described in the literature for numerical calculations of SAR distributions. Although experiments must be done to verify the accuracy of any theoretical procedure, the numerical methods allow a detailed modeling of the anatomically relevant human inhomogeneities that are not easy to model experimentally.

For SAR calculations the most promising techniques at present are the finite-difference time-domain (FDTD) method (Spiegel, 1984; Spiegel et al., 1985; Sullivan et al., 1987, 1988) and the sinc-basis fast Fourier transform (FFT)-conjugate gradient (CG) method (Borup et al., in press) which is a modified version of the traditional methods of moments (Chen and Guru, 1977; Hagmann et al., 1979). For lower frequencies at which quasistatic approximations may be made, the impedance method has been found to be highly efficient numerically and has been used for a number of applications (Armitage et al., 1983; Gandhi et al., 1984; Gandhi and DeFord, 1988; Orcutt and Gandhi, 1988).

The method of moments (MOM) has been extensively used to calculate SAR distributions in block model representation of the human body (Livesay and Chen, 1974; Chen and Guru, 1977; Gandhi et al., 1979; Hagmann et al., 1979; Hagmann and Gandhi, 1979). In this procedure the fields $\mathbf{E}$ (r) within the body are solved from the solution of the electric field integral equation (EFIE) obtained from Maxwell's equations. Assuming that the electric field is constant over the volume of the individual cells, the EFIE can be expressed in the form of the matrix equation:

$$A \cdot E = E^i$$

where A is a 3Nx3N matrix that describes coupling between the various cells, $\mathbf{E}$ is a vector of 3N unknown field components, 3 for each of the N cells into which the body is divided, and $\mathbf{E}^i$ is a vector of 3N known field components that would be incident at the same N cells if the body did not exist. Solution of this matrix equation by conventional methods requires order $(3N)^2$ storage and order $(3N)^2$ to $(3N)^3$ computation steps, both being requirements that expand rapidly as the number of cells N is increased. Consequently, it has not been possible to apply the traditional MOM to block model representation of the human body with N larger than 1,132 (DeFord et al., 1983).

Perhaps the most successful and the most promising of the numerical methods for SAR calculations is the FDTD method. This method was first proposed by Yee (1966) and later developed by Taflove and colleagues (Taflove and Brodwin, 1975a,b; Taflove, 1980; Umashankar and Taflove, 1982), Holland (1977), and Kunz and Lee (1978). Recently it has been extended for calculations of the distribution of electromagnetic fields in a human model for incident plane waves (Spiegel, 1984; Spiegel et al., 1985; Sullivan et al., 1987, 1988) and for exposures in the near field (Chen and Gandhi, 1989; Wang and Gandhi, 1989).

As previously described in Sullivan et al. (1987, 1988), the model for the human body has been developed from information in the book *A Cross-section Anatomy*, by Eycleshymer and Shoemaker (1911). This book contains cross-sectional diagrams of the human body that were obtained by making cross-sectional cuts about 1 inch apart in human cadavers.

For low-frequency dosimetry problems in which quasistatic approximations may be made the impedance method has been found to be highly efficient numerically (Armitage et al., 1983; Gandhi et al., 1984; Gandhi and DeFord, 1988; Orcutt and Gandhi, 1988). In this method the biological body or the exposed part thereof is represented by a 3-D network of impedances whose individual values are obtained from the complex conductivities $\sigma + j\omega\varepsilon$ of the various body regions. The effect of the incident electric field is to introduce currents into the model given by $j\omega\varepsilon_0 E_{ext}$, where E_{ext} is the external electric field (incident + scattered) at the surface of the body. The contribution from the **H** field is to set up electromotive forces (emfs) in the circuit loops given from Faraday's law of induction by $j\omega\mu_0 H_n \delta^2$, where H_n is the magnetic field normal to the plane of the loop of area δ^2. Applying Faraday's law to each cell face and requiring the current continuity for each cell in the network, a system of equations is obtained that can be solved for the induced currents. From these, SAR can be calculated for individual cells and for different organs, based on the knowledge of tissue percentages in each cell.

Presently, it can be stated that the numerical models for SAR calculations have reached a level of sophistication that would have been unthinkable a few years ago. These models are consequently being applied to some very realistic situations both in assessing RF safety and for medical applications.

With the anatomically based human models the impedance method has been also used for evaluating SAR distributions for RF magnetic fields representative of magnetic resonance (MR) imagers.

Persons undergoing magnetic resonance imaging (MRI) examinations are subjected to three different magnetic fields, namely a static field, a time-varying gradient field, and a radiofrequency field. In order to assess biological effects resulting from the simultaneous exposure to these fields, it is necessary that the effects of each of them be well understood. As regards radiofrequency fields, the interaction mechanism with a living body is well known, the primary effect being the heating of tissues, with consequent biological effects related to the thermally induced changes in the body. The distribution of RF power inside the body is not uniform, due to the different electrical characteristics of tissues. The pattern of energy absorption can be evaluated through measurements on phantoms, or by numerical techniques which make use of realistic models of human body.

In the frequency region currently used in MRI diagnostic procedures, the absorption of RF energy is almost completely associated with the magnetic field. The energy dissipation depends on frequency, strength of the incident field, coupling efficiency between the RF coil and the specimen, and on several properties such as the mass density, size, and orientation of the specimen relative to the field polarity.

In the impedance method (DeFord and Gandhi, 1985; Orcutt and Gandhi, 1988), SAR is calculated from these data, using a realistic model of the human body composed of a given number of cells, each one fully characterized, from the electrical point of view, by its impedance. The latter is calculated from the known permittivity and conductivity of the tissues composing the cell. In this way, the body is replaced by a three-dimensional network of impedances.

Grandolfo et al. (in press) have applied the impedance method for the evaluation of SAR in a human torso subject to MR tomography in different procedures.

For modelling purposes, the torso was enclosed in an ideal parallelepiped divided into $18 \times 12 \times 23 = 4,968$ cubic cells of a 2.6 cm edge. It was considered as composed by 13 different tissues, whose distribution inside each cell was obtained through a digital scanning of anatomic diagrams of cross sections of the torso (DeFord and Gandhi, 1985).

The complex impedances were calculated from permittivity and conductivity data found in the literature. It is to be noted that for the lung it was assumed 33% lung tissue and 67% air, and consequently the values for this tissue were obtained dividing the literature data for deflated tissue by a factor of 3.

The numerical calculations give SARs resulting from exposure to continuous waves. Since in all MRI procedures pulsed fields are employed, the data are to be scaled for the duty cycle, D, of the specific pulse sequence. Imagers can generally be programmed to produce several pulse sequences, and the patterns of energy deposition strongly depend on the chosen sequence, which can vary greatly in the number and type of pulses.

For the sake of simplicity, it was assumed that one pulse, at an angle θ, dominates within the chosen sequence. If the pulse sequence contains a number N_p of θ pulses, the duty cycle is multiplied by N_p. In the case of multi-slice imaging, a further correction for the number N_s of slices simultaneously scanned is needed, so that the general expression for the duty cycle is:

$$D = N_p N_s t T^{-1} \qquad (2)$$

where t is the time lenght of the pulse and T is the repetition period of the pulse sequence.

As an example of application of the impedance method, results are shown relative to an exposure similar to that hypotized by Bottomley and Edelstein (1981). These authors estimated the surface absorption of RF energy for a human head and torso imaged by nuclear magnetic resonance. The torso was approximated by a homogeneous cylinder of radius R. The permittivity (ε), conductivity (σ), and mass density (ρ) of the tissue inside the cylinder were assumed equal to those of muscle. The calculation was performed over a circular loop with radius R intersected by a spatially uniform RF magnetic field perpendicular to the loop and parallel to the torso axis.

For this model, it can be shown (Bottomley and Edelstein, 1981) that in the case of $\pi/2$ pulses of sinusoidal waves of frequency f the surface SAR is given by:

$$SAR = 6.80 \times 10^{-26} \sigma f^2 R^2 \rho^{-1} t^{-1} T^{-1} \qquad (3)$$

Applying the impedance method, it was assumed a torso of maximal cross section 46.8x31.2x59.8 cm along the x-, y-, and z-axis corresponding to the right-to left, front-to-back, and up-to-down directions in the torso, respectively. To obtain results for realistic situations, $\pi/2$ pulses of duration 5 ms and repetition period 100 ms were assumed, i.e. a duty cycle 5×10^{-2}, as frequently used in fast imaging. The magnetic field strength B_{rf} was obtained by the basic equation:

$$2\pi \gamma B_r t = \theta \qquad (4)$$

where γ is the gyromagnetic ratio (for protons $\gamma = 42.58$ MHz/T), and $B_r = 0.5\ B_{rf}$ is the magnitude of the rotating component of the RF field.

As already stated, the impedance method allows to evaluate the SAR distribution among the individual tissues and organs. Data shown in Fig. 2 for the particular frequency of 30 MHz allow a comparison with the results of Bottomley and Edelstein for the same frequency, polarization and duty cycle. The average SAR in the skin is 11 mW/kg, which compares with a value of 20 mW/kg given for the surface SAR by these authors. The discrepancy can be accounted for by the different dimensions of the realistic torso and the simplified model adopted by Bottomley and Edelstein; the radius of the latter equals in fact the maximal cross dimension used in applying the impedance method, allowing higher values of SAR in outer tissues.

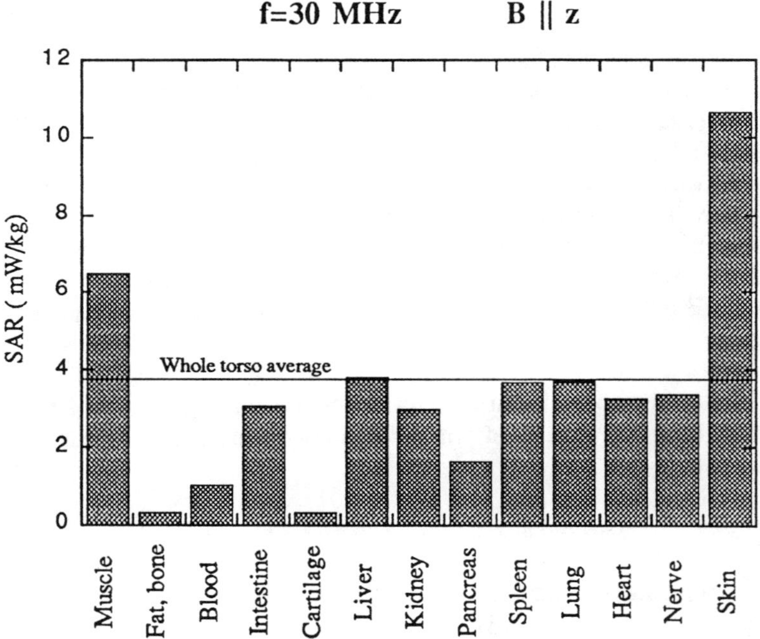

Fig. 2 - Average SARs in individual tissues and in the entire torso during a fast imaging procedure at 0.7 T, 30 MHz.

The dependence of average SAR on the polarization of RF fields has been proved for exposure to plane waves on simple models such as prolate spheroids. It is clearly seen that field polarization affects not only the values of average SAR in the torso, but also its relative magnitude in different tissues. In all cases, the distribution of energy deposition within each cross section perpendicular to the B-field shows a general increase with the distance from the center, as expected (DeFord and Gandhi, 1985) due to the larger current loops. That increase also explains the high value of SAR in the skin compared with all other tissues.

The average SAR in the whole trunk was calculated for each frequency under consideration, and to better describe the frequency dependence of average SAR, a best-fit analysis was performed. A f^k dependence, with k = 2.079, 2.081, and 2.080 for the x-, y- and z-polarization, respectively, has been found. The deviation of the exponents from the 2.0 value expected from the induction law is mainly due to the frequency dependence of conductivity.

From the results, the following general equations were derived that allow to calculate average SAR values in the whole trunk for any frequency (f in MHz) between 0.45 and 64 MHz:

$$SAR = 8.14 \times 10^{-10} f^{2.079} \theta^2 N_p N_s t^{-1} T^{-1} \qquad B//x$$

$$SAR = 1.29 \times 10^{-9} f^{2.081} \theta^2 N_p N_s t^{-1} T^{-1} \qquad B//y \qquad (5)$$

$$SAR = 6.40 \times 10^{-10} f^{2.080} \theta^2 N_p N_s t^{-1} T^{-1} \qquad B//z$$

To further verify the reliability of the method, the theoretical results were also compared with the experimental data of Roeschmann (1987) on the effective penetration depth and absorption of RF frequency in the human body as a function of frequency. These data were mainly derived from bench measurements of the quality factor and of the resonance frequency shift due to the loading of the different types of coils under investigation.

According to Roeschmann's experimental data, the SAR in the human torso can be calculated, in the frequency range 10-220 MHz, from the phenomenological equation (f in MHz):

$$SAR = 3.70 \times 10^{-10} f^{2.15} \theta^2 N_p N_s t^{-1} T^{-1} \qquad (6)$$

It is worth noting that, if the best-fit analysis of data on the average SAR in the torso is limited to the frequency range 13.56-64 MHz, which overlaps that considered by Roeschmann, a f^k dependence with k = 2.158, 2.161, and 2.155 for the three polarizations is obtained, in fair agreement with the dependence deduced from the experiment.

To assess the potential hazard of MRI examinations, calculations for some realistic procedures were made, corresponding to different pattern of RF excitation. The results, relative to z- polarization (magnetic field parallel to the body axis) are reported in Table 3 together with those obtained from Roeschmann's phenomenological equation under the same conditions.

Table 3. Whole-torso averaged SAR (W/kg) for some realistic MRI procedures

MRI procedure	Grandolfo et al. (In press)	Roeschmann (1987)
Spin echo, 1 echo, 1 π pulse t = 1 ms, T = 1 s B = 1.5 T, f = 64 MHz	0.038	0.028
Fast imaging, 1 echo, 1 $\pi/2$ pulse t = 5 ms, T = 0.1 s B = 0.7 T, f = 30 MHz	0.0037	0.0027
Inversion recovery, 4 echoes 5 π pulses, 12 slices t = 0.5 ms, T = 1 s B = 1.5 T, f = 64 MHz	4.56	3.35

The examples reported in Table 3 and Fig. 2 show the great potential of the impedance method for the estimation of energy deposition in different organs. That is of special importance in risk assessment, since it has been recognized that in certain cases localized SARs could be higher than the recommended limits, especially at higher fields, and in the case of multi-slice and multi-echo pulse sequences.

References

Armitage, D.W.; Leveen, H.H.; Pethig, R. (1983) Radio-frequency induced hyperthermia: computer simulation of specific absorption rate distributions using realistic anatomical models, Phys.Med.Biol., 28:31-42.

Borup, D.T.; Gandhi, O.P.; Johnson, S.A. (In press) A novel sinc basis FFT-CG method for calculating EM power deposition in anatomically-based man models, IEEE Trans. Microwave Theory Tech.

Bottomley, P.A.; Edelstein, W.A. (1981) Power deposition in whole-body NMR imaging, Med. Phys., 8:510-512.

Chatterjee, I.; Gandhi, O.P.; Hagmann, M.J. (1982) Numerical and experimental results for near-field electromagnetic absorption in man, IEEE Trans. Biomed. Eng., 30:2000-2005.

Chatterjee, I.; Gu, Y.G.; Gandhi, O.P. (1985) Quantification of electromagnetic absorption in humans from body-mounted communication transceivers, IEEE Trans. Vehicular Tech. 34:55-62.

Chatterjee, J.; Hagmann, M.J.; Gandhi, O.P. (1980) Electromagnetic energy deposition in an inhomogeneous block model of man for near-field irradiation conditions, IEEE Trans. Microwave Theory Tech., 28:1452-1459.

Chen, J.Y.; Gandhi, O.P. (1989) Electromagnetic deposition in anatomically-based model of man for leakage fields of a parallel-plate dielectric heater, IEEE Trans. Microwave Theory Tech., 37:174-180.

Chen, K.M.; Guru, B.S. (1977) Internal EM field and absorbed power density in human torsos induced by 1-500 MHz EM waves, IEEE Trans. Microwave Theory Tech., 25:746-755.

DeFord, J.F., Gandhi, O.P.; Hagmann, M.J. (1983) Moment-method solutions and SAR calculations for inhomogeneous models of man with large number of cells, IEEE Trans. Microwave Theory Tech., 31:848-851.

DeFord, J.F.; Gandhi, O.P. (1985) An impedance method to calculate currents induced in biological bodies exposed to quasi-static electromagnetic fields, IEEE Trans. Electromag. Comp., 27:168-173.

Duchêne, A.S.; Lakey, J.R.A.; Repacholi, M.H. (Eds) (1991) IRPA guidelines on protection against non-ionizing radiation, Pergamon Press, New York.

Durney, C.H.; Iskander, M.F.; Massoudi, H.; Allen, S.G.; Mitchell, J.C. (1980) Radiofrequency radiation dosimetry handbook, 3rd ed:, Pub. No. SAM-TR-80-32, USAF School of Aerospace Medicine, Aerospace Medical Division (AFSC), Brooks Air Force Base, Tex.

Durney, C.H.; Johnson, C.C.; Barber, P.W.; et al. (1978) Radiofrequency radiation dosimetry handbook, 2nd ed. USAF School of Aerospace

Medicine (RZP), Aerospace Medical Division (AFSC), Brooks Air Force Base, Tex., Contract No. SAM-TR-78-22.

Durney, C.H.; Johnson, C.C.; Massoudi, H. (1975) Long wavelength analysis of plane-wave irradiation of a prolate spheroid model of man, IEEE Trans. Microwave Theory Tech., 23:246-253.

Durney, C.H.; Massoudi, H.; Iskander, M.F. (1986) Radiofrequency radiation dosimetry handbook, 4th ed. Pub. No. USAF SAM-TR-85-73, USAF School of Aerospace Medicine, Aerospace Medical Division (AFSC), Brooks Air Force Base, Tex.

Eycleshymer, A.C.; Schoemaker, D.M. (1911) A croos-section anatomy, D. Appleton & Co., New York.

Franceschetti, G.; Gandhi, O.P.; Grandolfo, M. (Eds) (1989) Electromagnetic biointeraction: Mechanisms, safety standards, protection guides, Plenum Press, New York and London.

Gandhi, O.P. (1974) Polarization and frequency effects on whole-animal absorption of RF energy, Proc. IEEE, 62:1166-1168.

Gandhi, O.P. (1980) State of the knowledge for electromagnetic absorbed dose in man and animals, Proc. IEEE, 68:24-32.

Gandhi, O.P. (Ed.) (1990) Biological effects and medical applications of electromagnetic energy, Prentice Hall, Englewood Cliffs, New Jersey.

Gandhi, O.P.; DeFord, J.F.; (1988) Calculation of EM power deposition for operator exposure to RF induction heaters, IEEE Trans. Electromagnet. Comp., 30:63-68.

Gandhi, O.P.; DeFord, J.F.; Kanai, H. (1984) Impedance method for calculation of power deposition patterns in magnetically-induced hyperthermia, IEEE Trans. Biomed. Eng., 31:644-651.

Gandhi, O.P.; Hagmann, M.J.; D'Andrea, J.A. (1979) Part-body and multibody effects on absorption of radio-frequency electromagnetic energy by animals and by models of man, Radio Sci., 14(Suppl.):15:21.

Gandhi, O.P.; Hunt, E.L.; D'Andrea, J.A. (1977) Electromagnetic power deposition in man and animals with and without ground and reflector effects, Radio Sci., 12(Suppl.):39-47.

Grandolfo, M.; Polichetti, A.; Vecchia, P.; Gandhi, O.P. (In press) Spatial distribution of RF power in critical organs during magnetic resonance imaging, Annals of the New York Academy of Sciences, Special Issue on Biological Effects and Safety Aspects of Nuclear Magnetic Resonance Imaging and Spectroscopy.

Hagmann, M.J.; Gandhi, O.P. (1979) Numerical calculation of electromagnetic energy deposition in models of man with grounding and reflector effects, Radio Sci., 14(Suppl.):23-29.

Hagmann, M.J.; Gandhi, O.P.; Durney, C.H. (1979) Numerical calculation of electromagnetic energy deposition for a realistic model of man, IEEE Trans. Microwave Theory Tech., 27:804-809.

Hill, D.A. (1984) The effect of frequency and grounding on whole-body absorption of humans in E-polarized radiofrequency fields, Bioelectromagnetics, 5:131-146.

Holland, R. (1977) THREDE: a free-field EMP coupling and scattering code, IEEE Trans. Nucl. Sci., 24:2416-2421.

IRPA/INIRC (1985) Review of concepts, quantities, units and terminology for non-ionizing radiation protection, Health Physics, 49:1329-1362.

IRPA/INIRC (1988) Guidelines on limits of exposure to radiofrequency electromagnetic fields in the frequency range from 100 kHz to 300 GHz, Health Physics, 54:115-123.

Justesen, D.R. (1975) Toward a prescriptive grammar for the radiobiology of non-ionizing radiation, J. Microwave Power, 10:343.

Kunz, K.S.; Lee, K.M. (1978) A three-dimensional finite-difference solution of the external response of an aircraft to a complex transient EM environment. The method and its implementation. IEEE Trans. Electromagn. Comp., 20:328-332.

Livesay, D.E.; Chen, K.M. (1974) Electromagnetic fields induced inside arbitrary-shaped bodies, IEEE Trans. Microwave Theory Tech., 22:1273-1280.

Michaelson, S.M.; Lin, J.C. (1987) Biological effects and health implications of radiofrequency radiation, Plenum Press, New York and London.

Orcutt, N.; Gandhi, O.P. (1988) A 3-D impedance method to calculate power deposition in biological bodies subject to time varying magnetic fields, IEEE Trans. Biomed. Eng., 35:577-583.

Roeschmann, P. (1987) Radiofrequency penetration and absorption in the human body: Limitation to high-field whole-body nuclear magnetic resonance imaging, Med. Phys., 14:922-931.

Spiegel, R.J. (1984) A review of numerical methods for predicting the energy deposition and resultant thermal response of humans exposed to electromagnetic fields, IEEE Trans.Microwave Theory Tech., 32:730-746.

Spiegel, R.J.; Fatmi, M.B.E.; Kunz, K.S. (1985) Application of a finite-difference technique to the human radio-frequency dosimetry problem, J. Microwave Power, 20:241-254.

Stuchly, M. (1983) Dosimetry of radiofrequency and microwave radiation: theoretical analyses. In: Biological Effects and Dosimetry of Nonionizing Radiation: Radiofrequency and Microwave Energies, M. Grandolfo, S.M. Michaelson, and A. Rindi (Eds). Plenum Press, New York and London, pp. 163-177.

Stuchly, M.A.; Kraszewski, A.; Stuchly, S.S.; Hartsgrove, G.W.; Spiegel, R.J. (1987) RF energy deposition in a heterogeneous model of man near-field exposures, IEEE Trans. Biomed. Eng., 34:944-950.

Stuchly, M.A.; Spiegel, R.J.; Stuchly, S.S.; Kraszewski, A. (1986) Exposure of man in the near-field of a resonant dipole: comparison between theory and measurements, IEEE Trans. Microwave Theory Tech., 3426-31.

Stuchly, M.A.; Stuchly, S.S. (1986) Experimental radio and microwave dosimetry. In: Handbook of Biological Effects of Electromagnetic Fields, C. Polk and E. Postow (Eds). CRC Press Inc., Boca Raton, Florida, pp. 229-272.

Stuchly, S.S.; Kraszewski, A.; Stuchly, M.A.; Hartsgrove, G.W.; Spiegel, R.J. (1987) RF energy deposition in a heterogeneous model of man: far-field exposures, IEEE Trans. Biomed. Eng., 34:951-957.

Stuchly, S.S.; Stuchly, M.A.; Kraszewski, A.; Hartsgrove, G. (1986) Energy deposition in a model of man: frequency effects, IEEE Trans. Biomed. Eng., 33:702-711.

Sullivan, D.M.; Borup, D.T.; Gandhi, O.P. (1987) Use of the finite-difference time-domain method in calculating EM absorption in human tisues, IEEE Trans. Biomed. Eng., 34:148-157.

Sullivan, D.M.; Gandhi, O.P.; Taflove, A. (1988) Use of the finite-difference time-domain method in calculating EM absortion in man models, IEEE Trans. Biomed. Eng., 35:179-186.

Taflove, A. (1980) Application of the finite-difference time-domain method to sinusoidal steady-state electromagnetic-penetration problems, IEEE Trans. Electromagnet. Comp., 22:191:202.

Taflove, A.; Brodwin, M.E. (1975a) Computation of the electromagnetic fields and induced temperatures within a model of the microwave irradiated human eye, IEEE Trans. Microwave Theory Tech., 23:888-896.

Taflove, A.; Brodwin, M.E. (1975b) Numerical solution of steady-state electromagnetic scattering problems using the time-dependent Maxwell equations, IEEE Trans. Microwave Theory Tech., 23:623-630.

Umashankar, K.; Taflove, A. (1982) A novel model to analyze electromagnetic scattering of complex objects, IEEE Trans. Electromagnet. Comp., 24:397-405.

Wang, C.Q.; Gandhi, O.P. (1989) Numerical simulation of annular phased arrays for anatomically-based models using the FDTD method, IEEE Trans. Microwave Theory Tech., 37:118-126.

Yee, K. (1966) Numerical solution of initial boundary value problems involving Maxwell's equations in isotropic media, IEEE Trans. Antennas Prop., 14:302-307.

MEASUREMENT OF RADIOFREQUENCY FIELDS[a]

John A. Leonowich, Ph.D

Health Physics Department
Pacific Northwest Laboratory
Richland, Washington 99352, U.S.A.

1.0 Introduction

We are literally surrounded by radiofrequency (RFR) and microwave radiation, from both natural and man-made sources. The identification and control of man-made sources of RFR has become a high priority of radiation safety professionals in recent years (Allen et al., 1983; DeCaro, 1987; Hankins, 1987; ILO, 1985; Repacholi, 1983; WHO, 1981). For the purposes of this paper, we will consider RFR to cover the frequencies from 3 kHz to 300 MHz, and microwaves from 300 MHz to 300 GHz, and will use the term RFR interchangeably to describe both.

Electromagnetic radiation and fields below 3 kHz is considered Extremely Low Frequency (ELF) and will not be discussed in this paper. Unlike x- and gamma radiation, RFR is *non-ionizing*. The energy of any RFR photon is insufficient to produce ionizations in matter. The measurement and control of RFR hazards is therefore fundamentally different from ionizing radiation. The purpose of this paper is to acquaint the reader with the fundamental issues involved in measuring and safely using RFR fields.

2.0 Fundamentals of Electromagnetics Related to Hazard Analysis

The information in this section is a summation of the basic information required to understand the basic physics of RFR. More in-depth treatment of the subject can be found in a number of excellent references (Cheung et al., 1985; Marshall et al., 1987).

(a)   **Pacific Northwest Laboratory is operated by Battelle Memorial Institute for the Department of Energy under Contract DE-AC06 76RLO 1830.**

Electromagnetic energy is propagated through space in the form of a wave as shown in Figure 1. For plane waves travelling in free space, the electric field vector **E** and the magnetic field vector **H** are mutually orthogonal and are in phase - i.e. maxima and minima occur at the same point in time and space. The units of **E** are volts/meter (V/m) and for H amperes/meter (A/m).

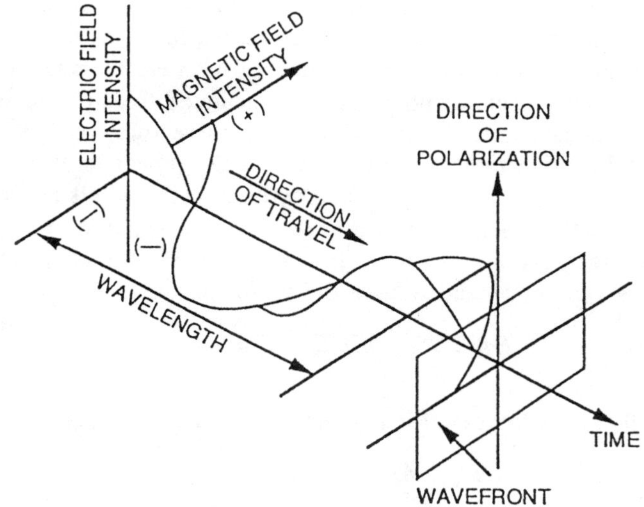

Figure 1: Electromagnetic Radiation Wave

Wavelength and frequency of any wave are related by the well known relationship:

$$Eq.\ 2.1 \qquad v = \lambda v$$

The frequency (f or v), is the number of wave cycles per second passing a fixed point along the direction of propagation. One cycle is represented as the period in which the magnitude of the electric field vector varies from zero, through its maximum value, back through zero to its minimum value, and finally back to zero. The unit of frequency is the *hertz*, or 1 cycle per second. The wavelength (λ), is the distance separating consecutive maxima or minima in the electric or magnetic field. The velocity of wave propagation, v, represents the speed at which the wave advances. In free space v is at right angles to both **E** and **H**. In a

vacuum, all electromagnetic waves travel at the speed of light c
(3×10^{10} cm/sec in free space).

It is important to realize that an electromagnetic wave carries energy.
The energy per photon is given by the classic relationship:

$$Eq.\ 2.2 \qquad E = h\nu$$

where h is Planck's constant, or 6.62×10^{-34} joule-sec. The energy carried
in an electromagnetic wave is usually expressed in terms of the energy
passing through a fixed area per unit time. For an electromagnetic wave
this *power density* W at any point may be calculated from the vector product of the electric and magnetic field strength vectors, i.e. $\mathbf{E \times H = P}$.
P is called Poyntings Vector and represents the power density and the
direction of energy propagation. The Poynting vector points in the direction in which the electromagnetic wave is travelling, thus transporting
energy in that direction. The magnitude of the power density propagated
in the wave can be calculated from the vector product:

$$Eq.\ 2.3 \qquad |ExH| = |E||H|\sin\Theta$$

For $\Theta = 90°$, as is the case for a plane wave in free space:

$$Eq.\ 2.4 \qquad |ExH| = |E||H|$$

Note that if **E** has dimensions of V/m and **H** is in units of amps/m the
dimensions of **P** and W are watts/m^2. More typically in hazard analyses
W is expressed in milliwatts/cm^2 (mW/cm^2).
<u>*HINT:*</u> $W(mW/cm^2) = W(W/m^2)/10$.

Polarization is another important property of electromagnetic waves and
is characterized by the oscillatory behavior and orientation of the electric
field vector. A wave referred to as being linearly polarized means that
the electric field vector varies in amplitude in only one direction as it
travels. It is conventional to describe polarization in terms of the electric
field only, not the magnetic field. An electromagnetic wave may exhibit
linear, circular, elliptical, or random polarization (such as in a light bulb).
A receiver of electromagnetic radiation must have the same sense of
polarization as the incoming wave for it to be detected <u>most efficiently</u>.

Suppose we could "freeze" the motion of an electromagnetic wave travelling in free space that we have previously discussed and take a snapshot of its **E** and **H** fields. The ratio of the electric field strength to the magnetic field strength is known as the characteristic impedance (or wave impedance Z_o) of free space. Therefore:

$$Eq.\ 2.5 \qquad Z_o = |E/H|$$

Note that Z_o is the ratio of **E** to **H** and is independent of their absolute magnitudes. It can also be shown using Maxwell's equations that:

$$Eq.\ 2.6 \qquad Z_o (free\ space) = \sqrt{\mu_o/\epsilon_o}$$

Where μ_o is the permeability of free space and has a value of 1.257×10^{-6} H/m and ϵ_o is the permittivity of free space and has a value of 8.855×10^{-12} F/m. The characteristic impedance of free space has a value of 377 ohms. This relationship is true for "well-behaved" electromagnetic waves in the so-called far field of an RFR emitter. Close to an emitter, in the near field, the 377 ohms relationship does not hold. This has important ramifications for how we measure RFR fields. This will be discussed further in Section 3.0. The characteristic impedance of other media can be calculated if their μ and ϵ are known. Free space, of course, has no mechanical resistance to motion, but it does have impedance to electromagnetic wave propagation. If space behaved like an electrical short circuit (i.e., no resistance), no radiation could leave an antenna and 100% reflection would result. Radio and other types of communication would not exist.

As we have previously stated, an electromagnetic wave represents a flow of energy in the direction of propagation. The rate at which energy flows through a unit area in space is called the power density of the wave, usually expressed in watts per square meter (W/m^2), or, milliwatts per square centimeter (mW/cm^2). If a source radiates power at a constant rate uniformly in all directions, the total power flowing through any spherical surface with its center at the source will be uniformly distributed over the surface and will equal the total power radiated. Such a source exists in theory only and is referred to as an isotropic radiator. If a source radiates a fixed amount of power (P) isotropically, then the power density (W) at a distance r from the source will be the total radiated power divided by the area of a sphere with its center located at the source and having a radius r:

$$Eq.\ 2.7 \qquad W = P/A_s = P/4\pi r^2$$

Examination of Eq. 2.7 shows that power density decreases as the distance from the source increases, and that the power density is inversely proportional to the square of the distance from the source. This is the inverse square law of radiation, observed experimentally for all forms of electromagnetic waves in free space. In obtaining this result it has been assumed that the source radiates uniformly in all directions. The same result will be obtained if the source were to radiate power preferentially in a certain direction or directions as happens in practice with directional antennas. We shall later see that antenna gain is a measure of the directionality of the power radiated. This is true whenever propagation occurs through a uniform medium in which the velocity of propagation is the same regardless of the direction of propagation.

The power density of the electromagnetic field is related to the electric and magnetic intensities in the same way that power in an electric circuit is related to the voltage and current, that is the product of the two. Since it is the average power density which is usually of interest, it is computed just as in AC circuits by multiplying the effective values of the E and H fields. The effective or *root mean square* (RMS) value is the maximum amplitude of the field multiplied by $1/\sqrt{2} = .707$. Thus:

$$Eq.\ 2.8 \qquad W = H_o/\sqrt{2} \times E_o/\sqrt{2} = E_oH_o/2$$

again where W is the power density in W/m^2 when E_o is expressed in V/m and H_o in A/m. Voltage (V) and current (I) in AC circuits are related to resistance (R) by Ohm's Law (V = IR). Similarly, we have seen that the electric and magnetic field intensities are related through the characteristic wave impedance of space Z_o, which has a value of 377 ohms, a value which also holds in air. Therefore:

$$Eq.\ 2.9 \qquad W = E^2/377 = 377\,H^2$$

the effective values ($E_o/\sqrt{2}$ and $H_o/\sqrt{2}$) in V/m and A/m.

Finally, by combining Eq. 2.7 with Eq. 2.9 at distances r and 2r we obtain:

$$Eq.\ 2.10 \qquad W_r/W_{2r} = (E_r^2/377)/(E_{2r}^2/377) = (E_r/E_{2r})^2$$

and:

$$Eq.\ 2.10 \qquad E_r/E_{2r} = 2$$

Thus while the power density of an electromagnetic wave falls off as inverse square, the field intensity falls off inversely proportional to the distance.

Finally, we can easily show from the above that power density and electric field strength are related in the far field according to the formula:

$$Eq.\ 2.11 \qquad W(mW/cm^2) = E^2/3770\ (Volts/meter)$$

Similarly, power density and magnetic field strength are related according to the formula:

$$Eq.\ 2.12 \qquad W(mW/cm^2) = H^2 x 37.7\ (Amps/meter)$$

All currently available broadband instrumentation measure either electric or magnetic field strength. If the meter face actually reads in power density instead of E or H, this power density is based on the above conversions. In the near field equations 2.11 and 2.12 do not hold, and therefore the power density is only equivalent to what it would be for that particular electric or magnetic field strength in the far field. All current standards for RFR are given in terms of these two basic quantities. Biologically, the amount of energy which an RFR field imparts to tissue can be quantified in terms of the specific absorption rate (SAR) in terms of watts/kg. SAR is difficult to measure outside of a laboratory, hence electric and magnetic field strength levels are kept below levels which would limit SAR to below physiologically acceptable levels. The interrelationship between the various quantities discussed is shown in Figure 2.

3.0   Hazard Analysis of RFR Systems

In order to understand how RFR emitters can emit electromagnetic radiation, and how there hazard can be controlled, a brief discussion of their component parts is in order. All RFR emitters have three basic components: a transmitter, a transmission line, and an antenna. Figure 3 illustrates the basic RFR emitter configuration.

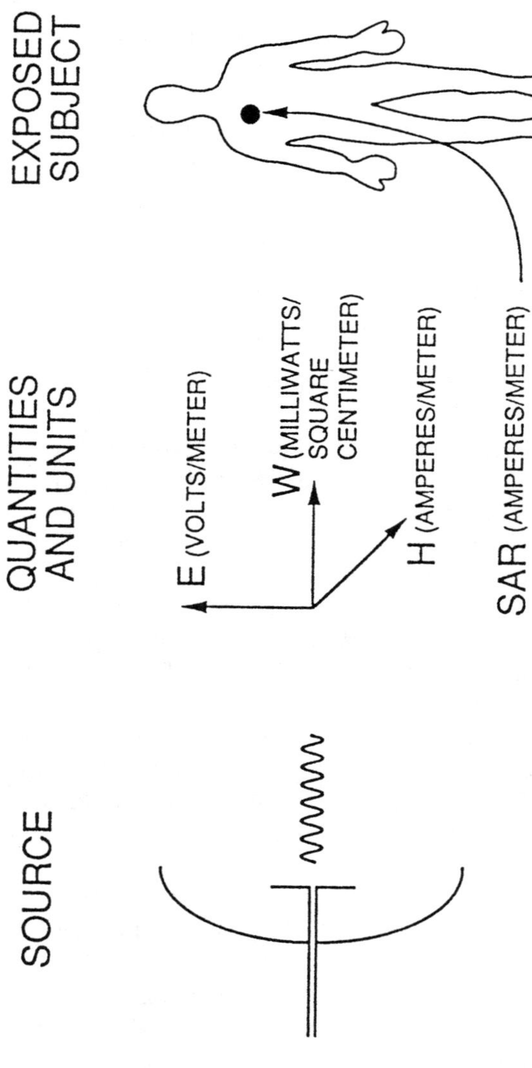

Figure 2: Quantities and Units Associated with Exposure of Man to RFR Electromagnetic Field (adapted from Guy, 1987 with permission)

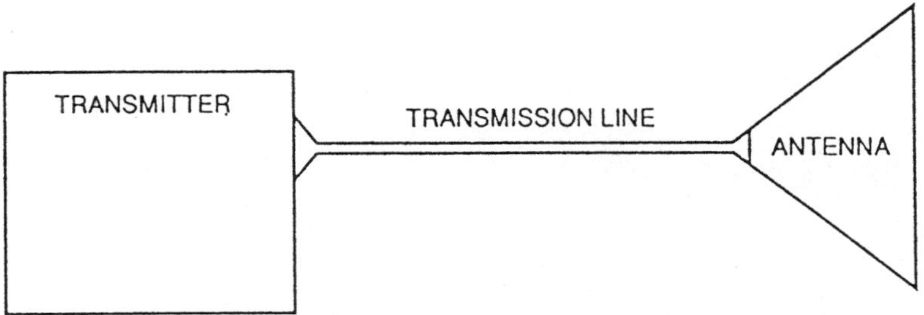

Figure 3: Basic RFR Emitter

There are four basic types of man-made devices for generating RFR: power-grid tubes; linear beam tubes such as klystrons; cross-field devices such as magnetrons and amplitrons; and solid-state devices. Once microwave energy is generated, it can be transmitted to an antenna through a waveguide or coaxial transmission line. An antenna is simply a means of radiating or receiving RFR. Radiating antennas such as parabolic reflectors are used to transmit RFR energy through free space or through a dielectric material.

The power rating of a transmitter must be known to evaluate an RFR system. All emitters transmit with either a continuous wave (CW) or a pulsed RFR signal. A CW system is one which is designed to produce its output in continuously successive oscillations (continuous waves). Rated output is normally *average power*. For pulsed systems, transmitter power is usually expressed as *peak power*. A system designed to produce its energy in short pulses or bursts, repeated at regular intervals, is therefore known as a pulsed system. In a pulsed system, average power is the peak power multiplied by the duty cycle (DC). Mathematically, the duty cycle is the product of the pulse width (PW) multiplied by the pulse repetition frequency (PRF). Alternately, it can be though of as the ratio of the pulse width to the pulse period (i.e., 1/PRF) of a periodic pulse train. The amount of time that each output pulse or burst of RFR energy is on is the PW, while the PRF is the number of output pulses per unit time, usually expressed in hertz (sec^{-1}).

$$Eq.\ 3.1 \qquad DC = PW(sec) \times PRF(sec^{-1}),$$
$$a\ dimensionless\ quantity\ less\ than\ 1$$

$$Eq.\ 3.2 \qquad P_{average} = P_{peak} \times DC$$

These concepts can be seen graphically in Figure 4.

(a)

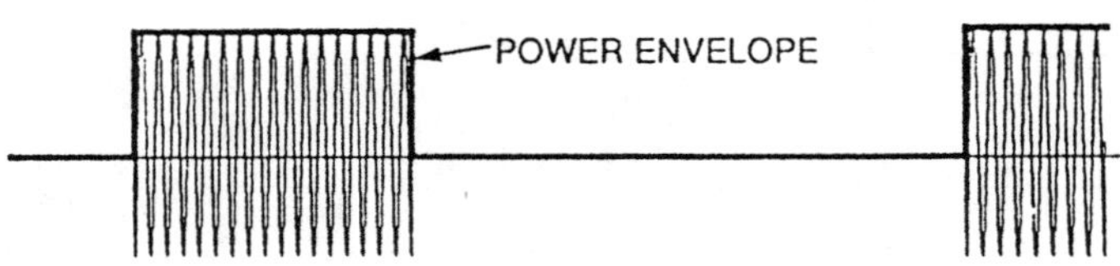

(b)

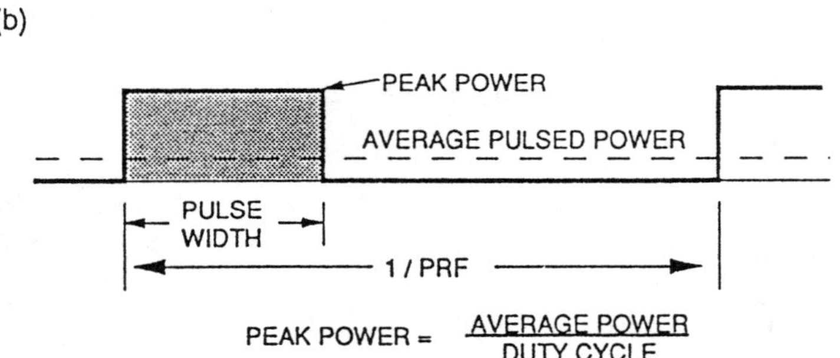

Figure 4: Relationship Between Peak Power,
Average Power, and Duty Cycle

Transmitter power can be specified in many units. Typically it is given in terms of watts (W), of kilowatts (kW). Another common unit is that of the dBm, which is a unit of power referenced to 1 milliwatt (mW). 0 dBm is equivalent to 1 mW. Other power levels are related to 1 mW by:

$$Eq.\ 3.3 \qquad P(dBm) = 10\ \log_{10}(P_1)/(1\ mW)$$

$$Eq.\ 3.4 \qquad P(mW) = \log^{-1}[P(dBm)/10]$$

Thus an output power of 1 kW (1000 W), would be equivalent to 60 dBm. There are two basic types of RFR transmission lines: one conductor and two conductor. A one conductor transmission guide propagates RFR through a hollow tube called a waveguide at microwave frequencies. There are numerous types of two conductor transmission lines, the most common being single coaxial line and parallel wires. These type of conductor lines are most commonly encountered at frequencies less than 1 GHz. It is important to be cognizant of the hazards associated with transmission lines. In particular, leaky or broken waveguides can propagate RFR like an antenna. The size of a waveguide break is the most important factor in determining its ability to behave like an antenna. For example, if the largest dimension of the break is smaller than one-half the wavelength, the break will not effectively emit RFR. On the other hand, if the break's greatest dimension is greater than one-half the system wavelength, the break could effectively emit RFR.

As was mentioned earlier, an antenna is a basic component of any RFR system. Regardless of the systems application all antennas have basic properties that can be well defined. Of these, gain, size, accessibility, and radiation pattern are of principle interest in the evaluation of RFR hazards.

Antenna gain is the ratio of the power gain of an antenna referred to a standard antenna, which is usually an isotropic emitter of RFR energy. An isotropic antenna is a hypothetical antenna radiating or receiving equally in all (4π) directions. Figure 5 illustrates an isotropic emitter.

In the case of electromagnetic waves, isotropic antennas do not exist physically but represent convenient reference antennas for expressing directional properties of actual antennas. An isotropic antenna would have a gain of one. Gain, therefore, can be thought of as a measure of directionality of an RFR emitter. It may be expressed as a pure number, or more commonly, in terms of decibels. It is therefore important to briefly review the decibel system. The decibel (dB), is a unitless quantity used to express a numerical ratio. For power considerations the decibel is equal to 10 times the logarithm of a power ratio expressed by the following:

$$Eq.\ 3.5 \qquad dB = 10 \log_{10}(P_1)/(P_2)$$

where P_1 and P_2 are two amounts of power. The ratio of P_1/P_2 is also known as the absolute gain, G(abs), of the antenna. Power ratios in

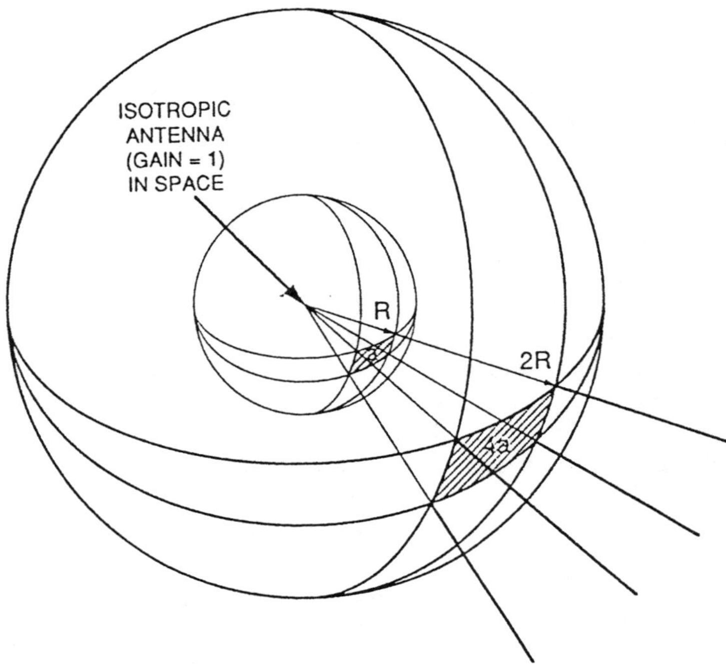

Figure 5: Isotropic Emitter

decibels can be added or subtracted like ordinary numbers (see the table below). The decibel scale is a convenient device for describing relative power magnitudes which may differ by orders of magnitude. P_2 may also be a particular reference level against which other power levels are compared. For example, dBw represents dB with respect to a reference level of 1 watt, while dBm represents dB with respect to a reference level of 1 mW. It is important to note that the dB scale represents a dimensionless ratio of two physical quantities, while the dBm scale represents an actual power level. The dB system may also be used to relate changes in voltage levels (or electric field strengths). Since the dB system is based on power ratios and since power depends on the square of the voltage, if we compare voltage levels we have:

$$Eq.\ 3.6 \qquad dB = 20 \log_{10}(V_1)/(V_2)$$

The following summarizes how various combinations of dB units can algebraically be combined:

Operation	Result	Physical Meaning	Allowed?
dB + dB	dB	product of 2 numbers	yes
dB-dB	dB	comparing 2 numbers	yes
$dB_m + dB_m$	XX	product of 2 powers	no
$dB_m - dB_m$	dB	comparing 2 powers	yes
$dB_m + dB$	dB_m	power amplification	yes
$dB_m - dB$	dB_m	power attenuation	yes

Practically, the range of antenna gains is from 2 dB for a dipole, to 70 dB for a very narrow aperture antenna.

The electromagnetic field radiated from an antenna varies greatly in configuration, depending on the distance from the emitter. Figure 6 details a two-dimensional representation of the regions around an aperture antenna.

The near field of an antenna can be divided into two distinct regions. In either region, the use of equations 2.11 or 2.12 is not appropriate. The reactive near field is that region of the field immediately surrounding the antenna where the reactive energy of the electromagnetic field is recovered and re-emitted during successive oscillations. True reactive near field conditions exist only to a distance of less than one-half wavelength of the emitted radiation from the radiator. Power density has no physical meaning in this region. Practically speaking, this region is important only at frequencies below 300 MHz, where independent measurement of E and H fields must be made to completely quantify the RFR field (ANSI, 1981). In the radiating near field region, the power density is not inversely proportional to the distance from the source. It is sometimes called the Fresnel region. In this region the power density increases irregularly with range to a maximum level then decreases at a near linear rate to the onset of the far field region. It is convenient and adequate from a personnel hazard viewpoint to consider the power density in the radiating near field to be constant with range, and equal to four times the average power density calculated at the antenna aperture

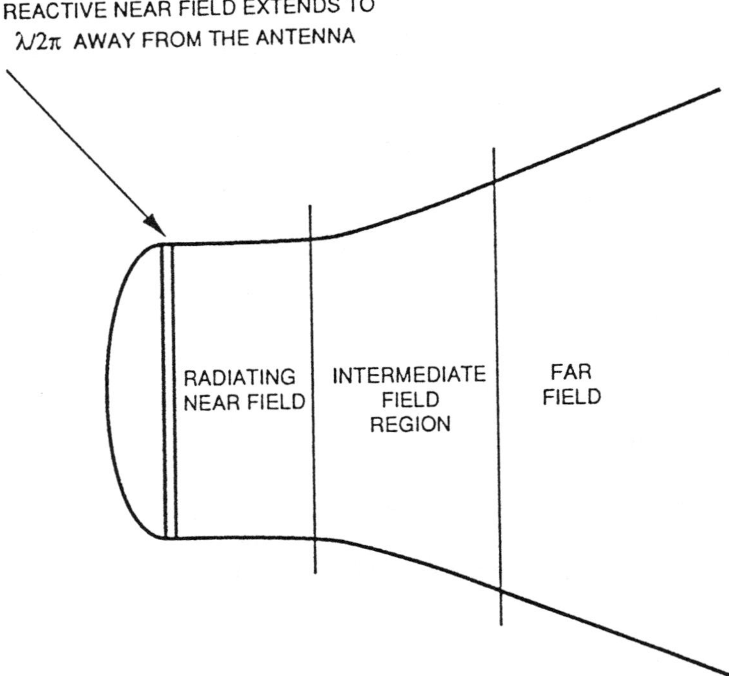

Figure 6: Antenna Regions

itself. Such a power density profile has proven accurate when compared to measured results. The intermediate field region is that portion of the Fresnel region of an antenna where the power density is decreasing at a near 1/r rate with range. It is sometimes ignored in hazard calculations. In the far field region (sometimes called the Fraunhofer region), power density is inversely proportional to the square of the distance from the source. In this region the electric and magnetic fields are perpendicular to each other, thus making it possible to make meaningful power density measurements. The impedance of free space in this region is 377 ohms, therefore measurement of either the E or H field is sufficient to completely characterize the RFR field.

The methodology presented in the remainder of this section will enable the health physicist or others concerned with RFR safety to make a very conservative estimate of the hazard distances involved which will then be useful as a starting point in making actual field measurements. It is based on basic antenna theory (Mumford, 1961; Hankins, 1986). The factor of conservatism will vary from as little as one, to as great as five, depending on how far from the antenna the far field actually begins. There are some small aperture antennae in the microwave frequency bands that have short near fields where these equations will yield very accurate predictions. The important point to remember is that any method other than actual measurements is only an estimate and/or prediction, and all are markedly influenced by a variety of factors, most of which are poorly understood. Actual measurements are always preferable, but estimates are useful as tentative numbers and for a starting place for any survey. The first step in every RFR hazard analysis is to compile a complete list of the systems characteristics. As a minimum, the following emitter information should be obtained if available:

1.   Operating frequency
2.   Transmitter Peak Power
3.   Pulse Width (PW), if any
4.   Pulse Repetition Frequency, (PRF), if any
5.   Antenna Gain in dB
6.   Antenna Dimensions
7.   Beam Width
8.   Scan or Rotation Rate, if any

The next step is to develop as complete a theoretical analysis as possible. In certain circumstances, due to the inaccessibility of the radiated beam, a theoretical analysis may be the only way to proceed. As was mentioned previously, this method of establishing hazard distances is based on a conservative, worst-case situation. The equations used in this section are all derived from basic antenna theory, and provide results consistent with measured power densities.

As was mentioned above, when an antenna is radiating into space, it is generally agreed that there are four distinct zones or regions wherein dissimilar behavior of the antenna's electromagnetic field is experienced. These zones include the reactive near field region, the radiating near field region, the intermediate and the far field region. The reactive near field region predominates over a very short range, usually less than 0.5 wavelengths (actually $\lambda/2\pi$) from the active antenna element. Because of the short wavelengths involved, this region is not usually significant in the microwave portion of the spectrum. The reactive near field can become important when dealing with resonant type antennas at frequencies below 300 MHz at power levels greater than 35 W. It is in the reactive near field region that separate measurements of E and H fields should be accomplished in order to produce a meaningful hazard survey. In the radiating near field, the energy is collimated in a beam having approximately the same size and shape as the far field beam. The radiating near field oscillates sinusoidally in amplitude with increasing range. The maxima are four times greater than the average power density W_o measured at the antenna aperture for an ideal antenna with 100% illumination. This average power density is given by:

$$Eq.\ 3.7 \qquad W_o = P_{ave}/A$$

Where P_{ave} = The average transmitter power that is available for radiation after transmission line losses are subtracted. In pulsed systems P_{ave} = Peak Power x Duty Cycle. A is the actual physical or effective area of the antenna aperture.

As was mentioned earlier, in the radiating near field it is convenient and adequate from a personnel hazard viewpoint to consider the power density in the radiating near field to be constant with range. The maximum radiating near field power density W_{nf} is:

$$Eq.\ 3.8 \qquad W_{nf} = \eta\, 4\, P_{ave}/A$$

where η is the antenna aperture efficiency, typically on the order of 0.5 to 0.75. In the intermediate field the power density is decreasing by a $1/r$ relationship with range, and can be represented as:

$$Eq.\ 3.9 \qquad W_{if} = W_{nf}(R_{nf}/R)$$

where R_{nf} is the extent of the near field and R is some range in the intermediate field. For a *circular* antenna, R_{nf} is given by:

$$Eq.\ 3.10 \qquad R_{nf} = D^2/4\lambda$$

where D is the antenna diameter and λ is the wavelength of the radiation.

In the far field an antenna has the characteristic that the power density W_{ff} decreases as the inverse square of the range. Equation 3.11 is a precise statement of the value of W_{ff} as a function of transmitter power, antenna gain, and range. This equation, the Friss free-space transmission formula, predicts the worst-case envelope of radiated power density from any antenna system. It is technically only accurate for plane-wave, far field conditions, though it can be used successfully as a worst-case predictor to zero range, where W would approach an infinite value. The formula is as follows:

$$Eq.\ 3.11 \qquad W_{ff} = P_{ave}\,G/4\,\pi\,R^2$$

where W_{ff} is the power density on axis; P_{ave} is the average power available for radiation; G is the absolute gain expressed as a power ratio; and $R > R_{ff}$, the distance which marks the beginning of far field conditions. For circular antennas Rff is given by:

$$Eq.\ 3.12 \qquad R_{ff} = 0.6\,D^2/\lambda$$

At the above distance, the power density falls to below 10 dB of its maximum value in the radiating near field as given by equation 3.8, and true far field conditions are reached.

If the gain of the antenna is not known, it can be closely approximated by the following:

$$Eq.\ 3.13 \qquad G = 4\,\pi\,A\,\eta/\lambda^2$$

where λ is the wavelength of the radiated energy, η is the antenna efficiency and A is the actual antenna aperture area. Combining equations 3.8 and 3.13 yields:

$$Eq.\ 3.14 \qquad W_{nf} = 16\,\pi\,P_{ave}\,\eta^2/G\lambda^2$$

Equation 3.11 can be solved for range as follows:

$$Eq.\ 3.15 \qquad R = \sqrt{(P_{ave}\,G/4\,\pi\,W)}$$

Once a power density is calculated using equation 3.11, the power density W_2 at any other distance R_2 in the far field is given by:

$$Eq.\ 3.16 \qquad W_1/W_2 = (R_2)^2/(R_1)^2$$

The above treatment was derived specifically for antennas with circular apertures. However, the above treatment can be extended to non-circular aperture antennas by representing them by a circular aperture of the same physical size and gain. In this case we can generalize our antenna equations as follows:

$$Eq.\ 3.17 \qquad R_{nf} = G\lambda/4\pi^2\eta$$

$$Eq.\ 3.18 \qquad R_{ff} = 0.6G\lambda/\pi^2\eta$$

The equations for W_{nf}, W_{if}, and W_{ff} and G are as before.

The effects on the above calculations if a system is rotating or "nodding" can now be shown. The power density produced at any point by an antenna which is rotating is given by:

$$Eq.\ 3.19 \qquad W = W_s x f$$

where W_s is the stationary power density and f is the so-called "rotational reduction factor." In the near field, the beam is considered to have a dimension in the plane of rotation equal to the length of the antenna axis, L in that plane. The near field rotational correction factor, f_{nf}, at R is given by:

$$Eq.\ 3.20 \qquad f_{nf} = L/R\Theta_s$$

where L equals the antenna dimension in the plane of rotation, R is the distance from a point in the near field to the antenna, and Θ_s is the scan angle in radians.

The power density calculated in the near and intermediate field using the above reduction factor is an overestimate, but is consistent with the conservative approach used in hazard calculations. The reduction factor f_{ff}, which Wff is multiplied by to determine the time-averaged power density found in the far field of a scanning antenna is given by:

$$Eq.\ 3.21 \qquad f_{ff} = \Theta_{1/2}/\Theta_s$$

where $\Theta_{1/2}$ is the half-power beam width of the antenna and Θ_s is the scanning angle of the emitter. *It is important to realize that f is independent of the scanning rate in revolutions per minute.*

It should be noted that experience has shown that only about 15% of the RFR emitters account for the bulk of the measurement problems encountered in managing a safety program. In some cases the calculations detailed in this section may not be required. There are a number of classes of emitters that may easily and promptly be removed from these calculations, however. For example, hand-held transreceivers, commonly known as "bricks," which operate from 136 to 174 MHz and at the 510 MHz regions of the spectrum are considered to be nonhazardous to personnel if they emit less than 7 watts (ANSI, 1982). *(Note that there will be some changes in the 7 watt exclusion in the new ANSI/IEEE RFR Standard when it comes into force These were not definitively available at press time).* As another example, many high-powered emitters have main beams which are normally inaccessible to personnel. Hazard calculations performed on these systems would be for academic interest only. A large class of emitters have characteristics which have been validated after a large number of surveys performed on them. Finally, it is important to note that there is a wealth of survey information on many "off-the-shelf" systems performed by many federal and state agencies. This data, coupled with interaction with the manufacturer of the system in question, can prove invaluable.

Finally, although the above treatment will provide reasonably accurate information for safety professionals when determining the hazards around a single emitter, calculations in a multi-emitter environment are much more difficult without the use of sophisticated computer programs. One such program is the Ohio State University Electromagnetic Reflector Antenna Code (NEC-REF) (Rudduck et al, 1982). This code is capable of calculating the power density in both the near and far field zones for any parabolic reflector antenna or combinations thereof. Typical output of the program is shown in Figure 7. The characteristic oscillations of the power density in the radiating near field are clearly shown in the figure. It is also important to note the extent to which the radiating near field extents for such a large antenna ($> 10^5$ ft.). Programs such as the NEC-REF can be powerful predictors of the radiation hazards of RFR emitters. Their disadvantages are that they are difficult to use and require reasonably sophisticated computers to run.

4.0   Conducting an RFR Hazard Survey

An onsite survey is absolutely imperative in order to have a total understanding of the RF/microwave (RFR) field conditions in the area of interest. Moreover, when conducted by an experienced surveyor, it can yield highly accurate and reproducible results. On the other hand, a

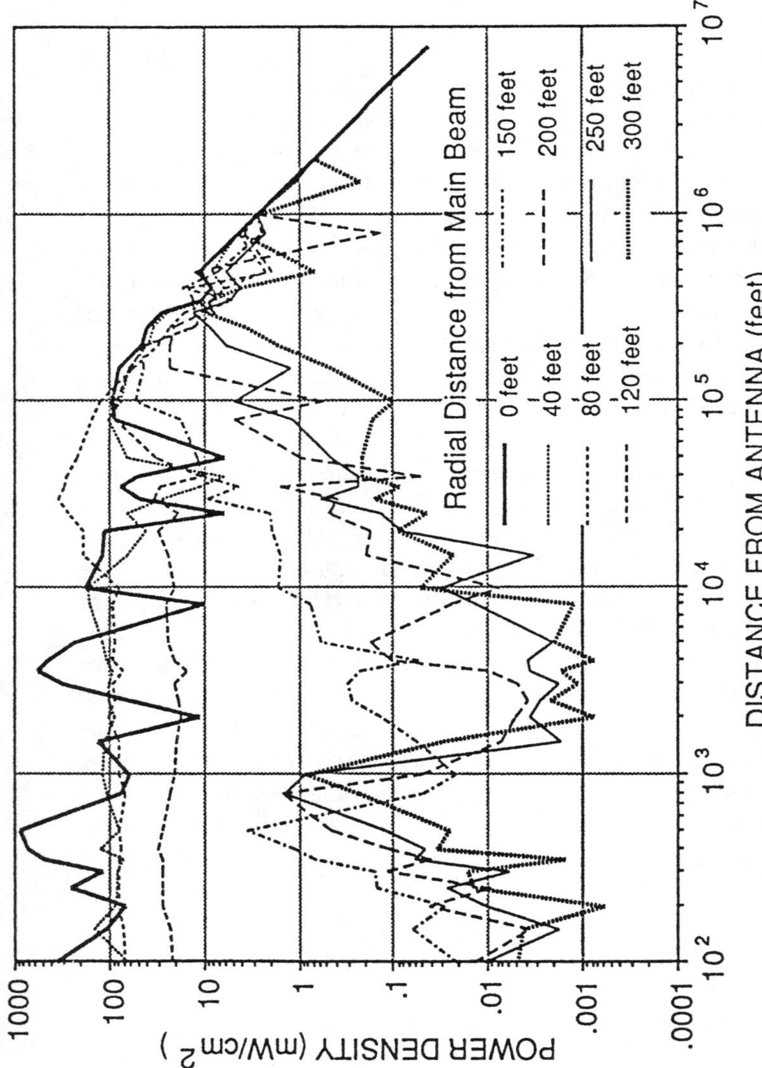

Figure 7: Power Density Along Main Beam of a 1 MW 70-m Diameter Antenna

survey accomplished by an inexperienced person can lead to erroneous data and false conclusions; possibly even to the damaging of expensive monitoring equipment. Theoretical calculations do not possess the reliability achieved through authentic measurements and should only be applied as a starting point and general guide for conducting a "hands on" survey. At the present state of the art, the onsite survey should remain as the primary basis for insuring compliance with appropriate regulatory guidance.

In general, a survey should entail consideration of the following:

a.   Emitter characteristics and operating parameters

b.   Purpose and use of the emitter

c.   Site configuration and terrain

d.   Procedures followed in all phases of operation,including maintenance and testing

e.   Magnitude of power densities created by the emitter

f.   Extent of potentially hazardous and hazardous areas, both on and off site; including the location and use of petroleum, oil, and lubricant (POL) or explosives

g.   Presence of ionizing radiation attributable to microwave generating equipment such as klystrons, thyratrons, traveling wave tubes, etc.

h.   Control techniques which will effectively reduce potential hazards. In this regard, application of the ALARA principle "as low as reasonably achievable" should be paramount in the surveyors mind

i.   Any other situations which may create a hazardous area [noise, chemicals, ventilation, high voltage, etc.]. Although you may not be an expert in these fields, you should refer any questionable situations to the appropriate individuals in your organization

Control techniques to reduce unnecessary personnel exposure will generally fall into one of two categories:

(1)   limiting the traverse of the radiated beam so it does not radiate into occupied areas

(2)   limiting the access of personnel into areas where hazardous power densities exist

If an evaluation of the RFR hazards of an installation reveals the existence of limited occupancy or denied occupancy areas, the following measures may be used to reduce unnecessary exposures:

a.   All potentially hazardous and hazardous areas should be conspicuously posted with appropriate warning signs

b.   Where operations allow, equipment positioning should be restricted in order to minimize the extent of exposure areas, thereby reducing unnecessary hazards. These restrictions can be implemented through the installation of cut-off devices in the electrical or mechanical components of a system which will automatically terminate transmission when the antenna is pointed in a predetermined direction. Such cut-off devices are much more desirable than simply instructing personnel not to point a system in a certain direction, even though these instructions may be in a standard operating procedure

c.   During test or maintenance procedures requiring free space radiation, the use of appropriate antenna positioning restrictions are necessary if the power densities exceed the permissible exposure limit (PEL)

d.   The use of dummy loads to absorb microwave energy output is recommended when free space transmission is not necessary. These dummy loads should also be checked for leakage

e.   The use of barriers, interlocks and visible/audible warnings is required to prevent ingress by personnel into denied occupancy areas

f.   Where an antenna is not permanently installed, the antenna may be relocated to reduce power densities in exposure areas to acceptable levels

g.   In situations where operations would be unduly restricted by implementation of the above methods, suitable attenuation of power density levels may be accomplished in the irradiated areas by shielding. The RFR attenuation factors of wire mesh, window glass, and concrete block are well known. This method is *only* recommended in the most. extreme circumstances

h.   The use of microwave protective clothing and eyewear can be considered, again in the most extreme circumstances only

Upon the completion of a survey, the results should be written up as a formal report. Items contained in the report should include but are not limited to:

a.   theoretical calculations carried out prior to the survey

b.   PEL hazard radius and height above the ground

c.   all areas in which the PEL is exceeded

d.   RFR power density levels at work stations and in "normally accessible" areas

e.   any "hot spots"

f.   adequacy of warning signs and access-limiting devices; they should be consistent with ANSI Standard C-95.2 (ANSI, 1981-2)

g.   adequacy of any standard procedures use to limit or avoid exposure

h.   personnel attitudes [if appropriate]

i.   conclusions and recommendations

j.   diagrams/photographs are often worth a thousand words

Remember - documentation is important. In the case of legal difficulties, the contents of your work will be scrutinized very carefully!

In conclusion, it should be stressed that even the most elaborate protective procedures and devices will be rendered ineffective unless all personnel concerned are cognizant of the potential hazards to which they may be exposed and of the measures employed to ensure their safety. There are many misconceptions as to the nature, magnitude, and extent of hazards attributable to RFR radiation. The importance of an adequate personnel instruction and training program cannot be overemphasized. This should be considered a prerequisite for any comprehensive policy designed to minimize these hazards.

5.0   Instrumentation To Conduct An RFR Hazard Survey

This section will review the capabilities of past and present RFR instrumentation and how their characteristics affect hazard measurements.

The average user is usually unaware of design limitations and compromises that different manufacturers have reached for their particular customer base or measurement philosophy. These compromises will dictate how well an instrument will perform in a given RFR environment. Approximately 20 years ago, the commercialization of microwave ovens generated a need for instrumentation operating at 915 and 2450 Megahertz (MHz) in order to obtain leakage information for manufacturers and repair organizations. Awareness of RFR energy and its possible effects lead to developments of broader frequency range monitors that at first were circularly polarized in an attempt to respond to all polarizations. These instruments were therefore not isotropic in their detection capability and their effectiveness was markedly affected by geometric considerations. Without a priori knowledge of the field to be measured it was therefore quite possible to make a totally erroneous conclusion on the amount of RFR energy present. These initial products were very broadband for their time, covering the spectrum from 1 to 14 Gigahertz (GHz) by the use of thermistor detectors. While the thermistor was very linear in its response, the receiving antenna design was not, necessitating multiple frequency calibrations to overcome polarization and frequency sensitivity errors of up to 10 decibels (dB). The next generation of these circularly polarized monitors incorporated thermocouple detectors and improved antenna designs which reduced frequency sensitivity errors to about 6 dB.

About 15 years ago the first isotropic detection probes came on the market. Electric field probes were made available covering the spectrum from 300 MHz to 18 GHz with a frequency sensitivity of only 3 dB, and a measuring range of 30 dB. During the early seventies there were also many advances in calibration methods and procedures for quantifying RFR fields from then United States National Bureau of Standards. Near field calculations and transverse electromagnetic (TEM cell) developments allowed for even higher calibration accuracies over a broad range of frequencies to uncertainties of ± 0.5 dB. Also in this time period the development of magnetic field probes was accomplished in part to measure the magnetic fields associated with high frequency (HF) communication systems. The impetus for the development of much of this isotropic instrumentation was the United States military, in particular the U.S. Air Force. The development of this broadband instrumentation overcame many of the problems associated with the earlier measurement equipment.

It is important now to discuss the ability of the present generation of broadband isotropic instrumentation to perform compliance measurements. The 1974 American National Standards Institute (ANSI) Radiofrequency/Microwave Exposure Standard did not include frequency dependent criteria, nor did it differentiate between partial or whole body exposure. In 1982 the ANSI Standard (ANSI, 1982) was extensively revised to include frequency dependent exposure criteria. One of the most challenging changes to the standard for equipment manufacturers was the inclusion of these frequency dependent levels combined with the use of spatial and time averaging of exposure. Additionally, in certain situations separate measurements of electric and magnetic fields were required.   Clearly, past instrumentation with limited dynamic and frequency ranges would have to be updated to meet new challenges, not only for standards compliance but also to accommodate the proliferation of industrial, medical, scientific, and communications applications of RFR energy.

Presently, the instrumentation available must be able to meet the broadband needs of major users such as the military, but also be applicable to the narrowband customer whose needs must be met with a cost-effective solution. Therefore there has been a proliferation of narrowband specialized products for 50/60 Hz fields, video display terminals, industrial heat sealers and induction heaters (operating at 27 MHz), and broadcast facilities. Broadband equipment has not gone without changes either.

Instrumentation currently available on the market have traditionally utilized either diode or thermocouple based electric field detection.

Uncompensated diode circuits, while providing higher dynamic ranges with excellent overload capabilities are nonetheless subject to large modulated signal and multiple emitter errors ranging from 1 to 30 dB. Early thermocouple based detectors exhibited excellent accuracy in complex, modulated field environments but were limited by overload specifications that a careless operator could exceed, thereby damaging or destroying these probes. Today there is at least one manufacturer providing higher frequency probes that operate to 100 GHz based on thermocouple detection with 1000 percent overload specifications to guard against failure or modulation errors. At lower frequencies, between 300 kHz and 1000 MHz, probes employing compensated diode detection circuitries that all but eliminate signal and modulation errors are now on the market.

New technology has not been limited to electric field probes either, as recent developments in magnetic field probes evidence. Early magnetic field probes of certain designs were not truly isotropic. A phenomena known as "spatial shadowing" existed wherein one or more of the three orthogonally mounted detector loops did not allow the same flux lines to pass through all loops. This was caused by the three loops not having a common vertex. Later designs have corrected this source of error. Nonetheless, these probes were still difficult, if not impossible, to use in multiple emitter applications because of their erroneous response to signals above their operating frequencies. Away from the controlled laboratory environment, where emitters are present throughout the frequency spectrum, large measurement errors are present when an out-of-band transmitter exists nearby a survey area. The 1982 ANSI Standard requested manufacturers to provide out-of-band response data; however, that only served to further complicate field measurements and to increase calibration costs. Newly designed magnetic field probes are now available that which greatly enhances the accuracy and operator confidence in magnetic field measurements.

Currently a new revision to the 1982 ANSI Standard has been drafted (ANSI/IEEE, 1990) and will probably be adopted as a joint IEEE/ANSI document in 1992. Major changes include allowing the use of temperature probes and thermography to measure SAR's of electromagnetic energy directly rather than the use of calculations based on external field strengths. There are also provisions for measuring induced body currents for radiofrequency fields below 30 MHz by standing an individual on a conductive plate electrode and monitoring current flow to ground with an RF ammeter. Another change that reflects on the direction that new instrumentation will take is the use of time and/or spatial averaging. Although the use of such averaging was already permitted in the 1982 document, it has often been neglected in practice. The new revision will emphasize that in certain conditions a much more useful picture of the actual exposure situation will be obtained by using these techniques. Therefore, in external field measurements where highly localized fields exist, averaging techniques should be applied. In continuous field applications, like those found around RF induction sealers, measurements may be performed at different portions of the body and averaged together manually. In the case of a rotating radar where field levels are varying constantly, a time averaging module is the only way to truly measure, and average, the total exposure. There are currently modules available that can perform this averaging automatically, assisting in determining time averaged exposure. There is also one system which can perform spatial averaging over the body or workplace area. This system has a "pause" feature which allows horizontal and vertical scan averages to be summed so that one can determine averages in multiple planes or for cubic areas. There is no doubt that continued advances in microprocessor technology will produce modules which will be able to perform all these functions, and because of their small size will produce minimal perturbations in the ambient electromagnetic field.

All of these newly available systems require comprehensive calibrations. The wide frequency ranges of modern probes requires the use of multiple point calibrations with traceability to a primary standard, both in and out of their frequency operating range. With the possibility of even more restrictive standards on the horizon, systems must not overestimate, nor underestimate, these critical safety measurement levels. The latest ANSI recommended practice document (ANSI, 1981-1) lists various methods that have been approved to certify survey instruments that all manufacturers should be following. Users must know whether their units have been calibrated to respond correctly in a real world environment, or simply the controlled laboratory area. An example of an improved method of calibrating a meter and probe used in the environs of an AM radio station would be to understand what is occurring when the instrument is brought into the survey field. The potential field effect predominant in these low frequency applications requires that the meter and probe be physically calibrated together, as they would be used in the field, rather than simply calibrating the probe in a TEM cell without the meter present. Performance testing of survey instruments, similar to that which is now standard in the ionizing radiation community, is still only in it infancy for RFR survey devices. As this area of radiation safety grows and matures, this type of independent testing by third parties will no doubt become standard.

In conclusion, users must be aware of not only the strongpoints of a particular system but also its weak points. No single broadband instrument currently available will meet the needs of all users (Mantiply, 1988). Even though products might be specified similarly their operation in a given environment may be significantly different. It is therefore important that the radiation protection professional be aware of these differences when choosing one for a particular application.

6.0 References

Allen, S. G., and Harlen F., *Sources of Exposure to Radiofrequency and Microwave Radiation in the UK,* National Radiological Protection Board (UK): (1983).

ANSI/IEEE C95.1-1990, *Draft American National Standard Safety Levels with Respect to Human Exposure to Radio Frequency Electromagnetic Fields, 3 kHz to 300 GHz.*

ANSI C95.1-1982, *American National Standard Safety Levels with Respect to Human Exposure to Radio Frequency Electromagnetic Fields, 300 kHz to 100 GHz.*

ANSI C95.2-1981-2, *American National Standard Radio Frequency Radiation Hazard Warning Symbol*; Reaffirmed in 1989.

ANSI C95.3-1973, *American National Standard Techniques and Instrumentation for the Measurement of Potentially Hazardous Electromagnetic Radiation at Microwave Frequencies.* Reaffirmed in 1979.

ANSI C95.5-1981-1, *American National Standard Recommended Practice for the Measurement of Hazardous Electromagnetic Fields---- RF and Microwave.*

Cember, H., *Introduction to Health Physics*, Chapter 14, New York: Pergamon Press, (1983).

Cheung, W. S., and F. H. Levien, Eds., *Microwaves Made Simple: Principles and Applications*, Dedham, MA: Artech House, (1985).

DeCaro, C., "The Zap Gap," *The Atlantic*, 259, No.3, p. 24, (March 1987).

Department of the Air Force, *A Practical R-F Guidebook for Bioenvironmental Engineers*, Brooks AFB, TX: USAF Occupational and Environmental Health Laboratory (USAF OEHL) Report 80-42, (1980).

Department of the Air Force, *Radio Frequency Radiation (RFR) Measurements in Operational Settings*, Brooks AFB, TX: USAF OEHL Report 85-028CV111ARA, (1985).

Department of the Air Force, *Base Level Management of Radio Frequency Radiation Protection Program*, Brook AFB, TX: AFOEHL Report 89-023RC0111DRA, (1989).

Department of the Army, *Radiofrequency, Microwave and Ultrasound Hazards Course Manual*, Chapters 2 and 3, Aberdeen Proving Grounds, MD: US Army Environmental Hygiene Agency, (1984).

Guy, A. W., Dosimetry of exposure to VLF to microwaves, *Health Physics* 53, 6 pp. 569-584 (1987).

Hankin, N. N., *The Radiofrequency Radiation Environment: Environmental Exposure Levels and RF Emitting Sources*, USEPA 520/1-85-014: (July 1986).

International Labour Office, Occupational Safety and Health Series No. 53: *Occupational Hazards from Nonionising Electromagnetic Radiation*, Geneva: (1985).

Mantiply, E. D., *Characteristics of Broadband Radiofrequency Field Strength Meters*, Proceedings of IEEE Engineering in Medicine and Biology Society, 10th Annual International Conference, New Orleans, LA (1988).

Marshall, S. V., and G. G. Skiter, *Electromagnetic Concepts and Applications*, Englewood Cliffs, NJ: Prentice Hall, (1987).

Mumford, W. W., "Some Technical Aspects of Microwave Radiation Hazards," *Proc. IRE*, 49 pp. 427-447, (1961).

Repacholi, M. A., "Sources and Applications of Radiofrequency (RF) and

Microwave Energy", contained in *Biological Effects and Dosimetry of Nonionizing Radiation - Radiofrequency and Microwave Energies*, M. Grandolfo, S. M. Michaelson, and A. Rindi, editors, Plenum Press, New York: (1983).

Rudduck, R. C., and Y. C. Chang, *Numerical Electromagnetic Code Reflector Code, NEC - REF (Version 2)*, Technical Report 712242-16, Ohio State University ElectroScience Laboratory, Department of Electrical Engineering, Ohio State University, (1982). (NOTE: Distribution limited to U.S. Government agencies and their contractors by Export Limitations Act.)

World Health Organization, *Environmental Health Criteria No. 16: Radiofrequency and Microwaves*, Geneva: (1981).

INTRODUCTION TO ULTRASOUND

William D. O'Brien, Jr.

Department of Electrical and Computer Engineering
University of Illinois
1406 West Green Street
Urbana, IL 61801 USA

I. WHAT IS SOUND?

Sound is the rapid motion of molecules. These molecular vibrations transport energy from a transmitter, a sound source like our voice, to a receiver like our ear. Sound travels in waves that transport energy from one location to another. When the molecules get closer together, this is called compression, and when they separate, this is called rarefaction. This mechanical motion, the rapid back and forth motion, is the basis for calling sound a mechanical wave or a mechanically propagated wave. Sound requires a medium in order to propagate. There are three types of medium: gas, liquid and solid. In a vacuum (such as outer space) sound cannot propagate - there is no medium.

This contribution is an introduction to the physical considerations of medical ultrasound and includes (1) a discussion of matter to understand issues of ultrasonic propagation in tissue and a means to quantify properties of matter, (2) a discussion of wave propagation and a means to quantify the temporal, amplitude, transmission and reflection characteristics of ultrasound waves and (3) concludes with a discussion of resolution trade-off issues.

II. TYPES OF ACOUSTIC WAVES

The classification of sound waves is based on the type of motion that is induced in the medium by the propagating sound wave. For purposes of ultrasonic physics, the lowest level of organization within material is called a particle (Kinsler et al., 1982). The particle is represented in the Fig 1 as dots and can be thought of as a volume of material. Each of these dots consists of millions of molecules and yet each has dimensions of a fraction of an ultrasonic wavelength. The assumptions of the particle are

$$a \ll (\Delta V)^{1/3} \ll L \tag{1}$$

where (1) the typical separation between molecules, a, is very small compared to the typical length of the particle, $(\Delta V)^{1/3}$, and (2) the typical length of the particle, is very small compared to the typical separation between maxima and minima of stresses, L, in the media. The former assumption assures that the forces experienced by the particle are an average over a large number of molecules and the latter assumption assures that such forces experienced by the particle are uniform.

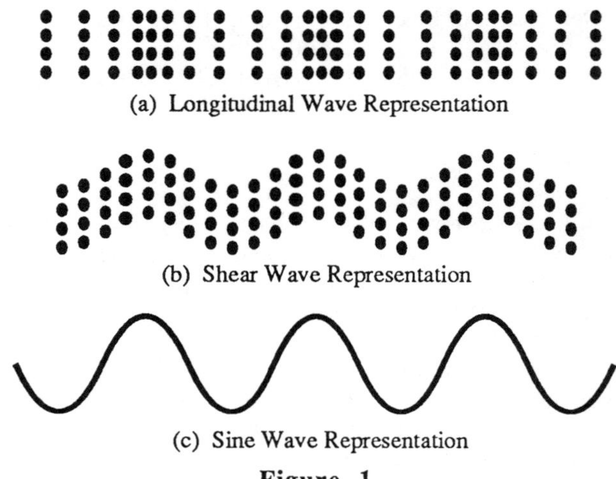

(a) Longitudinal Wave Representation

(b) Shear Wave Representation

(c) Sine Wave Representation

Figure 1

When an ultrasonic wave is propagated within material, the type of wave is classified in terms of (1) the direction the ultrasonic energy is traveling and (2) the direction the particle is moving. A longitudinal wave occurs when the particles move back and forth (that is, left to right and back - horizontally) relative to the direction of the wave energy, as demonstrated in Fig 1a. Propagated longitudinal waves travel through all kinds of materials: gases, liquids and solids.

In the case of shear waves, the particles move at right angles to the direction of the wave propagation as shown in Fig 1b. In this figure, the particles are moving vertically up and down while the wave energy is moving horizontally. Shear waves exist only in solid materials, not in liquids or gases, nor do they exist in soft tissues because soft tissues are approximated as a liquid. Shear waves do, however, travel in harder biological materials such as bone.

III. COMPOSITION OF MATTER

Since acoustic waves requires a material medium, an understanding of the physical properties of matter provides a basis for studying how ultrasound propagates. Physical properties of matter include: volume (it occupies space), mass (a quantitative measure of matter), weight (the way we measure mass on earth) and inertia (any body resists change in motion).

Tissue is matter. Since matter is very complex, simple models are used for illustrative purposes, namely, the three states of gas, liquid and solid. There are many physical properties which are similar with gases and liquids and because of these similarities, both are referred to as fluids.

Matter is composed of molecules. These molecules are held together by forces which, for modeling purposes, can be thought of as tiny springs or rubber bands. This model can be used to describe what happens when a sound wave moves through matter. In Fig 1, the particles are interconnected by the (invisible) springs. When the sound wave, in this case the driving force, interacts with the first group of particles, the interaction causes the particles to be pushed towards the adjacent particles. Through this process, the sound wave sets up a chain reaction, but each subsequent particle moves a little less than its neighbor due to the fact that there is friction in the system. If the sound wave, which is the driving force, changes direction, then the particles also change the direction of their movement. This movement occurs over a very short distance of several micrometers or less.

Considering only the longitudinal wave, the alternate compression and decompression action shown in Fig 1a corresponds to the crest (positive deflection) and trough (negative deflection) of the sine wave form shown in Fig 1c. Note that the location where the sine wave is a maximum corresponds to where the particles are very close together; and where the sine wave is a minimum the particles are spread apart. The correspondence between the sine wave and the particle density is important since it is far more common to diagram this with a sine wave.

Later, the ideas of material stress and strain will be discussed (Feynman, et al., 1965). Such media, whether gases, liquids or solids, are termed elastic media. An elastic medium is homogeneous when its physical properties are not dependent upon the location in the medium. Consider, for example, that the molecular composition and density of a very small volume element within a material are measured at many different locations. If the composition and density are the same at all points, the material is said to be homogeneous. Otherwise, the material is said to be inhomogeneous.

An elastic medium is isotropic when its physical properties do not influence differently the direction that a wave may travel through it. If the physical properties which affect the transport of energy (such as heat, electricity, light, sound) are the same in all directions, the material is said to be isotropic. Otherwise, the material is said to be anisotropic.

IV.   DENSITY, ELASTICITY AND SPEED

Density is a property common to all matter, but also is a property that makes different types of matter unique (Feynman, et al., 1965). It is an important material property which has a direct affect upon ultrasonic properties such as speed. Density is defined in two ways. The one with which we deal in ultrasonics is called mass density, and is defined as,

$$\rho = \frac{m}{V} \tag{2}$$

where m is the mass and has the unit of kilogram (kg) and V is the volume and has the unit of meter cubed (m^3). The other, weight density, is defined in terms of the object's weight divided by its volume.

If one tries to change the size or shape of a solid by applying a force, the object will resist the attempt by trying to return to its initial condition once the deforming force is removed. That is, if the solid is deformed, it will return to its original shape and size when the cause of the deformation is removed. Elasticity is the property of recovering size and shape when the forces producing deformations are removed.

Elasticity is quantified by relating force to deformation and this ratio of "force" to "deformation" is called an elastic modulus (Feynman, et al., 1965). The deforming force, called stress, is represented as force per unit area and has the unit of newton per meter squared (N/m^2) or pascal (Pa):

$$STRESS = \frac{F}{A} \tag{3}$$

The deformation, called strain, is represented in terms of a relative change in dimensions when subjected to a stress, that is,

$$STRAIN = \frac{\Delta L}{L} \tag{4}$$

The ratio of stress to strain is called the elastic modulus

$$ELASTIC\ MODULUS = \frac{STRESS}{STRAIN} \tag{5}$$

All substances exhibit the property of elasticity and hence each can be quantitatively described in terms of an elastic modulus. There are a number of different types of elastic moduli, each based on how the force is applied to the object. The more common types of elastic moduli include: Young's modulus, shear modulus and bulk modulus. Bulk modulus is one of the elastic moduli that is most often associated with fluid media, that is, liquids and gases (Kinsler et al., 1982). The reciprocal of bulk modulus is called compressibility and, thus, has the unit of reciprocal pascals (Pa^{-1}). It is more common to refer to fluids, which are quite compressible, in terms of their compressibility instead of their elastic modulus.

The speed of sound depends on both the density and elasticity of materials, that is (Kinsler et al., 1982),

$$c = \sqrt{\frac{\text{ELASTIC MODULUS}}{\text{DENSITY}}} \tag{6}$$

In terms of the three physical states of matter (gas, liquid, solid), Table 1 quantifies and demonstrates how elasticity and density affect speed.

TABLE 1

Typical elasticity and density values for the
three states of matter.

	TYPICAL ELASTICITY (Pa)	TYPICAL DENSITY (kg/m^3)	CALCULATED SPEED (m/s)
GAS	10^5	1	316
LIQUID	10^9	1000	1000
SOLID	10^{11}	5000	4472

V. TISSUE AS MATTER

From an acoustics point of view, the physical properties of tissue can be classified as a quasi-liquid or a quasi-solid (Dunn and O'Brien, 1976). The prefix quasi means "seeming like a" and is appropriate because the acoustical behavior of tissue sometimes behaves like a liquid and sometimes like a solid. To the touch, tissue appears to be a solid and yet some of its acoustical properties are very similar to those of water, a liquid.

The physical properties of tissue are influenced by and composed of water, ions, macromolecules and cells and are a consequence of the chemical structures of fibrous and nonfibrous components. Tissues are divided into various kinds, including epithelial, muscular, connective, nervous, blood, etc. Each of these has different physical properties. Common to all tissues is a large amount of water. Selected physical properties of pure water at 37°C (98.6°F) are listed in Table 2 (Nyborg, 1975). The physical properties of tissue depend strongly upon water because water makes up almost three-quarters of the entire mass of the human body. The water concentration varies from tissue to tissue with vitreous humor quite high at around 99%, liver at 70%, skin at 60%, cartilage at 30% and adipose as low as 10%.

TABLE 2

Selected physical properties of pure water at 37°C.

compressibility: 4.4×10^{-10} Pa^{-1}
bulk modulus: 2.3×10^{9} Pa
density: 990 kg/m^3
speed: 1527 m/s

Ions affect the physical properties of tissue. When sodium chloride (NaCl), an electrolyte, is added to water, the density of the solution increases and the compressibility decreases (elasticity increases). The addition of electrolytes causes the water molecules to form a hydration sheath of relatively high density and high elasticity around each ion. At physiological concentrations, the density is about 0.6% greater and the elasticity is about 2% greater than for distilled water. Since elasticity increases at a greater rate than density, from the discussion in Section IV, it can be seen that speed increases.

Collagen is an important macromolecule which influences the physical properties of tissue. Collagen is a high tensile strength, insoluble fiber found in most connective tissues, such as the connective tissues of cartilage, tendon, bone, skin and muscles. It is the most abundant protein in the human body. It constitutes twenty-five to thirty-three percent of the total protein, and therefore, about six percent of the total body weight. Collagenous fibers exhibit an elastic modulus approximately 1000 times greater than that of other tissues.

Fat is an almost water free tissue. Total body water is dependent upon the total amount of body fat. At least ten percent of the body weight is comprised of lipid (fat), the most abundant of which are the triglycerides, which are found throughout the body, as well as in certain specialized connective tissues, namely the adipose tissue.

VI.   FREQUENCY, WAVELENGTH AND SPEED

We have many perceptions of the nature of sound. The idea of pitch refers to our perception of frequency, that is, the number of times a second that air vibrates in producing sound that we hear. Voices are classified according to pitch in which the lowest frequency is a bass voice and the highest frequency is a soprano voice. This description of frequency, however, is limited to the frequency range, or spectrum, over which humans can hear sounds. There are sound frequencies below and above what humans can hear. The acoustic spectrum is shown in Fig 2a.

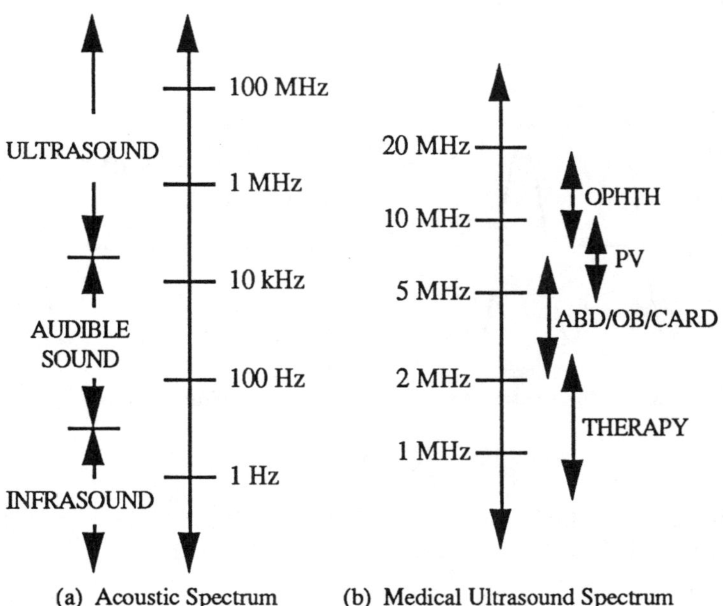

(a) Acoustic Spectrum   (b) Medical Ultrasound Spectrum

Figure 2

The lowest frequency classification in the acoustic spectrum is infrasound which has a frequency range below 20 Hz. Audible sound is what we hear and has an approximate frequency range of 20 Hz to 20 kHz. The ultrasound frequency range starts at a frequency of 20 kHz. Examples of devices that emit frequencies at the lower frequency end of the ultrasonic spectrum are a dog whistle and industrial ultrasonic cleaners.

Medical ultrasound equipment operates in the ultrasonic frequency range between 1 and 15 MHz (Fig 2b). Therapeutic (physical therapy) applications operate around 1 MHz. For most diagnostic applications in abdominal and OB-GYN ultrasound and in echocardiography, the frequency range is between 2.25 and 7.5 MHz. For very superficial body parts, such as the thyroid and the eye, and peripheral vascular applications where ultrasound does not have to penetrate very deeply into the body, higher ultrasonic frequencies in the range of 7.5 to 15 MHz can be used because ultrasonic attenuation increases with increasing frequency..

Ultrasound travels in waves that emanate from a source. The high crests and low troughs represent specific amplitude values of the wave and correspond to peak compressional and peak rarefactional values. The distance from one crest to the next, or from one trough to the next, has a particular distance associated with it and is called the wavelength and denoted by λ in Fig 3a.

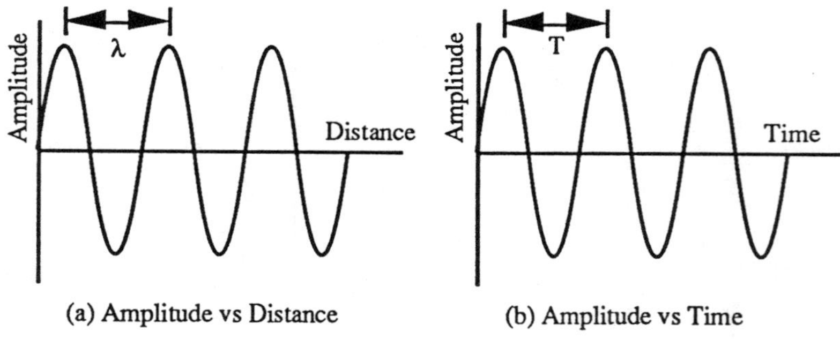

(a) Amplitude vs Distance   (b) Amplitude vs Time

Figure 3

The time that it takes for one cycle to occur is called the period (Fig 3b). The period (T) is the reciprocal of frequency (f), that is,

$$f = \frac{1}{T} \qquad (7)$$

In the diagnostic ultrasound frequency range, for example, the period for a frequency of 5 MHz is 0.2 μs (200 ns).

As demonstrated in Fig 3, the horizontal axis can illustrate either distance (Fig 3a) or time (Fig 3b). This is an important concept in diagnostic ultrasonic instrumentation. Distance information can be converted to time values, and time converted to distance information. Ultrasonic instruments are constantly performing these conversions in order to display sonographic images. The space (or distance) over which one cycle travels is called the wavelength and the time which one cycle occupies is called the period, that is, wavelength is "distance/cycle" and period is "time/cycle." Speed is the constant that relates wavelength (λ) to period:

$$c = \frac{\lambda}{T} = \lambda\, f \tag{8}$$

where, for medical applications, the tissue's propagation speed, c, is assumed to be constant at 1540 m/s. In the diagnostic ultrasound frequency range, for a frequency of 3.5 MHz, the wavelength is 0.44 mm (440 μm).

Propagation speed is very important in the proper design of diagnostic ultrasound systems. It must be known in order for the instrumentation to convert time values into distance or depth information, because the diagnostic system keeps track of only time. In order for the ultrasound instrumentation to perform this function, the speed must be set at a constant value. Although different tissues have different speeds, all of the propagation speeds in soft tissues fall within a rather narrow range. This narrow range allows for the use of an average speed which causes only a small margin of error when calculating distances or ranges in the body.

The propagation speed which has been most accepted for soft tissue is 1540 meters per second (1540 m/s, or 1.54 mm/μs). Another way to look at propagation speed is in terms of its reciprocal speed, which is 0.649 μs/mm or 6.49 μs/cm. This means that the wave travels a distance of 1 cm every 6.49 ms. If a structure is positioned 1 cm from the ultrasound source, then it would take 6.49 μs for the ultrasound wave to reach that structure and an additional 6.49 μs for an echo from that reflecting structure to return. Thus, the round trip reciprocal speed is 12.98 μs/cm and, for convenience, is usually quoted as 13 μs/cm. For example, if an object is located 10 cm from the source, then it would take 130 μs for the ultrasound wave to reach the reflecting structure and return to the source.

VII.   TEMPORAL CHARACTERISTICS

There are two basic generation modes of ultrasound used in medical ultrasound (Fig 4). Generation mode means the way in which the ultrasonic wave is "shaped" when it is transmitted from the ultrasonic transducer, that is, the waveform's temporal characteristics. One way is to continuously excite the ultrasonic transducer with an electrical sine wave at a constant amplitude. This produces a continuous ultrasonic wave at the same frequency as that of the electrical frequency and is termed continuous wave ultrasound (CW mode or CW ultrasound), as shown in Fig 4a.

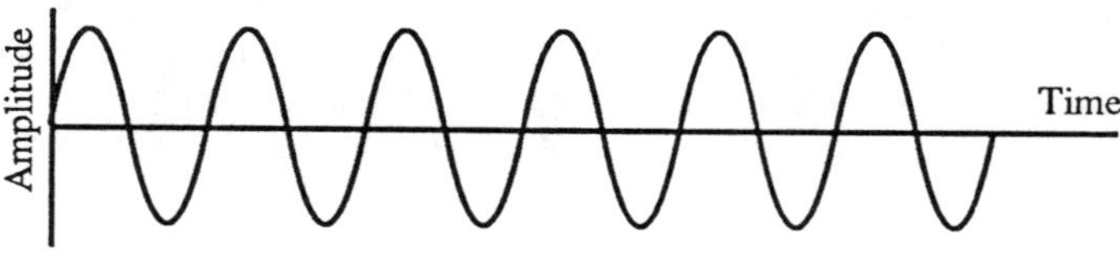

(a) Continuous Wave Representation

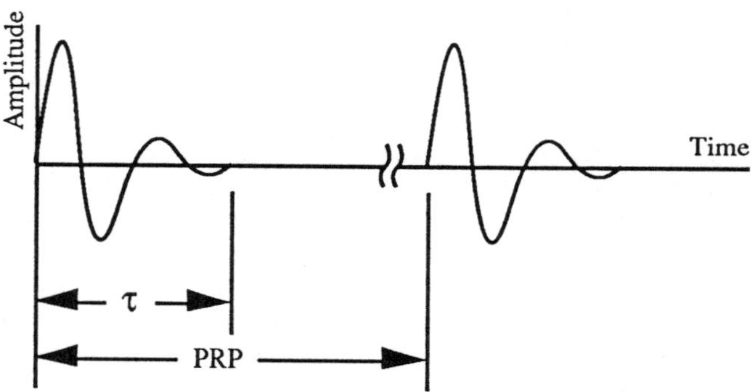

(b) Pulsed Wave Representation

Figure 4

Another way is to turn on the ultrasound for a very short period of time and turn it off for a much longer period of time and then to repeat this process. This is accomplished by exciting or shocking the ultrasonic transducer with very short electrical signals, waiting for some time and repeating the electrical shocking. The ultrasonic waves that are generated are termed pulse wave ultrasound (PW mode or PW ultrasound), as shown in Fig 4b.

Figs 3 and 4a show a CW ultrasound wave. To describe quantitatively a CW waveform, only two quantities are required, that is, amplitude and frequency (or period - see Eq 7).

To quantify a waveform of a single pulse (one of the pulses shown in Fig 4b), an additional piece of information is the time during which the pulse is on, termed the pulse duration (τ). If the number of cycles per pulse is N, then the pulse duration is

$$\tau = N\,T \tag{9a}$$

and from Eq 7

$$\tau = \frac{N}{f} \tag{9b}$$

To quantify a waveform of repeated pulses (Fig 4b), in addition to amplitude, frequency and pulse duration, the rate at which pulses are repeated is required, and quantified by either the pulse repetition frequency (PRF) or its reciprocal, the pulse repetition period (PRP).

The ratio of the pulse duration to the pulse repetition period is called the duty factor (DF), that is,

$$DF = \frac{\tau}{PRP} = \tau\,PRF \tag{10}$$

For example, if the pulse duration is 1 µs and the pulse repetition period is 1 ms (PRF = 1 kHz), then the duty factor is 0.001.

To summarize, the four quantities required to quantify a repeated pulse are: amplitude of the pulse, frequency of the ultrasonic signal in the pulse, pulse duration and pulse repetition frequency. Although these four quantities are sufficient, other combinations of these four quantities can also be used to quantify a repeated pulse waveform.

VII. WAVEFORM QUANTITIES IN A MEDIUM

A necessary concept to understand axial (or range) resolution is the distance one cycle (and hence one pulse) occupies in a medium. The distance one cycle occupies in a medium is the wavelength, λ (Eq 8).

For a repeated pulse waveform, as shown in Fig 5, the distance one pulse occupies in a medium is called the spatial pulse length (SPL), that is, the number of wavelengths per pulse where

$$SPL = N\,\lambda \tag{11a}$$

and from Eqs 9a and 7, respectively,

$$SPL = \frac{\tau}{T}\lambda = \tau\,f\,\lambda \tag{11b}$$

For example, at an ultrasonic frequency of 7.5 MHz, the wavelength in tissue is 0.205 mm and, for a three cycle pulse, the spatial pulse length is 0.615 mm.

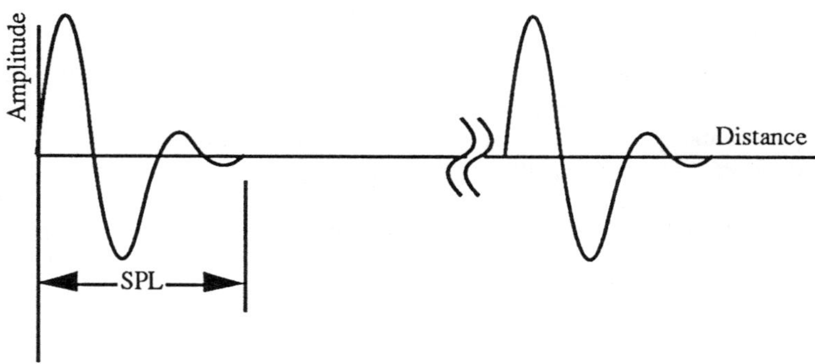

Figure 5

IX.   AMPLITUDE CHARACTERISTICS

In this section, amplitude quantities will be derived from basic principles under conditions in which the medium is assumed lossless and infinite in extent. It is the solution to the one-dimensional wave equation that yields the quantitative relations between the ultrasonic amplitude quantities (Kinsler *et al.*, 1982). The variables required to develop the lossless, one-dimension, acoustic wave equation are displacement, $\xi(x, t)$, density, $\rho(x, t)$ and pressure, $p(x,t)$ of a particle (Eq 1). Acoustic wave propagation, and the development of its wave equation, can be approached from the Equation of State which describes the change in density to the change in pressure, the Continuity Equation which relates particle motion to the change in density by invoking the conservation of mass principle and the Equation of Motion which compares the change in pressure to particle motion through Newton's Second Law of Dynamics.

The Equation of State is

$$p = \rho_e \left(\frac{dP}{d\rho}\right)_{\text{at constant } \rho_o} \tag{12}$$

where the propagation speed c_o is the derivative term

$$c_o{}^2 = \left(\frac{dP}{d\rho}\right)_{\text{at constant } \rho_o} \tag{13}$$

and, where the development was done under the small-signal conditions,

$$P = P_o + p \qquad p \ll P_o \tag{14a}$$
$$\rho = \rho_o + \rho_e \qquad \rho_e \ll \rho_o \tag{14b}$$

where P_o and ρ_o are the ambient pressure and undisturbed density, respectively, and p and ρ_e are the acoustic pressure and excess density, respectively. For a perfect gas, under adiabatic conditions where there is no heat transfer, that is, $pV^\gamma = \text{constant}$, where γ is the ratio of specific heats, Eq 13 yields

$$c_o{}^2 = \frac{\gamma P_o}{\rho_o} \tag{15}$$

The Continuity Equation expresses mass conservation as

$$\rho_e = -\rho_o \frac{\partial \xi}{\partial x} \tag{16}$$

which, in turn, expresses the fraction change in position of the particle displacement to the fraction change in density as

$$s = -\frac{\partial \xi}{\partial x} = \frac{\rho - \rho_o}{\rho_o} = \frac{\rho_e}{\rho_o} \tag{17}$$

where s is called condensation.

The Equation of Motion is expressed as

$$\frac{\partial p}{\partial (x+\xi)} = -\rho \frac{\partial^2 \xi}{\partial t^2} \tag{18}$$

Combining Eqs 12, 16 and 18 and linearizing yields the wave equation

$$\frac{\partial^2 \xi}{\partial t^2} = c_o{}^2 \frac{\partial^2 \xi}{\partial x^2} \tag{19}$$

The solution to Eq 19 considers one-dimensional traveling wave characteristics of the particle displacement, $\xi(x, t)$, particle velocity $u(x, t)$, particle acceleration $a(x, t)$ and acoustic pressure $p(x, t)$ in the positive x direction in a lossless medium, infinite in extent for a single frequency function. Assuming a solution to the wave equation as

$$\xi(x, t) = \xi_o \, \mathrm{Cos}\,(\omega t - kx) \tag{20}$$

where ξ_o is the amplitude particle displacement, $\omega = 2\pi f$ and k is the wave number ($= \omega/c_o$) and from this assumed solution, the particle displacement and particle acceleration is obtained, where

$$u(x, t) = \frac{\partial \xi}{\partial t} = -U_o \, \mathrm{Sin}\,(\omega t - kx) \tag{21a}$$

and

$$a(x, t) = \frac{\partial u}{\partial t} = -A_o \, \mathrm{Cos}\,(\omega t - kx) \tag{21b}$$

where U_o and A_o are their respective amplitude terms. To determine the acoustic pressure, $p(x, t)$, Eqs 12, 13 and 16 are combine to yield

$$p(x, t) = -\rho_o \, c_o^2 \, \frac{\partial \xi}{\partial x} \tag{22}$$

thus yielding

$$p(x, t) = -p_o \, \mathrm{Sin}\,(\omega t - kx) \tag{23}$$

In summary, the relationships between these amplitude terms are

$$\xi_o = \frac{U_o}{\omega} = \frac{A_o}{\omega^2} = \frac{p_o}{\omega \rho_o c_o} \tag{24a}$$

$$U_o = \omega \xi_o = \frac{A_o}{\omega} = \frac{p_o}{\rho_o c_o} \tag{24b}$$

$$A_o = \omega^2 \xi_o = \omega U_o = \frac{\omega p_o}{\rho_o c_o} \tag{24c}$$

$$p_o = \rho_o c_o \omega \xi_o = \rho_o c_o U_o = \frac{\rho_o c_o A_o}{\omega} \tag{24d}$$

For example, in water at 20°C where $\rho_o = 998$ kg/m^3 and $c_o = 1481$ m/s, at an ultrasonic frequency of 1 MHz, if $\xi_o = 18.5$ nm (185 Å), then $U_o = 12$ cm/s, $A_o = 7.3 \times 10^5$ m/s^2 and $p_o = 180$ kPa (1.8 atm).

X. WAVE PROPAGATION PROPERTIES

The propagation properties generally used to describe quantitatively the propagation of ultrasound in materials are speed, impedance and attenuation. The propagation of ultrasound is assumed to be an adiabatic process, that is, a process in which heat conduction does not occur. Therefore, the speed at which ultrasonic energy propagates in an isotropic fluid is (Pierce, 1981; Kinsler et al., 1982; Hall, 1987)

$$c_o = \sqrt{\frac{B_{AD}}{\rho_o}} \tag{25}$$

where B_{AD} is the adiabatic bulk modulus. Compared to Eq 6, the elastic modulus for an isotropic fluid is B_{AD}. As such, ultrasonic waves propagated in fluids are longitudinal waves. In a gas,

$$B_{AD} = \gamma P_o \tag{26a}$$

and

$$c_o = \sqrt{\frac{\gamma P_o}{\rho_o}} \tag{26b}$$

which agrees with Eq 15.

For a liquid, the elastic modulus is

$$B_{AD} = \gamma B_T \tag{27a}$$

where B_T is the isothermal bulk modulus and, therefore,

$$c_o = \sqrt{\frac{\gamma B_T}{\rho_o}} \tag{27b}$$

In an isotropic solid, both longitudinal and shear waves are supported wherein their respective propagation speeds are

$$c_L = \sqrt{\frac{Y(1-\sigma)}{\rho_o(1+\sigma)(1-2\sigma)}} \tag{28a}$$

and

$$c_S = \sqrt{\frac{Y}{2\rho_o(1+\sigma)}} \tag{28b}$$

where Y is the Young's modulus and σ is the Poisson's ratio. Since σ is less than 0.5, c_L is greater that c_S.

The specific acoustic impedance of the wave is defined as the ratio of the acoustic pressure to particle velocity. For plane waves under the conditions of Eqs 21a and 23, the specific acoustic impedance is defined as

$$Z_S = \frac{p(x, t)}{u(x, t)} \tag{29a}$$

where

$$Z_S = \frac{- p_o \sin(\omega t - kx)}{- U_o \sin(\omega t - kx)} = \rho_o c_o \tag{29b}$$

For other than plane waves, Z_S is generally different, that is, Z_S depends upon both the medium and the wave type (plane, cylindrical, spherical, etc). The $\rho_o c_o$ product is encountered frequently in analytic acoustics and is called the *characteristic acoustic impedance* of the medium or simply the acoustic impedance. Only for a plane wave are these two impedances the same. For an isotropic fluid, combining Eqs 29 with Eq 25 yield the characteristic acoustic impedance as

$$Z = \rho_o c_o = \sqrt{\rho_o B_{AD}} \tag{30}$$

and is a property <u>only</u> of the medium. The unit of the acoustic impedance is the rayl (kg/m^2s), after Lord Rayleigh. Table 3 summarizes the numerical ranges of ρ_o, c_o and Z for the various isotropic media.

TABLE 3

Typical density, propagation speed and characteristic
acoustic impedance values for isotropic media.

	ρ_o (kg/m^3)	c_o (m/s)	Z (rayl)
GAS	1	100-1000	100-1000
LIQUID	1000	1000-2000	1-2 x 10^6
SOLID	1000-10,000	2000-10,000(L)	10-100 x 10^6(L)
		1500-5000(S)	4-50 x 10^6(S)

where L represents a longitudinal wave and S a shear wave

XI. TRANSMISSION AND REFLECTION PHENOMENA

When an ultrasonic traveling wave impinges upon an acoustically discontinuous boundary, part of the energy is transmitted across the boundary into the second medium and part is reflected (Kinsler et al., 1982; Ensminger, 1988), as shown in Fig 6. If the incident wave in fluid medium 1 (in which the characteristic acoustic impedance $Z_1 = \rho_1 c_1$) impinges at an angle θ_1 relative to the normal of the boundary surface, then the reflected wave's angle is θ_1 and the transmitted wave's angle is θ_2 because of Snell's Law

$$k_1 \, \text{Sin} \, \theta_i = k_1 \, \text{Sin} \, \theta_r = k_2 \, \text{Sin} \, \theta_t \tag{31}$$

where k_1 (= ω/c_1) and k_2 (= ω/c_2)are the wave numbers in the two media, and θ_i, θ_r and θ_t are, respectively, the incident, reflected and transmitted angles relative to the normal of the boundary surface. Therefore, θ_i and θ_r are both equal to θ_1 because they are propagating in the same media with the same propagation speed, c_1, and θ_t is the same as θ_2. Eq 31 is more traditionally written as

$$\frac{\text{Sin} \, \theta_1}{c_1} = \frac{\text{Sin} \, \theta_2}{c_2} \tag{32}$$

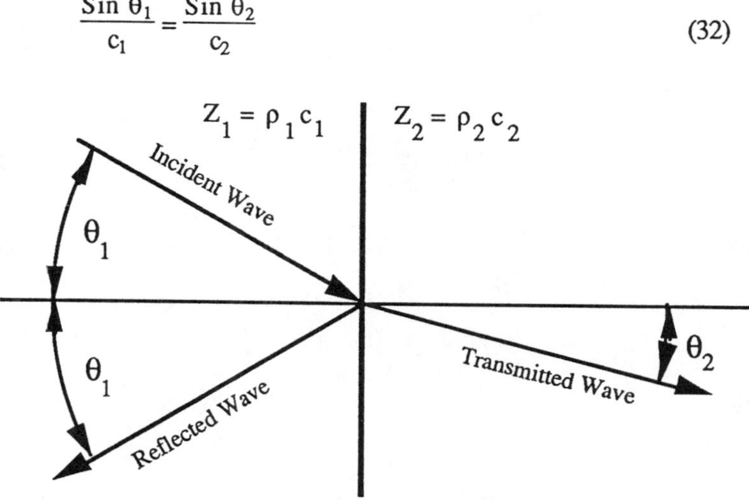

Figure 6

There are three conditions to evaluate Eq 32, that is, (1) normal incidence when $\theta_1 = 0$, (2) oblique incidence when c_1 is greater than c_2, and (3) oblique incidence when c_1 is less than c_2.

For normal incidence, $\theta_1 = 0$, and from Eq 32, $\theta_2 = 0$. For oblique incidence, if $c_1 > c_2$, then $\theta_1 > \theta_2$ which means the transmitted beam is bent towards the normal of the boundary surface as is shown in Fig 6.

For oblique incidence, if $c_1 < c_2$, then $\theta_1 < \theta_2$ which means the transmitted beam is bent away from the normal of the boundary surface. For this condition, a special case arises when the transmitted angle θ_2 is 90°. This condition is referred to as the *critical angle* for the incident angle, θ_c, that is,

$$\theta_c = \mathrm{Sin}^{-1}\left(\frac{c_1}{c_2}\right) \tag{33}$$

To evaluate the fractional amount of energy that is reflected from and transmitted across the boundary, as shown in Fig 6, two general quantities are used, one in term of the sound power reflected or transmitted and the other in terms of the sound intensity reflected or transmitted. The Sound Power Reflection (α_r) and Transmission Coefficients (α_t) are defined as, respectively,

$$\alpha_r = \frac{\text{Reflected Sound Power}}{\text{Incident Sound Power}} = \frac{W_r}{W_i} \tag{34a}$$

and

$$\alpha_t = \frac{\text{Transmitted Sound Power}}{\text{Incident Sound Power}} = \frac{W_t}{W_i} \tag{34b}$$

and the Sound Intensity Reflection (β_r) and Transmission Coefficients (β_t) are defined as, respectively,

$$\beta_r = \frac{\text{Reflected Sound Intensity}}{\text{Incident Sound Intensity}} = \frac{I_r}{I_i} \tag{35a}$$

and

$$\beta_t = \frac{\text{Transmitted Intensity Power}}{\text{Incident Sound Intensity}} = \frac{I_t}{I_i} \tag{35b}$$

Because of conservation of energy,

$$\alpha_r + \alpha_t = 1 \tag{36}$$

but the sum of β_r and β_t is not necessary unity.

In general, for oblique incidence,

$$\alpha_r = \left(\frac{Z_2 \, \text{Cos} \, \theta_1 - Z_1 \, \text{Cos} \, \theta_2}{Z_2 \, \text{Cos} \, \theta_1 + Z_1 \, \text{Cos} \, \theta_2} \right)^2 \qquad (37a)$$

and

$$\alpha_t = \frac{4 \, Z_1 \, Z_2 \, \text{Cos} \, \theta_1 \, \text{Cos} \, \theta_2}{(Z_2 \, \text{Cos} \, \theta_1 + Z_1 \, \text{Cos} \, \theta_2)^2} \qquad (37b)$$

and

$$\beta_r = \alpha_r = \left(\frac{Z_2 \, \text{Cos} \, \theta_1 - Z_1 \, \text{Cos} \, \theta_2}{Z_2 \, \text{Cos} \, \theta_1 + Z_1 \, \text{Cos} \, \theta_2} \right)^2 \qquad (37c)$$

and

$$\beta_t = \frac{4 \, Z_1 \, Z_2 \, \text{Cos}^2 \, \theta_1}{(Z_2 \, \text{Cos} \, \theta_1 + Z_1 \, \text{Cos} \, \theta_2)^2} \qquad (37d)$$

Note that $\alpha_r = \beta_r$ because the beam cross-sectional areas of the incident and reflected areas are equal whereas $\alpha_t \neq \beta_t$ because the beam cross-sectional areas of the incident and transmitted areas are, in general, not equal, except under normal incidence conditions.

At normal incidence where $\theta_1 = 0$, Eqs 37 become

$$\alpha_r = \beta_r = \left(\frac{Z_2 - Z_1}{Z_2 + Z_1} \right)^2 \qquad (38a)$$

and

$$\alpha_t = \beta_t = \frac{4 \, Z_1 \, Z_2}{(Z_2 + Z_1)^2} \qquad (38b)$$

Two cases under normal incident conditions of particular interest for medical ultrasound are (1) the acoustic impedances of the two media are very different and (2) the acoustic impedances are very similar.

For case (1) where either $Z_1 \gg Z_2$ or $Z_2 \gg Z_1$, Eqs 38 become

$$\alpha_r = \beta_r \approx 1 \qquad (39a)$$

and

$$\alpha_t = \beta_t \approx 0 \qquad (39b)$$

which means that virtually all of the incident energy is reflected and almost none is transmitted.

For case (2) where $Z_1 \approx Z_2$, Eqs 38 become

$$\alpha_r = \beta_r \approx 0 \tag{40a}$$

and

$$\alpha_t = \beta_t \approx 1 \tag{40b}$$

which means that virtually all of the incident energy is transmitted and almost none is reflected.

For medical ultrasound applications, it is very common that the characteristic acoustic impedances between two adjacent tissue types are within 1% or less of each other. For the case where the difference is 1%, that is, $Z_1 = 1.500$ Mrayles and $Z_2 = 1.515$ Mrayles, at normal incidence,

$$\alpha_r = \beta_r = 0.0000248 \tag{41a}$$

and

$$\alpha_t = \beta_t = 0.9999752 \tag{41b}$$

which means that only 0.00248% of the power (or energy) is reflected back from the boundary and 99.99752% is transmitted into the next medium. The reflected signal is about 46 dB down relative to the incident signal and for diagnostic ultrasound applications, tissue reflection coefficients range as low as 75 dB.

XII. RESOLUTION TRADE-OFFS AND CONCEPTS

The classical engineering trade-off of diagnostic ultrasound instrumentation is that between resolution and the depth of the image (depth of penetration). Both are directly affected by the ultrasonic frequency. As frequency is increased, resolution improves and penetration decreases. Resolution improves because the ultrasonic wavelength in tissue decreases (becomes a smaller number). Wavelength is inversely related to frequency; increase one and the other decreases.

As frequency increases, the ultrasonic attenuation also increases. Penetration is directly affected by the tissue attenuation coefficient which, in turn, is directly related to frequency. At an ultrasonic frequency of 1 MHz, an "average" attenuation coefficient for soft tissue is approximately 0.7 dB/cm whereas at 2 MHz, it is 1.4 dB/cm. Thus, the attenuation coefficient is directly related to frequency; increase one and the other increases. Thus, the attenuation coefficient can be normalized to frequency as 0.7 dB/cm-MHz.

Resolution is the ability to image or resolve discrete structures. Resolution is determined by many components and properties of the instrumentation and patient including transducer type, beam geometry, frequency and bandwidth; receiving and processing electronics; video monitor; and tissue attenuation and sound speed. For simplicity, it is easier to understand resolution by considering two types of resolution, *viz.*, axial resolution and lateral resolution.

Axial resolution (also termed range resolution or depth resolution) is the ability to resolve discrete structures along the beam axis. Quantitatively, it is represented as the minimum distance between two structures at different ranges at which both can just be discretely identified as two separate structures. The best axial resolution is represented by the expression

$$\text{best axial resolution} = \frac{\text{SPL}}{2} = \frac{N\lambda}{2} \tag{42}$$

where SPL is the spatial pulse length (see Eq 11). The transducer design affects the minimum number of cycles. More highly damped transducers (also referred to as low Q transducers) produce very few cycles of ultrasound when excited by the pulser voltage. If $N = 3$, at ultrasonic frequencies of 3.5 MHz ($\lambda = 0.44$ mm) and 7.5 MHz ($\lambda = 0.21$ mm), then, from Eq 42, the best axial resolutions are 0.67 mm and 0.32 mm, respectively. As the frequency increases, and other quantities remain constant, axial resolution improves.

The term "best axial resolution" has been employed because, in practice, the receiving and processing electronics affect axial resolution as does the quality of the video monitor. The electronics and monitor are often lumped into the term "system Q." Low-valued system Qs provide better axial resolution than do high-valued ones.

Lateral resolution is the ability to resolve discrete structures perpendicular, or lateral, to the beam axis. Quantitatively, it is represented as the minimum distance between two side-by-side structures at the same range at which both can just be discretely identified as two separate structures. The best lateral resolution is represented by the expression

$$\text{best lateral resolution} = \text{minimum beam width.} \tag{43}$$

The "best lateral resolution" term is employed here for the same reasons that the term "best axial resolution" was used.

Fig 7 shows the beam width of two, unfocused, plane piston source transducers (operating at the same frequency) of different radii, a_1 and a_2, where a_2 is greater than a_1. For the a_2 transducer, the distance of the near field is longer but the beam width is also wider. In other words, the best lateral resolution of a_2 transducer is worse than that of a_1 transducer in the near field. However, in the far field, at a sufficient range, the best lateral resolution is worse for the a_1 transducer demonstrating that lateral resolution is a function of imaging depth.

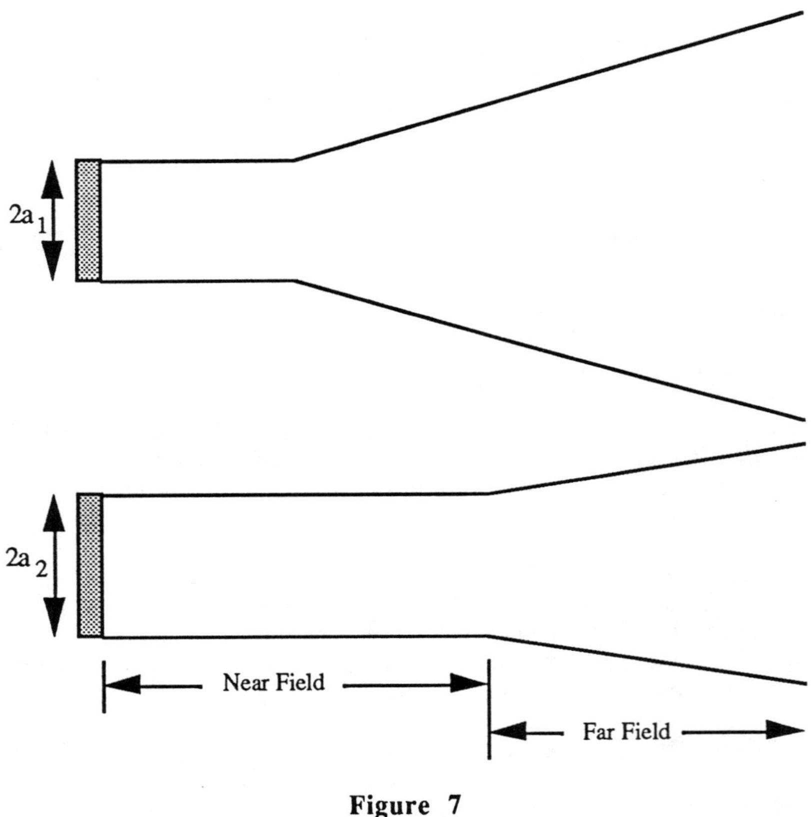

Figure 7

The range of the near field (also termed the Fresnel zone) is a function of transducer dimensions and wavelength, λ, by the expression

$$\text{near field range} = \frac{a^2}{\lambda} \qquad (44)$$

where a is the transducer radius. For an unfocused, plane piston source transducer, the best lateral resolution in the near field is affected mainly by transducer radius. Wavelength also affects lateral resolution here in terms of maintaining the same lateral resolution over the near field range.

When an ultrasonic field is focused, the focal range occurs in the near field of the transducer. Fig 8 shows the beam width from the same transducer operating at the same frequency but for two different focal lengths (focal length is the distance along the beam axis from the transducer to the focus). For the longer focus length, the minimum beam width is greater than that for the short focus case. The best lateral beam width at focus (BW) is directly proportional to wavelength (λ) and focal length (ROC, which stands for radius of curvature) and is inversely proportional to the transducer diameter (D), that is,

$$BW = \frac{1.4\ \lambda\ \text{ROC}}{D} \tag{45}$$

In imaging terminology, the term "f-number" or "f#" is often used to quantitate focusing where the lower the f-number value, the better is the focusing. The best lateral beam width at focus is related to the $f^{\#}$ by

$$BW = 1.4\ \lambda\ \frac{\text{ROC}}{D} = 1.4\ \lambda\ f^{\#} \tag{46}$$

where, at the same frequency, BW can be improved by decreasing the $f^{\#}$ (the ratio of ROC to D). Another way to improve lateral resolution at the focus is to decrease the wavelength (increase frequency).

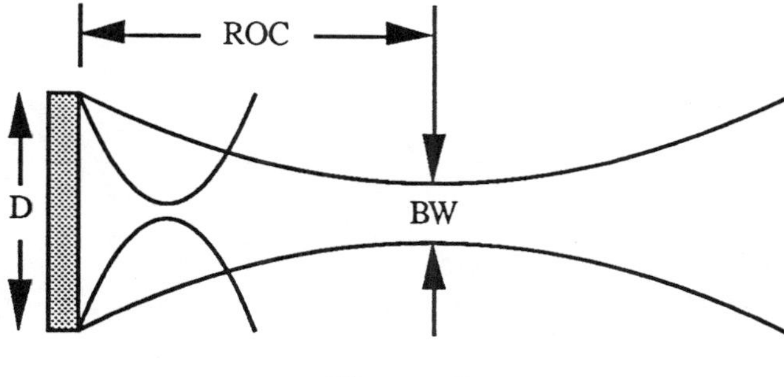

Figure 8

In summary, axial resolution is affected by the wavelength, number of cycles per pulse and system Q. As axial resolution improves, the wavelength decreases (frequency increases), the number of cycles per pulse decreases and the system Q decreases. Lateral resolution is affected by the wavelength, transducer size and geometry (focusing), and focal range. As lateral resolution improves, the wavelength decreases (frequency increases), transducer size increases and the focal range decreases. Away from the focal range axially, the lateral resolution quickly deteriorates.

XIII. REFERENCES

Dunn, F. and O'Brien, W. D., Jr. (eds) (1976). "Ultrasonic Biophysics," Dowden Hutchinson & Ross, Stroudsburg, PA.

Ensminger, D. (1988). "Ultrasonics: Fundamentals, Technology, Applications," Second Edition, Marcel Dekker, Inc., New York, NY.

Feynman, R. P., Leighton, R. B. and Sands, M. (1965). The Feynman Lectures on Physics," Addison-Wesley, Reading, MA.

Hall, D. E. (1987). "Basic Acoustics," Harper & Row, New York, NY

Kinsler, L. E., Frey, A. R., Coopens, A. B. and Sanders, J. V. (1982) "Fundamentals of Acoustics," Third Edition, Wiley, New York, NY.

Nyborg, W. L. (1975) "Intermediate Biophysical Mechanics," Cummings Publishing Co., Menlo Park, CA.

Pierce, A. D. (1981). "Acoustics: An Introduction to Its Physical Principles and Applications," McGraw Hill, New York, NY.

Temkin, S (1981). "Elements of Acoustics," Wiley, New York, NY

ULTRASOUND DOSIMETRY
AND INTERACTION MECHANISMS

William D. O'Brien, Jr.

Department of Electrical and Computer Engineering
University of Illinois
1406 West Green Street
Urbana, IL 61801 USA

I. INTRODUCTION

Ultrasonic biophysics is the study of mechanisms responsible for how ultrasound and biological materials interact. As shown in Fig 1, when one studies how ultrasound affects biological materials, this can be viewed as bioeffect studies or risk studies. On the other hand, the study of how tissue affects the ultrasound wave can be viewed as the basis for diagnostic ultrasound. Thus, an understanding of the interaction of ultrasound with tissue provides the scientific basis for understanding image production and risk assessment.

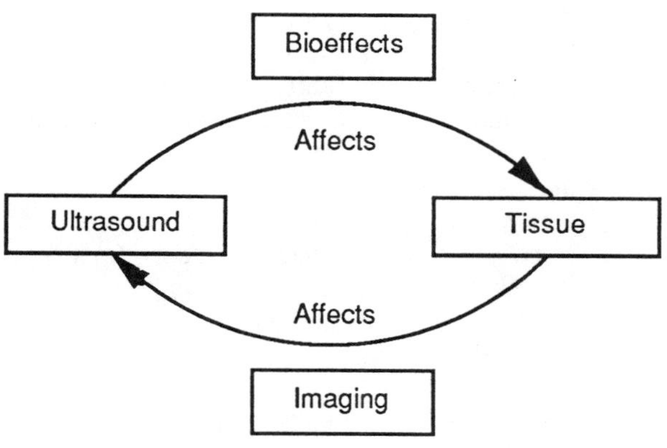

Figure 1

Ultrasonic dosimetry (O'Brien, 1978; O'Brien, 1986) is concerned with the quantitative determination of ultrasonic energy interaction with biological materials, that is, defining the quantitative relationship between some physical agent and the biological effect it produces. To understand more fully ultrasound dosimetry and interaction mechanisms, it is first appropriate to develop common nomenclature. Then, general dosimetric concepts will be presented since a large body of literature and history exists to quantitate the interaction of various propagated energies and biological materials. Ultrasonic dosimetry and its current status will be presented. To conclude, interaction mechanisms, both thermal and nonthermal, are discussed.

II. FIRST-ORDER AND SECOND-ORDER QUANTITIES

There are many buzz words to describe a general class of events such as the terms *first-order quantity* and *second-order quantity*. *Quantity* represents what is measured and *unit* represents the amount (see Table 1).

TABLE 1

Typical ultrasonic quantities and units

QUANTITY	UNIT
charge	coulomb (C)
current	ampere (A = C/s)
displacement	meter (m)
energy	joule (J = Ws)
energy density	joule per meter cubed ($J/m^3 = N/m^2$)
force	newton (N)
frequency	hertz (Hz)
intensity	watt per centimeter squared (W/cm^2)
length	meter (m)
mass	kilogram (kg)
power	watt (W)
speed	meter per second (m/s)
temperature	degree celsius (°C)
time	second (s)
ultrasonic pressure	pascal ($Pa = N/m^2$)
voltage	volt (V)
wavelength	meter (m)

First-order quantities are known as amplitude quantities and second-order quantities as energy quantities and are listed in Table 2. The basic ideas of first-order and second-order quantities are (1) both first-order and second-order quantities deal with the transport of energy, (2) all first-order quantities are directly proportional to each other, (3) all second-order quantities are directly proportional to each other, and (4) the product of any two first-order quantities is directly proportional to any second-order quantity, that is,

$$\text{any first-order quantity} \propto \text{any other first-order quantity} \qquad (1a)$$
$$\text{any second-order quantity} \propto \text{any other second-order quantity} \qquad (1b)$$
$$(\text{any first-order quantity})^2 \propto \text{any second-order quantity} \qquad (1c)$$

TABLE 2

List of first-order and second-order quantities used in ultrasound.

FIRST-ORDER QUANTITIES	SECOND-ORDER QUANTITIES
current	energy
particle acceleration	energy density
particle displacement	intensity
particle velocity	power
ultrasonic pressure	
voltage	

All ultrasonic amplitude quantities are directly proportional to each other. These quantities are (O'Brien, 1992)

$$\xi_o = \frac{U_o}{\omega} = \frac{A_o}{\omega^2} = \frac{p_o}{\omega \rho_o c_o} \qquad (2a)$$

$$U_o = \omega \xi_o = \frac{A_o}{\omega} = \frac{p_o}{\rho_o c_o} \qquad (2b)$$

$$A_o = \omega^2 \xi_o = \omega U_o = \frac{\omega p_o}{\rho_o c_o} \qquad (2c)$$

$$p_o = \rho_o c_o \omega \xi_o = \rho_o c_o U_o = \frac{\rho_o c_o A_o}{\omega} \qquad (2d)$$

where ξ_o, U_o, A_o and p_o are the particle displacement, particle velocity, particle acceleration and ultrasonic pressure, respectively, and ω is the angular frequency. In addition, the characteristic acoustic impedance is defined as

$$Z = \rho_o c_o \qquad (3)$$

where ρ_o and c_o are the undisturbed density and propagation speed, respectively.

Fundamentally, ultrasonic wave propagation transports energy. The total energy in a particle volume element is the sum of the particle's kinetic and potential energies, given by (Pierce, 1981; Kinsler et al., 1982; Hall, 1987; Ensminger, 1988)

$$\Delta E = \Delta KE + \Delta PE \tag{4}$$

The particle's kinetic energy represents the motion of the particle from

$$\Delta KE = \frac{1}{2} \, (\text{mass}) \, (\text{velocity})^2 \tag{5}$$

From the development of the continuity equation (Eq 16 in O'Brien, 1992), the undisturbed and disturbed incremental masses are, respectively,

$$M_u = \rho_o \, A \, \delta x \tag{6a}$$

$$M_d = \rho \, A \left(1 + \frac{\partial \xi}{\partial x} \right) \delta x \tag{6b}$$

which are the products, respectively, of the undisturbed and disturbed densities (ρ_o and ρ) and volumes $A \, \delta x$ and $A \left(1 + \frac{\partial \xi}{\partial x} \right) \delta x$. Since $M_u = M_d$ (conservation of mass), Eq 5 becomes

$$\Delta KE = \frac{1}{2} \, (\rho_o \, A \, \delta x) \left(\frac{\partial \xi}{\partial t} \right)^2 = \frac{1}{2} \, \rho \left(1 + \frac{\partial \xi}{\partial x} \right) A \, \delta x \left(\frac{\partial \xi}{\partial t} \right)^2 \tag{7}$$

where the particle velocity $u(x, t) = \frac{\partial \xi}{\partial t}$.

The particle's potential energy represents the compressed fluid state of the particle from

$$\Delta PE = - \int p \, dV \tag{8}$$

From the development of the continuity equation (Eqs 17 and 22 in O'Brien, 1992),

$$dV = - \frac{V_o}{\rho_o c_o^2} \, dp \tag{9}$$

which yields

$$\Delta PE = - \int p \left(- \frac{V_o}{\rho_o c_o^2} \, dp \right) = \frac{V_o}{\rho_o c_o^2} \frac{p^2}{2} + \text{constant} \tag{10}$$

where V_o $(= A \, \delta x)$ is the undisturbed particle volume. The constant is zero because under conditions in which the acoustic pressure is absent (the undisturbed condition), the particle's potential energy is zero. Therefore, the particle's total energy (Eq 4), from Eqs 7 and 10, becomes

$$\Delta E = \frac{1}{2} V_o \rho_o \left(u^2 + \frac{p^2}{\rho_o^2 c_o^2} \right) \tag{11}$$

The total energy per particle volume, or the total energy per volume, is the instantaneous energy density and is found from Eq 11 to be

$$\mathcal{E}(x, t) = \frac{\Delta E}{V_o} = \frac{1}{2} \rho_o \left(u^2 + \frac{p^2}{\rho_o^2 c_o^2} \right) \tag{12}$$

where u and p are the respective instantaneous values of particle velocity and acoustic pressure.

To evaluate the average energy density, the one-dimensional, harmonically varying particle velocity is assumed to be

$$u(x, t) = U_{op} \cos(\omega t - kx) + U_{on} \cos(\omega t + kx) \tag{13}$$

where U_{op} and U_{on} are the amplitude particle velocities for the positive and negative directed components, respectively, and the one-dimensional, harmonically varying ultrasonic pressure (from Sec IX of O'Brien, 1992) is

$$p(x, t) = p_{op} \cos(\omega t - kx) + p_{on} \cos(\omega t + kx) \tag{14}$$

where $p_{op} = \rho_o c_o U_{op}$ and $p_{on} = -\rho_o c_o U_{on}$. Therefore, the average energy density is

$$\langle \mathcal{E} \rangle = \frac{1}{T} \int_0^T \mathcal{E}(x, t) \, dt = \frac{\rho_o}{2} \left(U_{op}^2 + U_{on}^2 \right) \tag{15}$$

The same expression for $\langle \mathcal{E} \rangle$ results whether $\mathcal{E}(x, t)$ is averaged over either time or space.

Intensity is an extremely useful ultrasonic quantity which represents a measure of ultrasonic power flowing (time-averaged rate of flow of energy) at normal incidence to a specified unit area. The intensity concept is generally applied in connection with a traveling plane wave. Further, it is a vector quantity but, since the discussion herein is confined to an isotropic fluid and to the one-dimensional wave equation, vector notation is not used since the direction is known. The instantaneous intensity is defined as the dot product of the ultrasonic pressure and particle velocity. Its time-averaged representation is given by

$$I = \frac{1}{T}\int_0^T pu\,dt = \frac{\rho_o c_o}{2}\left(U_{op}^2 - U_{on}^2\right) \tag{16}$$

It should be noted that for a standing wave where $U_{op}^2 = U_{on}^2$, the time-averaged intensity is zero whereas the time-averaged energy density is not.

For a plane progressive ultrasonic wave propagating in only the $+x$ direction, $U_{on} = 0$, and Eqs 15 and 16 become

$$<\mathcal{E}> = \frac{\rho_o}{2}U_o^2 = \frac{1}{2\rho_o c_o^2}p_o^2 \tag{17}$$

and

$$I = \frac{\rho_o c_o}{2}U_o^2 = \frac{1}{2\rho_o c_o}p_o^2 = \frac{p_o U_o}{2} \tag{18}$$

and, combining these results yields

$$<\mathcal{E}> = \frac{I}{c_o} \tag{19}$$

which is an extremely useful expression in terms of measuring ultrasonic intensity and ultrasonic power with radiation force techniques.

III. DOSIMETRIC CONCEPTS

Dosimetry is the determination of a dose, or similar type of physical quantity, which characterizes the physical agent as to its potential or actual interaction with the biological material of interest. Ultrasonic dosimetry's objective is to relate magnitudes of specific quantities, such as intensity, acoustic pressure, particle displacement, etc., or perhaps some quantity yet to be developed, to the likelihood of occurrence of a biological alteration. To accomplish this, it is necessary (1) to quantify the output quantity or quantities of the source, (2) to determine the effect of the material on the propagating energy, *viz.*, reflections, refraction, scattering, absorption, etc., and (3) to relate quantitatively the first two items at the site of interest.

Typically, dose connotes something that is given or imparted in a quantitative manner. The history of other radiation forms has documented that defining dose, or dose-like concepts, is difficult, especially when the objective is to include all possible physical and biological variables. More commonly, however, special quantities are developed for the biological action under consideration. In ionizing radiation, for example, dose generally refers to the quantity *absorbed dose* which has been specifically defined as the energy imparted to matter by ionizing radiation per unit mass of irradiated material at the site of interest (ICRU, 1971). But other dose quantities have been defined for specific purposes such as genetically significant dose, cumulative dose, dose equivalent, threshold dose, etc (BEIR, 1972). In photobiology, dose sometimes refers to the quantity *dose of ultraviolet radiation* which has been defined as the energy per unit surface area applied to an object (Rupert, 1974). There has been much discussion regarding microwave dosimetry. Terms such as specific absorption rate, absorbed power density and specific absorption density and energy dose-rate have been used as a basic quantity to describe absorbed electromagnetic energy (Anderson, 1992; Bernhardt, 1992; Grandolfo, 1992; Leonowich, 1992; Sliney, 1992)

IV. SPECIFIC DOSIMETRIC EXAMPLE

For illustrative purposes, it is useful to examine in some detail the history of another radiation form with a view towards its dosimetry development. Ionizing radiation is chosen because it represents a well developed history (Spiers, 1956; Wang and Robbins, 1956; Spiers, 1964; Roesch and Attix, 1968; Taylor, 1971; Bushong, 1973). However, this examination is not meant to indicate that ultrasonic dosimetry should take this route. The mechanisms of interaction between ultrasonic and ionizing radiation are quite different. Rather, the rationale for ionizing radiation dosimetry, as with other radiations, is one of developing an acceptable and reasonable nomenclature by which researchers in various fields can compare results and from which radiation protection guidelines can be developed.

Knowledge of the energy deposition of the tissue site of interest is one of the critical elements in understanding the interaction between the radiation form and matter. One of the earliest ionizing radiation dosimetric concepts was the *skin unit dose*, more commonly known as the *skin erythema dose* or *threshold erythema dose*. One skin unit dose was the amount of ionizing radiation which just produced skin erythema within a period of about one week. The detector, human skin, was very imprecise but the skin unit dose was, nevertheless, used as a basis for the first radiation protection guideline in the mid 1920's. The *tolerance dose* was suggested to be a small fraction, around one percent, of a skin unit dose, averaged over a one month period.

Not until sensitive and reproducible measurement devices were developed was there a physical measurement of ionizing radiation. The concept and value of the unit *roentgen* (R) was established in 1928 and defined in terms of the ionization, or interaction, of x-ray radiation in air (ICRU, 1928). It was a special unit of exposure but no specific quantity was defined at that time for which the roentgen was its unit.

In an effort to relate the tolerance dose to a physical quantity, radiotherapists were polled as to the number of roentgens required to produce one skin unit dose. Based upon a rough value of 600 R for one skin unit dose, the tolerance dose worked out to be 6 R on a monthly basis, or 0.2 R/day. In the early to mid-1930's national and international organizations endorsed a tolerance dose of 0.2 R/day. Later this value was reduced to 0.1 R/day and remained at this level for 12 years. The term tolerance dose created many problems because it was impossible to predict just what level was tolerable over a long period of time. With the realization in the mid-1930's that ionizing radiation effects may not be threshold type reactions, the term *maximum permissible dose* was substituted for tolerance dose.

As a result of biophysical and biological effect studies with various types (qualities) of ionizing radiation, it was recognized that broader dosimetric concepts were required to define and describe quantitatively ionizing radiation fields. This was especially important when applying dosimetry to radiation protection in that the roentgen was inadequate because of its limitation to x- and γ-rays and because it was not a measure of absorbed energy. In the early 1930's, it was shown that the biological effect of ionizing radiation depended not only upon the exposure intensity and time but also upon the quality of radiation since differences between x-rays and γ-rays were observed in growth reduction and mortality studies (Failla and Henshaw, 1931). This was termed the *relative biological effectiveness* (RBE) concept and became even more important in the 1940's with the production and discovery of other ionizing radiation particles. In terms of the absorbed dose unit rad, which will be discussed shortly, this meant that the same number of rads of neutrons, for example, would produce a greater biological effect as compared x- or γ-rays.

In the late 1930's a unit, termed the *energy unit*, was suggested as dose of γ-rays delivered to tissue in terms of absorbed energy per gram tissue (Gray and Read, 1939). Also, around this time, another unit, the *gram-roentgen*, was suggested (Mayneord, 1940) as the amount of energy absorbed by one gram of air when irradiated to about 1 R. In the late 1940's another unit was suggested to describe energy absorption, *viz.*, the *rep* for *roentgen-*

equivalent-physical (Parker, 1948). Originally, one rep was defined as that dose of ionizing radiation which produced an energy absorption of 84 ergs/cm^3 in tissue. It was based upon the roentgen in that this was meant to be the energy absorbed by tissue when exposed to 1 R. That meant that the definition depended upon a calculation of energy absorption and upon other tissue parameters which were subject, in some cases, to wide uncertainty. These difficulties were reflected in redefining one rep from the original value to other values in order to reflect actual tissue absorption properties and to remove dependence of tissue density.

The rep concept led to what is currently the quantity *absorbed dose* with the unit *rad* for radiation absorbed dose. The advantage of the unit rad over the rep was that the one rad was arbitrarily defined as 100 ergs/gm and thus was independent of material properties (ICRU, 1954).

At the same time the rep was being suggested as a unit to describe dose, the unit *rem*, for *roentgen-equivalent-man*, was also being suggested for radiation protection purposes. The rem was defined as the product of energy absorption (in reps) and the relative biological effectiveness (dimensionless) of the energy under consideration. If there were energies of different RBEs, then the rem was the sum of each respective product (Parker, 1948). In the mid-1950's, the rem concept was adopted, using the rad instead of the rep. The quantity *RBE dose* in rems was equal to the product of absorbed dose in rads and the RBE and, in the case of multiple RBEs, the sum of each product (ICRU, 1956).

A few years later, RBE was changed to *quality factor*, QF, and assigned fixed values which were closely representative of actual RBE's for specific conditions and energies. This was done because RBE itself was dependent upon a large number of variables and for radiation protection purposes, the quality factor values chosen were representative of RBEs. Thus, the quantity *dose equivalent* was adopted, its unit the rem, and was equal to the product of absorbed dose, quality factor and other dose modifying factors to account for spatial and temporal dose distribution (ICRU, 1962).

In the history of ionizing radiation dosimetry is reflected the rationale for which national and international commissions have labored to develop concepts and define units and quantities. Initially, the threshold dose was defined as a monthly fraction of the skin unit dose and later, following proper instrumentation development, in terms of an exposure in roentgens. The term threshold dose was later called maximum permissible dose because of the realization that risk from ionizing radiation may not be represented by a threshold. Because of the desire to express the effect of ionizing radiation in terms of the interaction with or absorption by tissue at the site of interest, the absorbed dose concept was defined. Eventually the dose equivalent quantity evolved to embody both physical and biological quantities.

V. ULTRASOUND DOSIMETRY

By comparison, the field of ultrasonic dosimetry has not developed to the extent of ionizing radiation dosimetry. The most widely used quantity in ultrasonic bioeffect and biophysical studies is intensity in the unit of W/cm^2. The principal reason for the use of intensity is, perhaps, convenience since it is understood how it's measured (Lewin, 1992). However, intensity represents many of the same problems as does the ionizing radiation quantity "exposure" in that it is not a measure of dose. Yet the majority of bioeffect and biophysical reports use intensity as the measured physical quantity of the ultrasonic field. This extensive literature documents the actions of ultrasound but, in most cases, lacks the necessary characterization of the field at the site of interest. An ideal situation would be to know the instantaneous particle velocity, the instantaneous acoustic pressure and the phase between these two field parameters at the site or sites of interest (O'Brien et al, 1972; O'Brien, 1978; O'Brien, 1986).

There have been three ultrasonic dosimetric quantities which are noteworthy of comment in that they represent, in concept, the basic approach to dosimetry. The *cataract-producing unit*, CPU, was a quantity defined as the length of exposure necessary to produce a grossly observable cataract and expressed in units of seconds (Purnell et al, 1964). The dosimetric concept *damage ability index* with the unit second is a quantity intended to describe the effect of ultrasound on spinal cord hemorrhage (Taylor and Pond, 1972). It has been suggested (Johnston and Dunn, 1976) that a universal dosimetric response to ultrasonic exposure may exist for different tissues but the response has only been demonstrated, in a limited manner, in mammalian brain tissue. The response is in terms of *energy absorbed per unit volume* for histologically observable lesions at superthreshold levels as a function of the *delivered intensity*. It is shown that at two different ultrasonic frequencies, 3 and 4 MHz, identical constant volume curves result even though there are two different threshold levels (Dunn and Fry, 1971).

In another category of ultrasonic dosimetric studies, *in utero* ultrasonic intensity in both the gravid and nongravid human uterus have been estimated (summarized in Stewart and Stratmeyer (1982) and NCRP (1983)). In these early studies, a model of the tissue layers between the skin surface and fetal sac yielded a total attenuation in the range of 2 - 20 dB at frequencies between 2 and 5 MHz. The distances between the abdominal surface and the uterine cavity in early pregnancy ranged between 2 and 11 cm. In more recent work (Carson et al, 1989), similar distances were estimated to be 2.6

cm.

In very recent work wherein direct *in utero* intensity measurements were made (Daft et al, 1990; Siddiqi et al, 1991), the average attenuation was reported to be 6.2 ± 3.5 dB under full bladder conditions and 7.3 ± 4.9 dB under empty bladder conditions. Applying the fixed-attenuation tissue model (attenuation dependent upon frequency and independent of distance), and normalizing to the center frequency of 2.4 MHz, the attenuation coefficient was estimated to be 2.56 ± 1.47 dB/MHz for the full bladder condition whereas, with the overlying tissue model (attenuation dependent upon frequency and non-fluid distance), the attenuation coefficients were estimated to be 0.89 ± 0.71 dB/cm-MHz and 0.45 ± 0.32 dB/cm-MHz, respectively, for full and empty bladder conditions. The mean values for the fixed-attenuation tissue model's attenuation coefficient were about a factor of 3 greater than the values proposed to model the attenuation coefficient by Carson et al. (1989). FDA's Center for Devices and Radiological Health uses a value of 0.3 dB/cm-MHz as a derating factor for manufacturers in their 510(k) process to estimate ultrasonic intensity quantities in tissue (FDA, 1985; Harris, 1992). The measured mean values for the overlying tissue model's attenuation coefficient were a factor of 2 to 3 greater than the values used by FDA suggesting that FDA's values error on the side of safety.

In general, it is necessary to determine a firm data base from which various dosimetric modeling approaches can be explored. In one approach (Carson et al, 1989), a worst-case approach was employed to overestimate safety whereas, in another approach (Siddiqi et al, 1991), the mean values of the results and their distributions were reported so that the scientific community could make the appropriate safety judgements.

VI. ULTRASOUND DOSIMETRY RELEVANCE

Ultrasonic biological effect studies and biophysical research have shown that ultrasound can produce changes in living systems. The AIUM/NEMA Ultrasound Safety Standard for Diagnostic Ultrasound Equipment (AIUM/NEMA, 1983) and AIUM Acoustic Output Measurement and Labeling Standard for Diagnostic Ultrasound Equipment (AIUM, 1992) labeling requirements were based on the philosophy that there is a possible risk from diagnostic ultrasound exposure. The specific labeling requirements of these and other (Harris, 1992) safety standards were selected to include those quantities whose magnitudes are known or believed to be related to actual damage or to risk of damage to biological tissues as a result of ultrasonic irradiation. The Food and Drug Administration's 510(k) premarket notification requirements (FDA, 1985) have a similar basis, owing to the FDA's requirement to determine the safety and effectiveness of ultrasound equipment.

The basis for this rationale lies in an understanding of the mechanisms by which it is known that ultrasound can affect living systems. Such knowledge comes from fundamental laboratory studies (O'Brien, 1984; O'Brien and Withrow, 1985; O'Brien, 1991). These mechanisms can be classified and discussed in terms of whether heat is or is not believed to be the principal cause for the biological effect. The applicable ultrasonic exposure quantities will be identified during the course of this discussion. Both thermal and nonthermal mechanisms will be considered.

VII. THERMAL MECHANISM

Whenever ultrasonic energy is propagated into an attenuating material such as tissue, the amplitude of the wave decreases with distance. This attenuation is due to either *absorption* or *scattering*. Absorption is a mechanism that represents that portion of the wave energy that is converted into heat, and scattering can be thought of as that portion which changes direction. Since the medium can absorb energy to produce heat, a temperature rise may occur as long as the rate at which heat is produced is greater than the rate at which the heat is removed. In tissue, at the site where the ultrasonic temporal average intensity is I_{TA}, the average rate of heat generation per unit volume per unit time is given by the expression (Nyborg, 1981; Cavicchi and O'Brien, 1984)

$$Q = 2\alpha I_{TA} = \frac{\alpha p p^*}{\rho c} \tag{20}$$

where

$$I_{TA} = \frac{p p^*}{2\rho c} \tag{21}$$

where α is the ultrasonic amplitude absorption coefficient which increases with increasing frequency, p and p^* are the instantaneous ultrasonic pressure and its complex conjugate, respectively, ρ is density and c is sound speed. The product of p and p^* in Eq 21 is equal to the ultrasonic pressure amplitude (see Eq 2d) square, p_o^2, at the specific location in the medium where Q is determined and can be thought of as a temporal average quantity.

The temporal average intensity is not necessarily at the location where it is maximized, that is, at the spatial peak location. If it were, however, then the I_{TA} in Eq 21 would be I_{SPTA}, which would maximize Q for that tissue site. AIUM's Statement on Mammalian *In Vivo* Ultrasonic Biological Effects (see Table 3), sometimes referred to as the *100 mW/cm²*

Statement, is a generalization about the state-of-affairs with respect to an intensity (in terms of I_{SPTA})-time limit below which there have been no independently confirmed significant biological effects in mammalian tissues (AIUM, 1988).

TABLE 3

AIUM Statement on Mammalian *In Vivo* Biological Effects (AIUM, 1988)
(Approved August, 1976. Revised and approved October, 1987)

A review of bioeffects data supports the following statement:

In the low megahertz frequency range there have been (as of this date) no independently confirmed significant biological effects in mammalian tissues exposed *in vivo* to unfocused ultrasound with intensities[a] below 100 mW/cm^2, or to focused[b] ultrasound with intensities below 1 W/cm^2. Furthermore, for exposure times[c] greater than one second and less than 500 seconds for unfocused ultrasound, or 50 seconds for focused ultrasound such effects have not been demonstrated even at higher intensities, when the product of intensity and exposure time is less than 50 joules/cm^2.

[a] Free-field spatial peak, temporal average (SPTA) for continuous wave exposures, and for pulsed-mode exposures with pulses repeated at frequencies greater than 100 Hz.

[b] Quarter-power (-6 dB) beam width smaller than four wavelengths or 4 mm, whichever is less at the exposure frequency.

[c] Total time including off-time as well as on-time for repeated pulse exposures.

For a given I_{TA}, the maximum temperature rise, ΔT_{max}, under the assumption that no heat is lost by conduction, convection, or any other heat removal processes, is approximately described by

$$\Delta T_{max} = \frac{Q \Delta t}{C_h} \tag{22}$$

where Δt is the time duration of exposure and C_h is the medium's specific heat. This formula is valid only for short exposure times; for longer times, heat removal processes become significant. Nonetheless, as a "ballpark estimate," using the intensities from the *AIUM Statement* in Table 3 of $I_{SPTA} = 0.1$ and 1 W/cm^2 at an ultrasonic frequency of 5 MHz, from Eq 20, $Q = 0.05$ and 0.5 J/cm^3-s ($\alpha \approx 0.25$/cm at 5 MHz). Since the thermal properties of biological tissue can be approximated by water ($C_h = 4.18$ J/cm^3-C), the maximum time rate of change of temperatures, from Eq 22, are

$$\frac{\Delta T_{max}}{\Delta t} = 0.012 \text{ and } 0.12 \text{ °C/s} \tag{23}$$

which means that for a 1 second exposure, ΔT_{max} would be about 0.012 and 0.12 °C. If the exposure duration were longer than 1 second, the temperature would continue to rise but at a progressively slower rate, until the rate of heat generation was about the same as the rate of heat removal.

To estimate the temperature rise from a single pulse for clinical, diagnostic pulse-echo instrumentation, the local intensity of Eq 20 is considered to be the spatial peak value averaged over the duration of the pulse, that is, the spatial peak, pulse average intensity, I_{SPPA}. For typical instrumentation, a maximum value of I_{SPPA} may be as high as 500 W/cm^2. Thus, the maximum time rate of change of temperature is

$$\frac{\Delta T_{max}}{\Delta t} = 60 \text{ °C/s} \tag{24}$$

but with the duration of the pulse, Δt, of approximately 2 µs, the maximum temperature rise, $\Delta T_{max} \approx 120$ µ°C.

There have been several studies to calculate the temperature rise in mammalian tissue from ultrasonic exposure and some of them have shown to compare favorably with experimental results (Pond, 1970; Robinson and Lele, 1972; Lerner et al, 1973; NCRP, 1984; FDA, 1985; Nyborg and Steele, 1983; Cavicchi and O'Brien, 1985; AIUM, 1988). These demonstrate that selected aspects of the theory are reasonably well understood. But there are still many unanswered concerns in terms of being able to assess *in vivo* temperature rise.

VIII. NONTHERMAL MECHANISMS

The nonthermal mechanism that has received the most attention is acoustically generated cavitation. Cavitation, in a broad sense, refers to ultrasonically induced activity occurring in a liquid or liquidlike solid material that contains bubbles or pockets containing gas or vapor. These bubbles originate within materials at locations termed "nucleation sites," the exact nature and source of which are not well understood in a complex medium such as tissue. Cavitation can affect a biological system by virtue of a temperature rise, a mechanical stress, and/or free radical production. Even so, this is traditionally referred to as a nonthermal mechanism.

The discussion of cavitation will be less precise than that of the thermal mechanism owing to the fact that it has not been documented that cavitation occurs in biological tissue from diagnostic-like exposure conditions, whereas it is known that ultrasound can increase the temperature of tissue. So, in one sense, research continues to determine whether cavitation is a mechanism that needs to be addressed from the aspect of diagnostic, imaging equipment. In another sense, it is known that cavitation does occur in tissue at excessively high intensity levels and in "model systems" at quite low intensity levels. Excellent reviews of cavitation have been published (NCRP, 1984; Flynn, 1964; Nyborg, 1965; Nyborg, 1975; Coakley and Nyborg, 1978; Apfel, 1981; Flynn, 1982).

Cavitation can be discussed in two general categories termed transient cavitation and stable cavitation (Flynn, 1964). Transient cavitation connotes a relatively violent activity (bubble collapse) in which "hot spots" of high temperature and pressure occur in very short (of the order of microseconds) bursts at points in the sonicated medium. These bursts may be accompanied by localized shock waves and/or by the generation of highly reactive chemical species.

In contrast, a much less violent form is stable cavitation, which is associated with vibrating gaseous bodies. The nature of this form of cavitation consists of a micron-size gaseous body (at diagnostic ultrasonic frequencies) that remains spatially stabilized within but not necessarily because of the ultrasound field and, because of the ultrasound field, oscillates or pulsates. When such volumetric oscillations are established, the liquid-like medium immediately adjacent to the gas bubble flows or streams (termed microstreaming) (Nyborg, 1965). Microstreaming has been shown to produce stresses sufficient to disrupt cell membranes.

The occurrence of cavitation, and its behavior, depends on many factors, including: the ultrasonic pressure; whether the ultrasonic field is focused or unfocused, or pulsed or continuous; to what degree there are standing waves (i.e., energy reflecting back onto itself); and the nature and state of the material and its boundaries. Experimentally, since cavitation would probably affect only a single or a few cells, it would be extremely difficult to detect an adverse biological effect, unless the cavitation events were widespread among a large volume of tissue. The latter has been shown to be the case when mammalian nervous system tissue was exposed to ultrasonic levels in excess of I_{SPTP} of 1000 W/cm^2 for a duration of at least 1 ms; these conditions are outside of the diagnostic equipment range (Dunn and Fry, 1971).

A theory has recently been put forth that results in a derived formula that predicts the conditions that will produce transient ultrasonic cavitation on a scale of a single biological cell in one ultrasonic cycle (Apfel, 1986). For sufficiently high material viscosity, a threshold pressure, p_{th}, is proportional to ultrasonic frequency, the material's viscosity, and reciprocal ambient pressure. The p_{th} quantity is the peak rarefactional pressure, p_r, quantity. This approximate analytic expression agreed reasonably well with another acoustic cavitation theory calculation (Flynn, 1982) that predicted that microsecond pulses could cause cavitation nuclei, that is, gas bubbles.

An experimental study showed evidence for cavitation from microsecond pulses of ultrasound (Carmichael et al, 1986) that qualitatively agreed with earlier theoretical predictions of Flynn (1982). Aqueous solutions were exposed to short ultrasonic pulses (approximately 6 to 20 μs) and $^\bullet$OH radicals and $^\bullet$H atoms were detected by spin trapping and electron spin resonance. These findings were ultrasonically quantified with the I_m, which is believed to be an estimator of the maximum ultrasonic pressures, either p_r or p_c.

There has been no experimental evidence to suggest that cavitation occurs in mammalian parenchymal tissue from exposure-like conditions employed with diagnostic ultrasound equipment (Apfel and Holland, 1991). However, there have been some theoretical studies that suggest that under precise conditions, cavitation may be induced by microsecond-type pulses of ultrasound (Flynn, 1982). The comments here have shown that in terms of assessing the potential for cavitation and a mechanism responsible for producing a biological effect, the important exposure quantities seem to relate to the instantaneous, and, specifically, the maximum ultrasonic pressure only.

But, there has been a suggestion that some type of ultrasonically-induded bubble activity induces lung damage in mice. Child et al. (1990) have observed an approximate ultrasonic pressure threshold for lung damage in mice of 0.8 MPa at 1 MHz. The observations of Child et al (1991) are in good agreement with the frequency dependent, *in vitro* cavitation experiments of Apfel and Holland (1991).

IX. REFERENCES

AIUM (1988). Bioeffects Considerations for the Safety of Diagnostic Ultrasound, J. Ultrasound Med., 7(September): S1.

AIUM (1992). "AIUM Acoustic Output Measurement and Labeling Standard for Diagnostic Ultrasound Equipment," American Institute of Ultrasound in Medicine, Rockville, MD.

AIUM/NEMA (1983). "Safety Standard for Diagnostic Ultrasound Equipment," AIUM/NEMA Standards Publication No. UL 1-1981. J. Ultrasound Med., 2 (April):S1.

Anderson, L. (1992). Bioeffects of 50/60 Hz Fields. This Proceedings.

Apfel, R. E. (1981). Acoustic Cavitation. In "Ultrasonics: Methods of Experimental Physics Series,"P. D. Edmonds (ed), p 356, Academic Press, New York, NY.

Apfel, R. E. (1986). Possibility of Microcavitation from Diagnostic Ultrasound, IEEE Trans. Ultrasonics, Ferroelectrics, and Frequency Control, UFFC-33:139.

Apfel, R. E. and C. K. Holland (1991). Gauging the Likelihood of Cavitation from Short-Pulse, Low-Duty Cycle Diagnostic Ultrasound, Ultrasound Med. Biol., 17:179.

BEIR (1972). "The Effects on Populations of Exposure to Low Levels of Ionizing Radiation," National Academy of Sciences - National Research Council, Washington, DC.

Bernhardt, J. (1992). Bioeffects of Radiofrequency Fields. This Proceedings.

Bushong, S. C. (1973). "The Development of Radiation Protection in Diagnostic Radiology," CRC Press, Cleveland, OH.

Carmichael, A. J., M. M. Mossoba, P. Riesz et al. (1986). Free Radical Production in Aqueous Solutions Exposed to Simulated Ultrasonic Diagnostic Conditions, IEEE Trans. Ultrasonics, Ferroelectrics, and Frequency Control, UFFC-33:148.

Carson, P. L., J. N. Rubin and E. H. Chiang (1989). Fetal Depth and Ultrasound Path Lengths through Overlying Tissues. Ultrasound Med. Biol., 15:629.

Cavicchi, T. J. and W. D. O'Brien, Jr. (1984) Heat Generated by Ultrasound in an Absorbing Medium, J. Acoust. Soc. Amer. 70:1244.

Cavicchi, T. J. and W. D. O'Brien, Jr (1985). Heating Distribution Color Graphics for Homogeneous Lossy Spheres Irradiated with Plane Wave

Ultrasound, IEEE Trans. Sonics Ultrasonics, SU-32:17.

Child, S. Z., C. L. Hartman, L. A. Schery and E. L. Carstensen (1990). Lung Damage from Exposure to Pulsed Ultrasound, Ultrasound in Med. Biol., 16:817.

Coakley, W. T. and W. L. Nyborg (1978). Cavitation; Dynamics of Gas Bubbles; Applications. In "Ultrasound: Its Applications in Medicine and Biology," F. J. Fry (ed), p 77, Elsevier, New York, NY

Daft, C. M. W, T. A. Siddiqi, D. W. Fitting, R. A. Meyer and W. D. O'Brien, Jr. (1990). *In Vivo* Fetal Ultrasound Exposimetry, IEEE Trans. Ultrasonics, Ferroelectrics, and Frequency Control, 37:501.

Dunn, F. and F. J. Fry (1971). Ultrasonic Threshold Dosages for the Mammalian Central Nervous System, IEEE Trans. Biomed. Engr., BME-18:253.

Ensminger, D. (1988). "Ultrasonics: Fundamentals, Technology, Applications," Second Edition, Marcel Dekker, Inc., New York, NY.

Failla, G. and P. S. Henshaw (1931). The Relative Biological Effectiveness of X-Rays and Gamma Rays, Radiology, 17:1.

FDA (1985). "501(k) Guide for Measuring and Reporting Acoustic Output of Diagnostic Ultrasound Medical Devices, December, 1985," Center for Devices and Radiological Health, US Food and Drug Administration, Rockville, MD.

Flynn, H. G. (1964). Physics of Acoustic Cavitation in Liquids. In "Physical Acoustics," vol. lB, Mason, W. P. (ed), p 57, Academic Press, New York, NY.

Flynn, H. G. (1982). Generation of Transient Cavities in Liquids by Microsecond Pulses of Ultrasound, J. Acoust. Soc. Amer., 72:1926.

Grandolfo, M. (1992). Electromagnetic Field Dosimetry. This Proceedings.

Gray, L. H. and J. Read (1939). Measurement of Neutron Dose in Biological Experiments, Nature, 144:439.

Hall, D. E. (1987). "Basic Acoustics," Harper & Row, New York, NY.

Harris, G. R. (1992). Ultrasound Safety Standards. This Proceedings.

ICRU (1928). International X-Ray Unit of Intensity. International Commission on Radiological Units and Measurements. Report No. 2, Brit. J. Radiol, 1:363.

ICRU (1954). Recommendation of the International Committee for Radiological Units, International Commission on Radiological Units and Measurements, Report 7, Radiology, 62:106.

ICRU (1956). "International Committee for Radiological Units and Measurements," Report No. 8, National Bureau of Standards, Handbook 62, Washington, DC.

ICRU (1962). "Radiation Quantities and Units," International Committee for Radiological Units and Measurements, Report No. 10a, National Bureau of Standards Handbook 84, Washington, DC.

ICRU (1971). "Radiation Quantities and Units," International Commission on Radiation Units and Measurements, Report 19, Washington, DC.

Johnston, R. L. and F. Dunn (1976). Ultrasonic Absorbed Dose, Dose Rate, and Produced Lesion Volume, Ultrasonics, 14:153.

Kinsler, L. E., Frey, A. R., Coopens, A. B. and Sanders, J. V. (1982) "Fundamentals of Acoustics," Third Edition, Wiley, New York, NY.

Leonowich, J. (1992). Measurements of Radio Frequency Fields. This Proceedings.

Lerner, R. M., E. L. Carstensen and F. Dunn (1973). Frequency Dependence of Thresholds for Ultrasonic Production of Thermal Lesions in Tissue, J. Acoust. Soc. Amer., 54:504.

Lewin, P. (1992). Ultrasound: Measurements and Instrumentation. This Proceedings.

Mayneord, W. V. (1940). Energy Absorption, Brit. J. Radiol., 13:235.

NCRP (1983). "Biological Effects of Ultrasound: Mechanisms and Clinical Implications," NCRP Report No. 74, National Council on Radiation Protection and Measurement, Bethesda, MD.

Nyborg, W. L. (1975) "Intermediate Biophysical Mechanics," Cummings Publishing Co., Menlo Park, CA.

Nyborg, W. L. (1965). Acoustic Streaming. In "Physical Acoustics," vol.

2B, Mason, W. P. (ed), p 265, Academic Press, New York, NY.

Nyborg, W. L. (1981). Heat Generation by Ultrasound in a Relaxing Medium, J. Acoust. Soc. Amer., 70:310.

Nyborg, W. L. and R. B. Steele (1983). Temperature Elevation in a Beam of Ultrasound, Ultrasound Med. Biol., 9: 611.

O'Brien, W. D., Jr. (1978). Ultrasonic Dosimetry. In "Ultrasound: Its Applications in Medicine and Biology," F. J. Fry (ed), pp 343-391, Elsevier, New York, NY

O'Brien, W. D. Jr. (1984). Safety of Ultrasound with Selected Emphasis for Obstetrics, Seminars in Ultrasound, 5:105.

O'Brien, W. D., Jr. (1986). Biological Effects of Ultrasound: Rationale for the Measurement of Selected Ultrasonic Output Quantities, In Echocardiography, A Review of Cardiovascular Ultrasound, vol 3, pp 165-179, Futura Publishing Co., New York, NY.

O'Brien, W. D., Jr. (1991). Ultrasound Bioeffects Related to Obstetric Sonography. In "The Principles and Practice of Ultrasonography in Obstetrics and Gynecology, fourth edition," A. C Fleischer et al (eds), pp 15-23, Appleton & Lange, Norwalk, CT.

O'Brien, W. D., Jr. (1992). Introduction to Ultrasound. This Proceedings.

O'Brien, W. D., Jr., M. L. Shore, R. K. Fred and W. M. Leach (1972). On the Assessment of Risk to Ultrasound. In "1972 IEEE Ultrasonics Symposium," pp 486-490, IEEE Catalog No. 72 CHO 708-8SU, Institute of Electrical and Electronics Engineers, New York, NY.

O'Brien, W. D., Jr.and T. J. Withrow (1985). An Approach to Ultrasonic Risk Assessment and an Examination of Selected Experimental Studies. In "The Principles and Practice of Ultrasonography in Obstetrics and Gynecology, third edition," R. C. Sanders and A. E. James, Jr., (eds), pp 15-22, Appleton-Century-Crofts, Norwalk, CT.

ODS (1991). "Standard for Real-Time Display of Thermal and Mechanical Indices on Diagnostic Ultrasound Equipment," American Institute of Ultrasound in Medicine, Rockville, MD.

Parker, H. M. (1948). Health Physics, Instrumentation and Radiation Protection, Adv. Biol. Med. Phys., 1:223.

Pierce, A. D. (1981). "Acoustics: An Introduction to Its Physical Principles and Applications," McGraw Hill, New York, NY.

Pond, J. B. (1970). A Study of the Biological Action of Focussed Mechanical Waves (Focussed Ultrasound). Ph.D. Thesis, London University.

Purnell, E. W., A. Solollu, R. Torchia and H Taner (1964). Focal Chorioretinitis Produced by Ultrasound, Invest. Ophth., 3:657.

Robinson, T. C. and P. P. Lele (1972). An Analysis of Lesion Development in the Brain and in Plastics by High Intensity Focused Ultrasound at Low Megahertz Frequencies, J. Acoust. Soc. Amer., 51:1333.

Roesch W. C. and F. H. Attix (1968). Basic Concepts of Dosimetry. In "Radiation Dosimetry, Volume 1," F. H. Attix and W. C. Roesch (eds), pp 1-41, Academic Press, New York, NY.

Rupert, C. S. (1974). Dosimetric Concepts in Photobiology, Photochem Photobiol, 20:203.

Siddiqi, T. A., W. D. O'Brien, Jr., R. A. Meyer, J. M. Sullivan and M. Miodovnik (1991). *In Situ* Exposimetry: The Ovarian Ultrasound Examination, Ultrasound Med. Biol., 17:257.

Sliney, D. (1992). Measurements and Bioeffects of Infrared and Visible Light. This Proceedings.

Spiers, F. W. (1956). Radiation Units and Theory of Ionizing Dosimetry. In "Radiation Dosimetry," G. J. Hine and G. L. Brownell (eds), pp 1-47, Academic Press, New York, NY.

Spiers, F. W. (1964). The Historical Development of Radiation Dosimetry. In "Proc Inter School of Physics, Course XXX, Radiation Dosimetry," G. W. Reid (ed), pp 1-6, Academic Press, New York, NY.

Stewart, H. F. and M. E. Stratmeyer (eds) (1982). "An Overview of Ultrasound: Theory, Measurement, Medical Applications and Biological Effects," Health and Human Services Publication (FDA) 82-8190, US Government Printing Office, Washington, DC.

Taylor, K. J. W. and J. B. Pond (1972). A Study of the Production of Hemorrhagic Injury and Paraplegia in Rat Spinal Cord by Pulsed Ultrasound of Low Megahertz Frequencies in the Context of the Safety for Clinical Usage, Brit. J. Radiol., 45:343.

Taylor, L. S. (1971). "Radiation Protection Standards," CRC Press, Cleveland, OH.

Wang C. C. and L. L. Robbins (1956). Biological and Medical Effects of Radiation. In "Radiation Dosimetry," G. J. Hine and G. L. Brownell (eds), pp 125-152, Academic Press, New York, NY.

Ultrasound Sources and Human Exposures

S.H.P. Bly[a] and G.R. Harris[b]

[a]Non-Ionizing Radiation Section
Bureau of Radiation & Medical Devices
Health & Welfare Canada
Ottawa, Ontario K1A 0L2

[b]Center for Devices & Radiological Health
U.S. Food & Drug Administration
5600 Fishers Lane
Rockville, MD, USA 20857

People are exposed to ultrasound in a great variety of ways including medical exposures, both therapeutic and diagnostic, occupational exposures and exposures to the general public. This article reviews current ultrasound sources and the exposures of people to ultrasound.

1. Ultrasound Sources

Ultrasound in the megahertz (MHz) frequency range is used for diagnosis, physical therapy, hyperthermia, focal lesion work and in ultrasonic humidifiers. In this frequency range, the ultrasound is generated exclusively by piezoelectric transducers, normally ferroelectrics. The transducer converts applied electrical voltage oscillations into ultrasonic waves. Barium titanate and lead zirconate titanate are two of the more common ferroelectric materials (Benwell and Bly, 1987).

Diagnostic devices are focused, either geometrically or electronically, and scanned, either electronically or mechanically. Electronic scanning and focusing are performed through appropriate phase delays applied to the active elements of transducer arrays (Ramm and Smith, 1983). Electrical matching and tuning are required to provide efficient electrical energy transfer to the piezoelectric element. A heavy absorbing damping metal/epoxy backing is required for obtaining the short pulses required for high resolution, and 1/4 wavelength impedance matching epoxy layers are required for efficient transmission into the body.

Therapy devices consist of two basic components: a generator and an applicator. The generator contains a radio-frequency (RF) oscillator, along with various operator output and timer controls. The applicator is connected to the generator by a flexible electrical cable. The applicator contains an air backed piezoelectric crystal protected by either an epoxy or aluminum facing. In most cases the piezoelectric transducers in the various applicators are circular disks.

Devices used for hyperthermia and focal lesion work normally consist of air backed piezoelectric elements run in continuous wave mode. Some prototype hyperthermia devices may consist of several large aperture (8-16 cm diameter) transducers strongly geometrically focused and scanned to maintain a relatively uniform temperature distribution. Commercial hyperthermia transducers are plane disks with large apertures, 6-16 cm in diameter (Hynynen, 1990). Transducers for focal lesion work are air backed ferroelectrics, shaped geometrically to provide an extremely highly focused beam (Lizzi et al., 1984).

Among the at least eight manufacturers of extracorporeal shock wave lithotripter devices, three different shock wave generation techniques are used: spark-gap (SG) (also called electrohydraulic), electromagnetic (EM), and piezoelectric (PE) (Reichenberger, 1988; Coleman and Saunders, 1989; Davros et al., 1991). The aperture diameter for the various focusing systems typically lies in the 10-50 cm range. In SG lithotripters, a pair of underwater electrodes is placed at one focus of an ellipsoidal reflector. When a high voltage capacitor is discharged through these electrodes, the resulting expanding shock wave is focused by the reflector onto the ellipsoid's second focus, which is the point at which the treated stone is located. EM lithotripters generate an ultrasonic pulse by inducing movement in a thin metallic disk via an electrically-excited coil. This pulse is focused into a small volume using either an acoustic lens or a parabolic reflector. Piezoelectric generators use piezoelectric transducers to generate the ultrasonic pulse. The transducers can be mounted on the inside of a spherical dish, or the pulse can be focused with a lens.

The transducers used for low frequency ultrasonics, from 20-50 kHz, are often magnetostrictive (Benwell and Bly, 1987). Magnetostrictive transducers are particularly useful for the production of intense cavitation required in ultrasonic cleaners. They are also used for ultrasonic bathtubs.

2. Human Exposures

2.1 Diagnostic Exposures

Currently, in terms of population exposed, the greatest human exposure to ultrasound is during medical procedures, particularly diagnostic ones (ter Haar and Hill, 1989). Diagnostic ultrasound has been used to examine almost every part of the body, but the largest uses have been in obstetrics, cardiology, and in examining soft tissue structures of the abdomen. Obstetrics especially has seen a significant gain. In a U.S. study, Moore et al. (1990) found that between 1980 and 1987 the use of ultrasound in pregnancy more than doubled (33.5% vs. 78.8% of pregnancies). It has been estimated that in 1989 a total of 60-90 million ultrasound examinations were conducted worldwide (Merritt, 1989).

Ultrasound examinations can be divided into several categories depending on the mode of operation of the device (Kremkau, 1989). These modes can produce: (1) one dimensional echo patterns vs. distance (A-mode) or time (M-mode); (2) real-time two-dimensional grey-scale images (B-mode); (3) Doppler-derived information on moving structures, such as blood flow or fetal heart movement (pulsed or continuous wave (CW) Doppler); and, most recently, (4) two-dimensional color flow maps using Doppler techniques (color Doppler ultrasound, or CDU). In this last mode of operation the colour image of the flowing blood is superimposed on the grey scale B-mode image. In the A-, M-, and pulsed or CW Doppler modes, the ultrasound beam emitted from the transducer assembly is stationary in space. However, in the B- and CDU modes, the beam is swept or scanned through the image plane, usually in a sector or rectilinear imaging format. These various modes of operation can be used either singly or in combination. Most examinations use real-time B-mode imaging.

Diagnostic devices introduce ultrasonic energy into the body either continuously (CW Doppler) or in short (eg., 1-10 μs) pulses (A- and M-mode, pulsed Doppler, CDU, and all B-mode). The ultrasonic frequency is usually in the 2-10 MHz range, although in some applications such as ophthalmology higher frequencies are used. With the exception of fetal heart beat monitors, virtually all diagnostic beams are focused, and focal dimensions are typically less than a few millimeters.

Complementing diagnostic ultrasound's proven value as a diagnostic tool is its excellent safety record, in that there are no known instances of human injury due to its use. However, evidence does exist that supports at least a hypothetical risk from ultrasound exposure at diagnostic levels, primarily due to the two most widely studied potential bioeffects mechanisms, heating and cavitation (AIUM, 1988; O'Brien, 1992). For these reasons, the use of minimum practical exposure levels has been advocated widely, particularly with regard to fetal applications. This is particularly important since it is not clear that increasing the exposure level will always result in improved diagnostic information (Harris et al., 1989).

The possibility of risk of significant injury during diagnostic ultrasound examinations, particularly to the fetus, has generated considerable interest in the estimation of ultrasound exposure during diagnostic examinations. Historically, this has been mainly achieved by measuring ultrasound exposure quantities under standard conditions (Lewin and Schafer, 1992) in water and then, using a knowledge of tissue attenuation properties, estimating the in situ exposure. Exposure quantities used to describe diagnostic ultrasound fields generally are chosen based on the above-mentioned thermal and mechanical (cavitation) bioeffects mechanisms. These quantities include the peak compressional (positive) and rarefactional (negative) pressures, p_c and p_r, respectively; the temporal average ultrasonic power, W_o; the spatial-peak intensity, temporally averaged over the pulse repetition period (I_{SPTA}) or the pulse duration (I_{SPPA}); the center frequency, f_c; and the focal and source aperture dimensions.

Data on intensities, powers, pressures and other exposure quantities have recently been compiled for devices in Europe, Japan, the U.S. and Canada (Duck, 1989; Ide, 1989; Zagzebski, 1989; Bly et al., 1992; Environmental Health Directorate, 1989). Approximate ranges of exposures are shown in Table 1.

As can be seen, there is a wide range of exposure levels for diagnostic ultrasound devices. The typical time-averaged exposure levels for B-mode are relatively low, the spatial peak temporal average intensity being typically about 20 mW/cm^2. Peak pressures are typically 4 megapascal (MPa) for p_c and 2 MPa for p_r (Duck, 1989). Peak pressures in CDU modes are similar or tend to be lower than those during B-mode operation, but the repetition rates are higher and the pulses longer, which leads to considerably higher output powers and intensities.

The temporal-average exposure levels generally are higher in the stationary modes (A, M, pulsed and CW Doppler). Pulsed Doppler values can be especially high, because of the longer pulse durations and higher pulse repetition frequencies sometimes used. Typical I_{SPTA} values in pulsed

Doppler mode are about 1-2 W/cm^2 and typical ultrasonic powers are

Table 1. Table of exposure quantities for three modes of operation. The entries give the ranges of the exposure quantities for each mode. Adapted from Duck, 1989.

Exposure Quantity	Mode of Operation		
	B-mode	M-mode	Pulsed Doppler
W_o (mW)	-	0.5-360	3.2-500
I_{SPTA} (mW/cm^2)	0.3-166	6.6-340	110-9500
P_c (MPa)	0.8-8.8	-	0.2-6.3
P_r (MPa)	0.8-3.9	-	0.2-2.6

between 50-100 mW. In M-mode, the I_{SPTA} is typically about 100 mW/cm^2 and the ultrasonic powers are typically between 10 and 20 mW (Duck, 1989).

Although the samples in the surveys could be biased, the surveys do suggest that in recent years, more equipment is at higher output levels, limited only by regulatory requirements. No console/transducer/operating mode/intended use combination can currently receive fast-track (so called 510(k)) pre-market approval from the U.S. Food and Drug Administration (FDA) if the maximum estimated in situ temporal average intensity exceeds 720 mW/cm^2 (Nyborg, 1989).

Recently, there has been considerable progress in assessing human exposure in terms of "worst case" temperature elevations, also known as thermal indices, computed from measured exposure quantities. Thermal indices have been proposed by the U.S. National Council on Radiation Protection and Measurements (NCRP 1991). Bly et al. (1991, 1992) computed the number of devices in pulsed Doppler mode with $I_{SPTA} \geq 500$ mW/cm^2 as a function of the NCRP thermal indices for fetal exposure. These devices were either intended for peripheral vascular or cardiac use, but their focusing characteristics suggested that they could also be used in fetal examinations. For first trimester examinations the thermal indices for all 236 devices studied were less than 1.5°C. However, due to bone heating in the 2nd and 3rd trimesters, approximately half of the devices yielded thermal indices greater than 2°C, some values reaching 9°C.

A different set of thermal indices has been used in the draft output display standard (ODS) (Harris, 1992; AIUM 1991). The ODS thermal indices are based on a model which assumes a homogeneous tissue path in which the power decreases exponentially at a rate of α dB/cm-MHz (Stratmeyer, 1989), where a value of 0.3 is used for α. Devices complying with this standard will provide a real-time display of thermal and mechanical indices.

For diagnostic ultrasound alone, there is a wide range of exposures. Recent progress in measurement, estimation and disclosure of human exposure during diagnostic examinations should help to ensure the continuation of the remarkable safety record of this modality for the future.

2.2 Therapeutic Exposure

Ultrasound has been used in physical therapy since the 1950s for an assortment of ailments of the joints and soft tissues. The advantage of ultrasound over other diathermy methods is its ability to heat at moderate depth in a localized target tissue volume. The indicated conditions include arthritis, bursitis, muscle spasms, traumatic soft tissue injuries, and certain collagen diseases. Treatment objectives are relief of pain, decrease in soft tissue stiffness, and accelerated healing. Although heating of the tissues via ultrasound absorption is generally regarded as the principal rehabilitative mechanism, beneficial non-thermal reactions to ultrasound treatment have been described (Lehmann and Lateur, 1982).

Ultrasound therapy devices have been used in hospitals, clinics, private medical offices, nursing homes, spas, and athletic facilities. In the 1970s the number of treatments per year in the United Sates was estimated to be about 15 million (Stewart and Stratmeyer, 1982). In a 1985 survey in the U.K. it was estimated that 1.1 million treatments were performed in a year in the U.K. (ter Haar et al., 1987).

The frequency of the ultrasonic waves emitted by current therapy devices is usually about 1 MHz, but values range from 0.75 MHz to 3 MHz. In continuous-wave operation the amplitude of the ultrasound is constant; i.e., there is no (or very little) modulation of the ultrasonic waveform. In the pulsed (amplitude-modulated) mode, the ultrasound is either gated on and off, or has a sinusoidal modulation envelope. Pulse duration (PD) and pulse repetition rate or frequency (PRR or PRF) values usually lie in the 2-10 millisecond and 20-120 pulses-per-second range, respectively. Treatment times are typically 5-15 minutes.

Ultrasonic power and intensity are the two most important acoustical exposure quantities for ultrasound therapy. The total power radiated by the applicator in the form of ultrasonic radiation is known as the ultrasonic power (W). Intensity defines the spatial concentration of power, and is expressed in units of watts per square centimetre (W/cm^2). Intensity can vary both temporally (eg., continuous or pulsed) and spatially. The ultrasound therapy transducers produce a beam of ultrasound that is essentially collimated in the region near the applicator face, but eventually the beam begins to diverge. The ultrasonic intensity distribution within this beam is not uniform, and the highest intensity at a point in the beam is generally referred to as the spatial-maximum or spatial-peak intensity. The term "effective" intensity (I_{eff}) has been used to denote a spatial-average intensity obtained by dividing the ultrasonic power by the effective radiating area (ERA) of the applicator face. As a first approximation, this area can be thought of as the surface area of the piezoelectric transducer in the applicator. However, more precise definitions exist for purposes of standardization (Harris, 1992). The time-averaged I_{eff} is directly related to the heating effects of ultrasound. The temporal maximum I_{eff} in an amplitude modulated wave, the largest intensity, averaged over the period of the carrier wave, is more related to stable cavitation effects.

Both W and I_{eff} are indicated on the ultrasonic therapy generators. The temporal average quantity is indicated in continuous wave operation and the temporal maximum quantity is indicated in the pulsed mode. Typical indicated effective intensities during treatment are in the range of 0.2-1 W/cm^2 (ter Haar et al., 1987). The trend recently has been to this lower intensity range. In the past, most treatments were performed with indicated effective intensities from 1 to 2 W/cm^2 (Environmental Health Directorate, 1980). Effective intensities usually do not exceed 3 W/cm^2 and some modern devices do not exceed 1 W/cm^2. The typical range of indicated ultrasonic power has diminished from 5-10 watts to 1-5 watts. The full range of ERA values on modern equipment is from 0.5 to 10 cm^2 and the indicated ultrasonic power used can range from about 0.5 to 30 watts.

Some types of human exposure (ter Haar et al., 1987) are contraindicated for ultrasound therapy. For example, it is extremely important that the pregnant uterus not be irradiated inadvertently during treatment. The embryonic phase (first 8-10 weeks after conception) is particularly vulnerable to giving rise to serious birth defects after thermal insult. Unfortunately this is exactly when the patient and therapist may not be aware of the patient's pregnancy (Environmental Health Directorate, 1989).

2.3 Surgical Exposure

Extracorporeal Shockwave Lithotripsy (ESL) is a technique that has proven useful as a non-surgical means to treat urinary tract stones by fracturing the stones with focused acoustic shock waves (Neuhausel, 1987; Reichenberger, 1988). In the U.S., approximately 50,000 people per year are potential candidates for ESL. Since its first use in Germany in 1980, over 300,000 people worldwide have been treated with ESL for urinary tract stones.

ESL is also being investigated for use in the treatment of gallstones (Steinberg et al., 1989; Torres and Nelson, 1989; Davros et al., 1991). As many as 150,000 people annually in the U.S. may be eligible for the ESL procedure under current protocols (Brink et al., 1988; Steinberg et al., 1989).

During treatment the shock-wave source is coupled to the body, either by submersing the patient in water or using a coupling oil or gel. Positioning the stone at the focus of the shock wave source is accomplished either by fluoroscopic or ultrasound imaging. The strength of the shock wave can be varied by a generator output control, sometimes referred to as the kilovoltage control. Up to several thousand shocks per treatment may be required. Periodic evaluation, and, if necessary, realignment of the stone with respect to the focus are done using x-rays or ultrasound during the procedure.

To date a consensus has not been reached as to which shock-wave field characteristics are important in evaluating an ESL device's ability to fragment stones, in part because the exact mechanisms of stone destruction are not completely understood. However, the following pulse quantities have been mentioned as being of potential importance in the disintegration process: peak positive and negative pressure amplitudes, pulse rise time and width, focal region dimensions and volume, and energy per pulse (Coleman and Saunders, 1987; Davros et al., 1991).

The focal region (usually taken as the region where the pressure is greater than 50% of its maximum value) has cross-sectional dimensions on the order of a centimetre or less, and is about one to twenty centimetres in length, depending on the shock-wave source (Coleman and Saunders, 1989; Davros et al., 1991). Spark-gap (SG) sources commonly have the largest focal volumes, piezoelectric (PE) the smallest, and electromagnetic somewhere in between. Likewise, the energy in a single shock-wave pulse at the focus is usually greatest for SG types and smallest for PE types, with values ranging from 1-100 mJ (millijoules).

The pressure pulse at the focus of the shock-wave field is characterized by an initial positive half-cycle having a rapidly rising leading edge, followed by a more slowly varying negative half-cycle. After the negative pressure peak, the pressure amplitude gradually decreases, either monotonically or with damped oscillations. Because this slowly varying "tail" can persist for many microseconds, most of the energy in an ESL pulse lies at frequencies below a few megahertz. Positive and negative pressure amplitudes are in the range 20-100 MPa and 3-10 MPa, respectively. At the highest generator output settings, positive pressure rise times typically are less than 50 ns, and may be less than 10 ns. The pulse width, defined as the time between the half-amplitude points on the initial positive pressure half cycle, usually is several hundred nanoseconds in duration.

Improved understanding of the shock-wave characteristics required to fragment stones may, with appropriate exposure control, provide further reductions in the risk of tissue injury and pain for the patient.

Hyperthermia is relatively moderate ultrasonic heating of the tumour to 43°C for approximately 30 minutes. Hyperthermia exposures currently occur only in investigational procedures or clinical trials for cancer therapy and only about 500-1000 cancer patients worldwide have ever been treated using ultrasound (Hynynen, 1991). Frequencies are usually 1 MHz, but they can range from 0.5 to 6 MHz (Hynynen, 1990). Intensities for unfocused devices are only about 0.5-2 W/cm^2 but total power is about 40-100 watts. Intensities from focused and mechanically scanned devices are about 200-400 W/cm^2 and total power is about 40-150 watts (Hynynen et al., 1990; Hynynen, 1991).

Another ultrasonic surgical application is treatment of glaucoma via induction of focal lesions through ultrasonic heating. Intensities are approximately 10 kW/cm^2 (Burgess et al., 1986). The frequency is about 5 MHz (Lizzi et al., 1984) and maximum total powers are estimated to be about 75 watts.

2.4 Occupational Exposures

Industrial ultrasound is used in a variety of applications in both liquids and solids (Shoh, 1988). The largest use in liquids is ultrasonic cleaning. The largest use in solids is welding, particularly plastic welding.

Occupational exposures can occur through contact with the ultrasound emitting device or through a coupling medium. Direct contact could occur with an ultrasonic welder. Coupling medium contact can occur in devices such as ultrasonic cleaners. Ultrasonic cleaners and welders undoubtedly constitute the major devices potentially giving rise to contact exposures. The ultrasonic frequencies of the radiation from these devices are usually in the range of 20-50 kHz. The intensities within ultrasonic cleaners typically range from 1 to 6 W/cm^2 at the face of the active element. For welders, the intensity at the welding tip (where contact would occur), range from 100 W/cm^2 to 2000 W/cm^2 (Environmental Health Directorate, 1991). Contact exposures with industrial ultrasound must be avoided because of the high risk of significant tissue damage.

For human exposure to airborne ultrasound, exposure levels commonly are quantified by measuring the sound pressure level (SPL) in air, usually in third-octave bands (Herman and Powell, 1981; Acton, 1983; INIRC/IRPA, 1984). Values are expressed in decibels (dB), with the standard reference pressure being 20 micropascals (μPa).

There have been a number of reports and investigations into the exposure of workers to airborne ultrasound (Environmental Health Directorate, 1991). A wide variety of SPL values were obtained depending on the device, the location of measurement relative to the device, and whether the device is enclosed. Frequencies ranged from 20 to 31.5 kHz. Ultrasonic welders yielded SPL values at the operator position of 100 to 127 dB in the 20 kHz third-octave band. Ultrasonic cleaners yielded SPL values in the range of 80 to 105 dB depending on the third-octave band and the size of the cleaners. Exposure to industrial ultrasound was found to only rarely exceed 120 dB.

Some commercial devices are also encountered occupationally. Since 1976, a number of measurements of airborne ultrasound SPLs from such commercial devices have also been made by the Bureau of Radiation and Medical Devices (BRMD) in Canada and by the Center for Devices and Radiological Health (CDRH) of the Food and Drug Administration (FDA) in the U.S.A. (FDA, 1981; Herman and Powell, 1981; Environmental

Health Directorate, 1991). Exposures from commercial devices are usually from single pure tones (although sometimes frequency sweeps are also used) and tend to be at lower SPLs than exposures from industrial devices. Exposure levels from commercial devices designed to emit ultrasound, such as ultrasonic occupancy sensors or dog repellers, rarely exceed 110 dB. Typical SPL values for occupancy sensors are 93 dB. Commercial devices which incidentally emit ultrasound, such as VDTs, yield SPLs at the operator's ear of less than 70 dB and only rarely exceed 65 dB. Typical SPL values for VDTs are between 50 and 60 dB.

2.5 General Public Exposures

Until relatively recently, there were no deliberate non-medical contact exposures to ultrasound. This has changed with the advent of the ultrasonic bath which is used to clean people who have limited mobility. Intensities are in the range 20-80 mW/cm^2 at the patient, the maximum time-averaged intensities being approximately 300 mW/cm^2 near the transducer. The ultrasonic frequency is 30 kHz. These devices create stable cavitation in the bath water and the vibrating bubbles likely provide the cleaning action.

Ultrasonic humidifiers are also a potential source of damaging contact exposures but only by accident. These operate in the low MHz range with intensities of several W/cm^2 (Environmental Health Directorate, 1991).

The main source of exposure to airborne ultrasound for the general public is from ultrasonic occupancy sensors. With increasing concern about energy conservation, there may be an increase in office lighting/heating control systems and a subsequent increase in human exposure to ultrasonic occupancy sensors. Exposure levels are given in section 2.4.

Summary

There are a wide variety of sources and human exposures to ultrasound. The exposure may be mediated by air at low ultrasonic frequencies, 20-40 kHz, or by water over the full range of frequencies from 20 kHz to 20 MHz. Irradiation may be whole body or localized. Medical applications are diagnostic, therapeutic and surgical. Non-medical exposures are usually either accidental or incidental to the purpose for which the ultrasound is being used. Ultrasonic power and intensities may vary enormously over the full range of applications available. Through careful control of the ultrasound exposure via measurement, standardization, and an understanding of ultrasound's biological effects, this non-ionizing radiation should continue to have an excellent safety record.

References

Acton, W.I. (1983) Exposure to industrial ultrasound: hazards, appraisal and control, J. Soc. Occup. Med., 33:107-113.

AIUM (1988) Bioeffects considerations for the safety of diagnostic ultrasound, J. Ultrasound Med., 7(suppl):S1-S38.

AIUM (1991) Standard for real time display of thermal and mechanical indices of diagnostic ultrasound equipment, American Institute of Ultrasound in Medicine, Rockville, MD.

Benwell, D.A. and Bly, S.H.P. (1987), Sources and applications of ultrasound, In: "Ultrasound. Medical applications, biological effects, and hazard potential", Repacholi, M.H., Grandolfo, M., and Rindi A. (eds.), Plenum Press, 29-47.

Bly, S.H.P., Hussey, R.G., Mabee, P. and Vlahovich S. (1991) Risk assessment: Worst case fetal temperature elevations during discrete mode transabdominal pulsed doppler examinations, In: Proc. The Canadian Medical and Biological Engineering Society Conference, May 7-11, Banff, Alberta, 91-92.

Bly, S.H.P., Vlahovich S., Mabee, P.R. and Hussey, R.G. (1992) Computed estimates of maximum temperature elevations in fetal tissues during transabdominal pulsed doppler examinations, Ultrasound in Med. & Biol. (in press).

Brink, J.A., Simeone, J.F., Mueller, P.R., Richter, J.M., Prien, E.L., Ferrucci, J.T. (1988) Physical characteristics of gallstones removed at cholecystectomy: Implications for extracorporeal shock-wave lithotripsy, AJR, 151:927-931.

Burgess, S.E.P., Silverman, R.H., Coleman, D.J., Yablonski, M.E., Lizzi, F.L., Driller, J., Rosado, A. and Dennis, P.H. (1986) Treatment of glaucoma with high-intensity focused ultrasound, Ophthalmology, 93:831-838.

Coleman, A.J. and Saunders, J.E. (1987) Comparison of extracorporeal shock wave lithotripters (based on measurements in the acoustic field), In: Lithotripsy II. Textbook of second generation lithotripsy, Coptcoat, M., Miller, R., Wickham, J., eds., BDI Publ., London, 121-131.

Coleman, A.J. and Saunders, J.E. (1989) A survey of the acoustic output of commercial extracorporeal shock wave lithotripters, Ultrasound in Med. & Biol., 15(3):213-227.

Davros, W.J., Garra, B.S. and Zeman, R.K. (1991) Gallstone lithotripsy: Relevant physical principles and technical issues, Radiology, 178(2):397-408.

Duck, F.A. (1989) Output data from European studies, Ultrasound in Med. & Biol., 15(suppl 1):61-64.

Environmental Health Directorate (1980) Canada-wide survey of non-ionizing radiation emitting devices, Health and Welfare Canada, 80-EHD-53.

Environmental Health Directorate (1989) Safety Code-23. Guidelines for the safe use of ultrasound. Part I. Medical and paramedical applications, 1987 Update, Health and Welfare Canada, 88-EHD-59.

Environmental Health Directorate (1991) Safety Code-24. Guidelines for the safe use of ultrasound: Part II - Industrial and commercial applications, Health and Welfare Canada, EHD-TR-158.

FDA (1981) An evaluation of radiation emission from video display terminals, Center for Devices and Radiological Health, Food and Drug Administration, HHS Publ. (FDA) 81-8153, Rockville, MD, USA 20857.

Harris, G.R. (1992) Ultrasound Safety Standards, Proceedings of Non Ionizing Radiation: 2nd International IRPA Workshop, Vancouver, B.C., May 10-14, 1992.

Harris, G.R., Stewart, H.F., Leo, F.P. and Sanders, R.C. (1989) Relationship between image quality and ultrasound exposure level in diagnostic US devices, Radiology, 173(2):313-317.

Herman, B.A. and Powell, D. (1981) Airborne ultrasound: Measurement and possible adverse effects, HHS Publ. (FDA) 81-8163, FDA, Center for Devices and Radiological Health, Rockville, MD, U.S.A. 20857.

Hynynen, K. (1990) Biophysics and technology of ultrasound hyperthermia, In: "Clinical thermology, subseries thermotherapy. Methods of external hyperthermic heating", Gautherie, M. (ed.), Springer-Verlag Berlin Heidelberg, 92-103.

Hynynen, K. (1991) private communication.

Hynynen, K., Shimm, D., Anhalt, D., Stea, B., Sykes, H., Cassady, J.R. and Roemer, R.B. (1990) Temperature distributions during clinical scanned, focused ultrasound hyperthermia treatments, Int. J. Hyperthermia, 6:891-908.

Ide, M. (1989) Acoustic data of Japanese ultrasonic diagnostic equipment, Ultrasound in Med. & Biol., 15(suppl 1):49-53.

INIRC/IRPA (1984) Interim guidelines on limits of human exposure to airborne ultrasound, International Non-Ionizing Radiation Committee (INIRC) of the International Radiation Protection Association (IRPA), Health Physics, 46(4):969-974.

Kremkau, F.W. (1989) Diagnostic ultrasound: Principles, instruments, and exercises (3rd ed.), W.B. Saunders, Philadelphia.

Lehmann, J.F. and Lateur, B.J. (1982) Therapeutic heat, In: Therapeutic heat and cold, 3rd edition, Lehmann, J.F., ed., Williams and Wilkins, Baltimore.

Lewin, P.A. and Schafer, M. (1992) Ultrasound: measurement and instrumentation, Proceedings of Non-Ionizing Radiation: 2nd International IRPA Workshop, Vancouver, B.C., May 10-14, 1992.

Lizzi, F.L., Driller, J. and Ostromogilsky, M. (1984) Thermal model for ultrasonic treatment of glaucoma, Ultrasound in Med. & Biol., 10:289-298.

Merritt, C.R.B. (1989) Ultrasound safety: What are the issues?, Radiology, 173(2):304-306.

Moore, R.M., Jr., Jeng, L.L., Kaczmarek, R.G. and Placek, P.J. (1990), Use of diagnostic imaging procedures and fetal monitoring devices in the care of pregnant women, Public Health Reports, 105(5):471-475.

NCRP (1991) Exposure criteria for medical diagnostic ultrasound, Part I: Criteria based on thermal mechanisms, NCRP Publ., 7910 Woodmont Ave., Suite 800, Bethesda, MD. (in press).

Neuhausel, D.J. (1987) Lithotripsy, A survey, J. Clinical Engineering, 12(4):283-295.

Nyborg, W.L. (1989), Session 6. Statements, recommendations and guidelines for diagnostic equipment. Overview, Ultrasound in Med. & Biol., 15(1):85-86.

O'Brien, W.D. Jr. (1992) Ultrasound dosimetry and interaction mechanisms, Proceedings of Non-Ionizing Radiation: 2nd International IRPA Workshop, Vancouver, B.C., May 10-14, 1992.

Reichenberger, H. (1988) Lithotripter systems, Proc. IEEE, 76(9):1236-1246.

Ramm, O.T. von, Smith, S.W. (1983), Beam steering with linear arrays, IEEE Trans. Biomed. Eng., BME-30(8):438-452.

Shoh, A. (1988) Industrial applications of ultrasound, in: Ultrasound its chemical, physical, and biological effects, ed. Suslick, K., VCH, New York, 97-122.

Steinberg, H.V., Torres, W.E. and Nelson, R.C. (1989) Gallbladder lithotripsy, Radiology, 172(1):7-11.

Stewart, H.F. and Stratmeyer, M.E., eds. (1982) An overview of ultrasound: Theory, measurement, medical applications, and biological effects, Health and Human Services, HHS Publ. (FDA)82-8190, U.S. Government Printing Office, Washington D.C.

Stratmeyer, M.E. (1989) FDA model for regulatory purposes, Ultrasound in Med. & Biol. 15(suppl 1):35-36.

ter Haar, G., Dyson, M. and Oakley, E.M. (1987) The use of ultrasound by physiotherapists in Britain, 1985, Ultrasound in Med. & Biol.,

13:659-663.

ter Haar, G.R., Hill, C.R. (1989) Ultrasound, In: Non-ionizing Radiation Protection, Suess, M.J., Benwell-Morison, D.A., eds., WHO Regional Publ., European Series 25, Geneva.

Torres, W.E. and Nelson, R.C. (1989) Lithotripsy at top of list for treating cholelithiasis, Diagnostic Imaging, March 1989, 90-96.

Zagzebski, J.A. (1989) Acoustic output of ultrasound equipment summary of data reported to AIUM, Ultrasound in Med. & Biol., 15:(suppl 1):55-59.

ULTRASOUND: MEASUREMENT AND INSTRUMENTATION

Peter A. Lewin
Department of Electrical and Computer Engineering
and Biomedical Engineering and Science Institute,
Drexel University, Philadelphia, PA 19104

and

Mark E. Schafer
Sonic Technologies, Inc. 333 Byberry Rd, Hatboro, PA 19040

Introduction:

This paper discusses practical aspects of ultrasound field measurements and describes instrumentation needed to adequately characterize acoustic output of ultrasound imaging systems. Such adequate characterization requires information on both temporal and spatial behavior of the field and can be most conveniently obtained using calibrated piezoelectric hydrophone probes with active elements made of piezoelectric polymer, polivinylidene fluoride (PVDF). Accordingly, attention here is focused primarily on measurement systems relying on the use of calibrated ultrasonic PVDF hydrophones and computerized data acquisition systems. Fundamental limitations of such measurement systems are also discussed and possible improvement in measurement techniques by using fiber optic sensors is pointed out. Since other measurement devices such as radiation force balance, calorimeters, and thermocouples very often require customized design and are not capable of providing both temporal and spatial data, they are only briefly mentioned. A detailed description of these devices can be found in (Ziskin and Lewin, 1992, IEEE, 1989). In the following general requirements for ultrasonic field sensors are outlined and a succinct description of two most widely used hydrophone designs is given. This is followed by a brief discussion of the devices which are based on the measurement of radiation force and thermal energy. Finally, a versatile computerized system for measuring ultrasound fields is described.

1. Piezoelectric hydrophones

Ultrasound pressure sensors or hydrophones convert mechanical energy of the pressure wave traveling through a medium into electrical signals whose amplitude is proportional to the pressure measured. In general, hydrophones fall into two broad classes based on the type of piezoelectric material used. The older hydrophones are made from piezoceramic material such as lead zirconate titanate (PZT) or lead metaniobate (PMN), and a succinct review of these types of hydrophones can be found in (Lewin and Chivers, 1981) However, in general, PZT hydrophones have limited bandwidth, dynamic range, and linearity. Also, they suffer from radial resonances and severe variation in sensitivity with frequency (Lewin, 1983).

The availability of piezoelectric plastic materials such as the polymer material polivinylidene difluoride (PVDF) has dramatically improved the overall acoustic performance of piezoelectric hydrophones. At present calibrated PVDF hydrophone probes are not only preferred but are also recommended for precise field measurements (AIUM, 1981, FDA, 1985, IEC, 1991). Therefore, in the following, the attention is focused on the performance of piezoelectric polymer hydrophones which have been introduced in the last decade. PVDF material has certain characteristics that make it well suited to hydrophone design, such as good mechanical flexibility and an acoustic impedance close to that of water. Because of PVDF's low planar coupling coefficient, it provides a very wide uniform frequency response, without the spurious resonances characteristic of ceramic materials. In addition, unlike the conventional piezoelectric ceramics, PVDF material is extremely flexible and can be conveniently formed into the desired shapes. Several early designs of PVDF hydrophone probes which have been reported in the literature, have been summarized in (Lewin, 1981, Lewin and Schafer, 1986). At present, two basic designs are being used: the needle-type, often referred to as the Lewin-type hydrophone (Lewin, 1981), and the membrane design originally suggested in (DeReggi et al., 1977), and later described in (Harris, 1982, Preston et al., 1983).

1.1. Hydrophone design requirements

Prior to the description of measurement arrangements in which the ultrasonic hydrophone probe are used, it may be useful to summarize briefly basic requirements for hydrophones. Such summary will help to identify the technological limitations of the current designs. In the following, such parameters as size, angular response, frequency response, linearity, sensitivity, and radio-frequency (RF) shielding will be discussed.

Physical size and angular response: The dimensions of the hydrophone active area or, preferably, of the probe itself, should be as small as possible. There are two reasons for this: first, the hydrophone should not disturb the sound field being measured, and second, the smaller the probe, the better or more omnidirectional is the angular response of the hydrophone. This is particularly important during off-axis measurements of the field. In general, the smaller is the ratio of the ultrasonic hydrophone diameter to the source dimensions, the more accurate the ultrasonic field measurement. Also, the smaller this ratio, the more omnidirectional is the angular response or directivity pattern of the hydrophone probe. However, in practice, it is rather difficult to fabricate a piezoelectric hydrophone which exhibits omnidirectional angular response at the biomedical range of frequencies; it is sufficient to recall that the acoustic wavelength in water at 10 MHz is about 150 microns. Typically, the diameters of commercially available probes ranges from 0.4 - 1 mm. As noted below the most promising way to achieve further subminiaturization of the ultrasound field probes is to develop sensors based on fiber optic technology. However, at the time of this writing, such sensors are not commercially available and in comparison with the proven PVDF hydrophone technology, their characteristics and performance are not well documented.

Frequency response: A uniform frequency response of at least 15 MHz is desirable for recording the spectrum of the transmitted pulses generated at the output of medical ultrasound equipment under normal conditions. A frequency response well beyond 15 MHz is preferable, since the medical equipment in current use produces nonlinear propagation phenomena in water, leading to the generation of higher harmonics (Harris, 1988, Lewin and Goldberg, 1989).   In addition, ultrasound catheter transducers are operating typically at the frequency of 20 MHz or above. Also, in skin imaging applications the use of frequencies in the range 40-60 MHz has been reported (Foster et. al., 1989). Furthermore, a faithful reproduction of the shock waves, such as those generated by lithotripter devices for fragmentation of kidney and gall-bladder stones requires hydrophones with bandwidth on the order of 100 MHz (Lewin and Schafer 1991).

Linearity Another important hydrophone parameter is linearity. This is because of the wide range of pressure amplitude values encountered in measurement practice. Thus, peak compressional (i.e. positive) pressure of the lithotripter generated shock wave may be on the order of 100 MPa (Platte, 1986, Lewin et al., 1989) while the levels from specialized small-parts scanners or diagnostic ophthalmological units are usually two orders of magnitude lower. In addition, the hydrophone must have an adequate signal-to-noise ratio for the low-pressure (or off-axis) measurements, and it must not disturb the measurements taken in high pressure fields. Low pressure measurements, whether for a low power device or in the off-axis region of diagnostic scanning, raise the question of the hydrophone's voltage sensitivity. This sensitivity, stated as an end-of-cable response with a specified complex load (i.e. input impedance; typically 1 MOhms in parallel with 15-30pF), is expressed in dB re 1 V/µPa, or in nV/Pa. The sensitivity requirements can be estimated based on the expected range of pressure levels and on the minimum voltage required by the data acquisition system for adequate results. Finally, good RF shielding is a requirement because of the electrically noisy environment created by medical ultrasonic equipment. Diagnostic imaging systems often drive the transmitting transducer with high voltage shock excitation, which may produce a significant level of electromagnetic interference. In addition, this is also the case when the probe is used to determine field parameters of the lithotripter generated shock waves (Lewin et al., 1989).

1.2. Hydrophone configurations.

There are two well established sensor configurations which are widely used in measurement practice. Fig. 1 shows the design principle of both sensors.

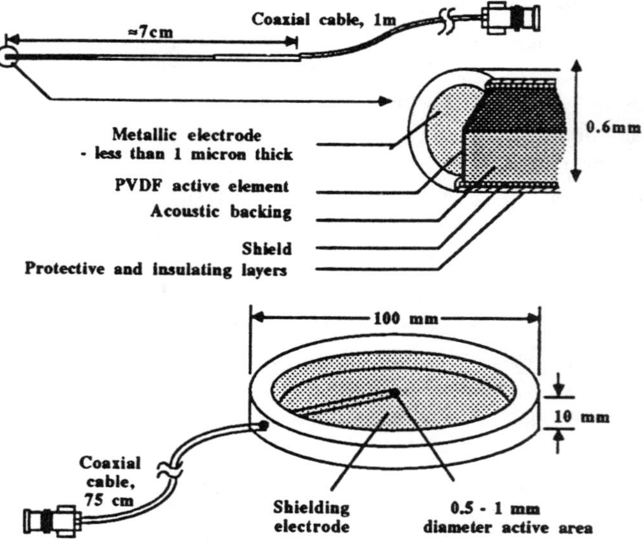

Fig. 1. Schematic construction of two PVDF hydrophone configurations:
a) Lewin - or needle-type (courtesy of The Danish Institute of Biomedical Engineering, Copenhagen-Brondby, Denmark),
b) membrane - type (courtesy of Sonic Technologies, Hatboro, PA).
Reprinted with permission (Lewin and Schafer, 1988), ©1988, IEEE.

Membrane design.   The first configuration (see Fig. 1b) uses the membrane approach originally suggested in (DeReggi, 1977).   In this configuration thin PVDF film is streched taut across an annular frame approximately 5-10 cm in diameter.   Only a selected portion of the PVDF film is made piezoelectrically active (spot poled) (Harris, 1982, Lewin and Schafer, 1986). Sensitive element radius ranges from 0.1 - 0.5 mm, and is located in the center of the membrane. Electrodes are usually made of gold and are vacuum deposited.   The PVDF polymer sheet is thin enough that is essentially transparent to the sound waves in the 1-15 MHz range (IEC, 1988). Polyvinylidene film is available in sheets as thin as 9 μm. For a sensor operating in a half wavelength thickness mode (as is clearly the case for membrane configuration), 9 μm corresponds to a resonant frequency of approximately 110 MHz (Lewin and Schafer, 1991).

An example of the frequency response of the membrane hydrophone is shown in Fig. 2. The frequency response of a well designed probe is generally flat below the resonance frequency and falls off beyond the resonance. Due to the complexity of the task, the absolute calibration is seldom carried out beyond 15 MHz.

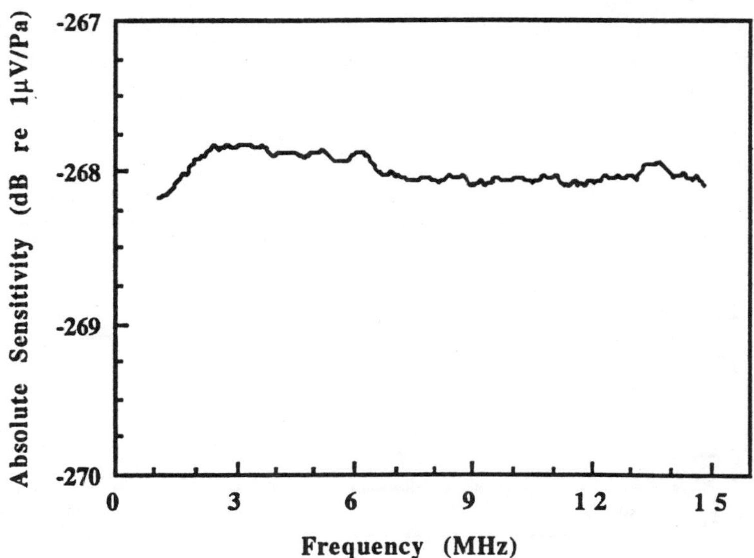

Fig. 2. Calibrated end-of-cable frequency response of the PVDF probe measured in water and terminated into 1 MOhm and 30 pF: 0.4 mm diameter, 9 μm thick PVDF membrane hydrophone. Coaxial cable length 0.5m. (courtesy of Sonic Technologies, Hatboro, PA, USA.),

Needle-type hydrophone The second configuration of the PVDF hydrophone is shown in Fig. 1a. It illustrates the needle-type sensor described in detail in (Lewin, 1981). Briefly, this hydrophone consists of a 16 μm thick circular PVDF film, typically 0. 5-1 mm diameter, attached to an insulating layer at the end of a hypodermic needle. (Fig. 1a). The backing material behind the element has a higher acoustical impedance than water and, as a result, the resonance corresponds roughly to a quarter wavelength thickness of the PVDF film (Lewin and Schafer, 1991). This construction is particularly useful for spatial field plotting and near field (CW measurements). Also, since, the physical dimensions of the hydrophone probe are virtually the same as the dimensions of the active sensor element (Shombert et al., 1982), the disturbance of the acoustic field is minimized. Typical frequency response of a Lewin-type needle probe is shown in Fig. 4.

1.3. Selecting and Using a Hydrophone.

The acoustic pressure-time waveform (see Fig. 3.), if measured with a calibrated hydrophone probe, contains the information required to determine most of the ultrasound exposure parameters. From the analysis of the waveform, the characteristic parameters of the field (Harris, 1992) can be determined. In addition, scanning of the field with the hydrophone probe allows the beam pattern distribution to be determined and subsequently to calculate total acoustic power (see also section 2 of this chapter).

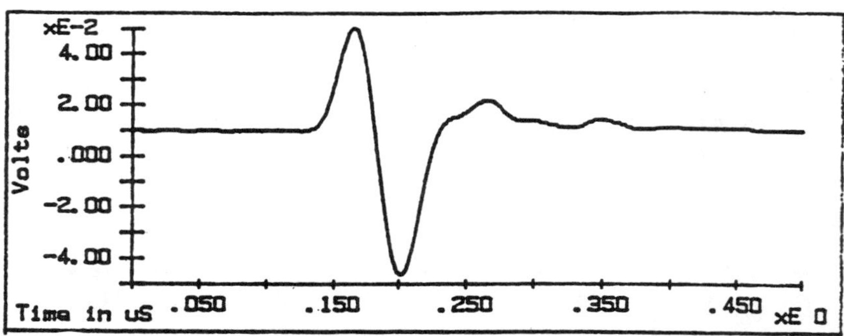

Fig. 3. Example of a pressure-time waveform recorded using a PVDF
hydrophone and transferred from a digital oscilloscope to a PC
computer for further analysis.

In ultrasonic field measurement practice the choice between the membrane- and the needle hydrophone configuration depends on particular application and on the limitations associated with the hydrophone parameters such as end-of-cable sensitivity, uniformity or flatness of frequency response, and angular response, which determines the effective diameter (Lewin and Schafer, 1986).

Effective diameter. As already mentioned, the smaller the element diameter, the more omnidirectional the hydrophone's angular response becomes. In practice, however, there is a difference between the geometrical and effective diameter of the probe. Since, in general, and in the case of membrane design in particular, the effective diameter is larger than the geometrical one (Harris, 1988, Shombert et al. 1982) it has to be determined separately. This can be conveniently done through a careful measurement of directivity pattern in connection with an absolute calibration of the hydrophone probes, as discussed below.

Characterization of hydrophones In general, for quantitative exposimetry work, the hydrophones should be absolutely calibrated i.e. the absolute sensitivity response of the hydrophone should be given as a continuous function of frequency (Fig.2). In this way the uniformity or flatness of the response can be verified.

An absolute calibration of of voltage sensitivity versus frequency can be carried out by primary or secondary methods (Bobber, 1970). Secondary methods are those in which a hydrophone, previously calibrated by a primary method, is used as a reference standard. In the megahertz frequency range, three primary calibration methods are commonly used: 1) the reciprocity method, 2) the radiation force method, and 3) optical methods using light diffraction by ultrasound or based on interferometric measurements (Haran, 1976, Baboux et al. 1988). A comprehensive discussion of those calibration methods can be found in review works (FDA, 1973, O"Brien, 1978) and (Schafer, 1992). However, all three methods suffer from one principal shortcoming: the calibration is usually carried out at discrete frequencies often separated by more than 1 MHz. A discrete frequency calibration procedure often overlooks rapid and significant variations in hydrophones frequency response, for example due to spurious mechanical resonances (Lewin, 1983, Pedersen and Lewin, 1988). This drawback can be overcome by combining the reciprocity technique with the TDS technique. The principles and applications of the technique are briefly outlined in (Lewin, 1983, Pedersen and Lewin, 1988, Ludwig and Brendel, 1988, Chivers, 1986).

It has been demonstrated that, when carefully performed such combination of the reciprocity technique and the TDS technique allows the free field, end-of-cable sensitivity and directivity pattern to be determined as a continuous function of frequency with an overall uncertainty of less than 9% (RMS error) (Preston et al., 1988) Typically, 0.4% is achievable in the low MHz range, increasing to +/-12% (1 dB) at 15 MHz (Preston et al., 1988). Frequency response of the hydrophone shown in Fig 2. was obtained using this technique. Fig. 4 compiles typical frequency responses and sensitivities of both membrane- and needle-type hydrophones for different film thicknesses and configurations. Lewin- or needle type hydrophone exhibits low Q resonance frequency between 25 and 40 MHz. Bilaminar hydrophone configuration using 50 µm thick PVDF film suffers from a severe resonance peak at approx. 20 MHz. The widest bandwidth is offered by a 9 µm thick film membrane design (Lewin and Schafer, 1991).

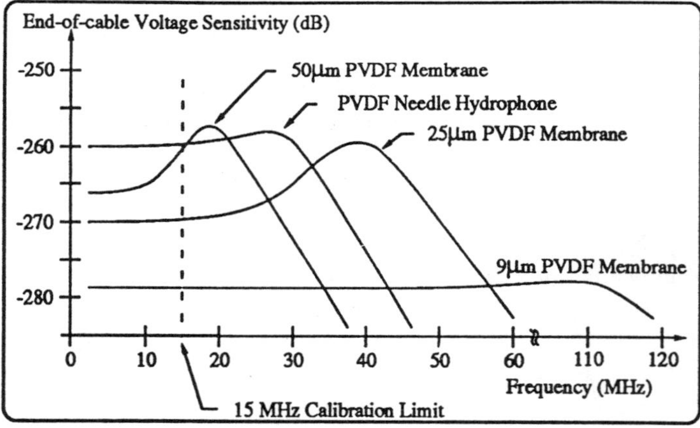

Fig. 4. Typical frequency responses of both membrane and Lewin-type hydrophone for different film thicknesses and configurations. dB scale refers to 1V per µPa.

Angular response. The time delay spectrometry technique can also be used to measure another important hydrophone parameter, namely, the angular sensitivity response or directivity pattern. A comprehensive comparison of angular responses for different hydrophone designs can be found in (Shombert et al., 1982). It appears, that the Lewin-type hydrophone directivity patterns are close to those predicted by a simple piston model. In contrast, the membrane-type designs exhibits accentuated side lobes and, in general, the membrane directivity patterns cannot be predicted using the simple uniform circular receiver model. This is due to the generation of membrane (Lamb) waves when the sound field is incident at the critical angle (approximately 50º), (Shombert et al., 1982).

Linearity: As already mentioned modern diagnostic imaging equipment is capable of generating instantaneous pressure amplitudes on the order of 10 MPa (Lewin and Goldberg, 1989). Such amplitudes are only an order of magnitude lower than those encountered in the focal region of the extracorporeal shock wave generators (lithotripters). Therefore, a faithful reproduction of these highly distorted pressure pulses is important for quantitative acoustic field measurements. Currently available evidence seems to indicate that the PVDF polymer material is linear up to extremely high pressures (100 MPa) (Lewin, 1982, Meeks and Ting, 1984). Additional evidence supporting linear response of the PVDF polymer material emerges from the recent testing of the specially designed PVDF hydrophones for underwater acoustic applications (Tancrell et al. 1985). Negligible changes in the frequency response of the hydrophones were observed up to hydrostatic pressures of 6900 MPa (Tancrell et al. 1985).

1.4. *Other hydrophone designs.*

Although the discussion above was focused primarily on the needle-and membrane type hydrophone probes, a few other designs have been reported. A multielement (array) membrane design can be made by spot poling a number of elements along the axis (DeReggi, et. al., 1977). Such construction has been suggested for rapid lateral electronic scanning of the field generated from diagnostic transducers however, it requires specially designed electronic circuitry for retrieving the signals from individual elements (Preston, 1988).

Yet another design of the needle-type probe has been suggested in (Platte, 1986). In this design the liquefied polymer material is corona polarized on a convex backing support and hence the probe shape corresponds to a fraction of a sphere. This construction reportedly exhibits further improvement in directivity patterns in comparison with the original needle design (Lewin, 1981) and bring them closer to the more desirable spherical characteristics. However, this is done at the expense of the uniformity of the frequency response.

Bilaminar hydrophone configuration (Preston et al., 1983) uses a fully metalized but piezoelectrically inactive, 24 μm thick, PVDF layer glued to piezoelectrically active film, also 24 μm thick. In this way, excellent RF shielding is achieved at the expense of bandwidth (see Fig. 4).   Another modification of the membrane design has been described in (Lewin and DeReggi, 1988).   In this design the PVDF sensor element is mounted on a heavily absorbing backing and protected from the environment by a thin layer of (usually gold) metalized film. This hydrophone also has a built-in preamplifier.   A comprehensive discussion of the advantages and disadvantages associated with the use of integrated preamplifiers is given in (Lewin et al., 1987).

1.5. Ultrasonic Hydrophone Probes: Conclusions

The advent of piezoelectric plastics such as PVDF polymers has markedly contributed to the recent advances in the development of wideband hydrophones for medical ultrasonics applications. Most common ultrasonic polymer hydrophones are the needle-and membrane type, which when properly designed have already proven suitable as reference hydrophones, with good spatial and temporal resolution. The useful bandwidth depends on the overall thickness of the construction and ranges from approx. 20 MHz (See Fig.4) to beyond 100 MHz for a single 9 μm PVDF membrane. This bandwidth ensures that the probes are uniquely suited for measurements and comprehensive characterization of ultrasonic fields generated by both CW and pulsed acoustic sources. The hydrophones feature well behaved directivity patterns, and excellent linearity over a wide range of pressures (up to approx. 100 MPa in biomedical ultrasonics range of frequencies), and last but not least, long term stability (Lewin, 1981).

The choice of particular configuration depends on the application or given measurement task. The embrane design exhibits wider bandwidth due to its pure half wavelength resonance configuration, however its angular response is inferior to the one achievable with a well designed needle-type hydrophone (Shombert, et al. 1982). A needle-type hydrophone may be preferable for measurements of CW fields such as those produced by a CW Doppler device. This is because standing waves may be generated by reflections from the membrane hydrophone. In addition, the membrane design is not well suited for CW fields, and wherever measurements have to be carried out in confined spaces. For in-vivo measurements, the classical membrane design is impractical and a needle design or its modification must be used (Lewin and Jensen, 1979, Daft et al., 1990). The most severe limitation of the currently available piezoelectric hydrophones is due to their finite aperture. Current experience indicates that fabrication of 0.2 mm effective radius hydrophone is feasible. Better spatial resolution can be achieved by using by fiber optic hydrophones at the expense of simplicity of the basic measurement arrangement (Chan et al., 1988).

2. Radiation Force Devices

This section describes devices which are based on the measurement of radiation force. Radiation force devices are attractive because they provide an absolute measurement of ultrasonic power. However, unlike hydrophones they are not suitable for temporal wave analysis and measurements of spatial beam distribution. In the following, only the most widely used implementations of the radiation force devices are discussed. A more comprehensive description of the different implementations of the radiation force technology including strain gauge, suspended ball (elastic sphere) and float radiometers can be found in (IEEE, 1989).

The radiation force devices utilize the fact that the incident acoustic energy acting on appropriately selected, either reflecting or absorbing target, transfers a momentum to this target. This momentum can be measured as a force which is directly proportional to the spatial and temporal-averaged acoustic power. Radiation force devices may be divided into two types depending on the size of the target. The "large" target devices use a balance to "weigh" the force exerted on the target, while the "small" target devices use other means, e.g. cathetometer, to determine this force. The use of a large target which covers the total effective beam cross section will result in a measurement of total power, while the small target, totally immersed in the field, will yield local intensity. It is worthwhile to note that while the principle of radiation force balance is simple and requires only accurate measurement of the static force produced by the ultrasonic energy, the implementation of the device is fairly complex. In practice, a careful customizing of the design is often required. A good quality analytical chemical microbalance is needed for the measurement of the low level acoustic power, such as that generated by diagnostic equipment. Typically, measurement of acoustic powers in the diagnostic range 1 - 50 mW requires sensitivity of the microbalance on the order of 0.01 mg.

For a perfect "large" reflecting target the force F acting in the direction of propagation of the acoustic wave can be expressed as: $F = (2P \cos 2\phi)/c$, where P is the time averaged acoustic power in Watts, ϕ is the angle of incidence, and c is the speed of sound in a medium in m/s. The radiation force balance is shown schematically in Fig. 5. In practical implementation

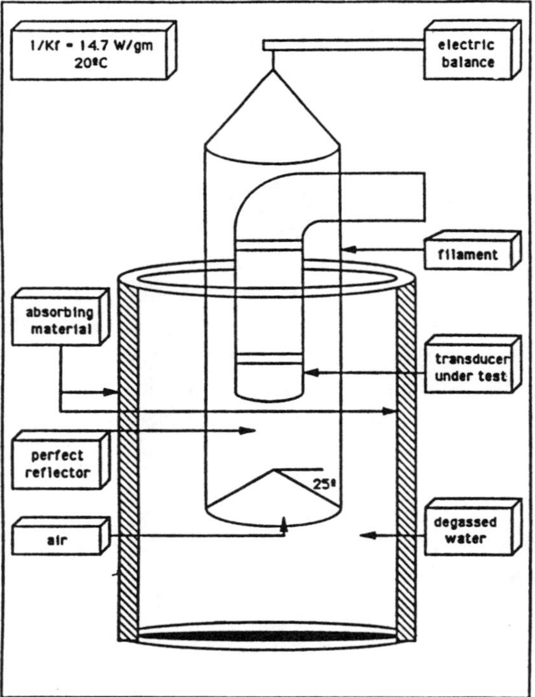

Fig. 5 The radiation force measurement principle. The measurements are performed in degassed water.

of the power measuring device an angle of incidence for a perfect reflector is selected to be 45 degrees, as shown in Fig.5. At room temperature (22^0+/-3^0C) the sensitivity of the radiation force balance is typically 14. 55 +/-0.05 W/g (IEEE, 1989).   The same sensitivity is valid for a perfectly absorbing target and normal or zero degree angle of incidence.

Both absorbing and reflecting targets have their advantages and disadvantages. In the case of perfectly reflecting 45^0 target, a 10% error in target angle leads to 3.5 % error in the measured power. While perfectly absorbing targets will minimize positioning error, because there is no dependence on cosine term, perfectly absorbing wideband target materials are not easily available.   Absorbing targets can also cause standing waves due to the imperfect absorption of acoustic energy. These standing waves can lead to fluctuations as large as +/-13% about the mean (Hussey and Bly, 1988).

It is appropriate to stress that there are relatively few commercial sources of complete radiation force balances. Most of the devices described in the literature are custom made prototypes utilizing a commercially available balance or electrobalance. The absorbing or refelecting targets, their suspension, and the water container with lined walls to minimize standing wave phenomena must all be individually designed and constructed. Sensitivity of such radiation pressure balances depends mainly on construction and the precision of the balance. In general, the higher sensitivity that is needed, the more precise (and expensive) the balance which must be used. Often, the sensitivity is sufficient for therapeutic range of acoustic power, i.e. 0.1-15W. However, diagnostic ultrasound levels are usually significantly lower and therefore, a special design is needed to measure average power on the order of microwatts (Rooney, 1983).

Tethered Float Wattmeters Buoyant and tethered devices (FDA,1973, IEEE, 1989) constitute interesting variations of the large target radiation force devices. These devices, often called wattmeters, are used to measure acoustic power ranging from 1.5 to 15 W. In this design, the ultrasonic transducer is aimed at a tethered buoyant float which is covered with an absorbing material. The radiation force due to ultrasound energy causes a downward displacement of the float from the equilibrium position. This displacement is proportional to the ultrasonic power (FDA,19723 IEEE, 1989).

Suspended sphere and strain gauge devices While radiation force balances gained largest dissemination, radiation force phenomena is also utilized in measurements of acoustic power with suspended sphere and strain gages. Elastic sphere or suspended ball radiometer is often used for measurements of acoustic intensity in the range 1W-10W, although spatial-peak temporal-average intensities as low as 10 mW/cm^2 have been determined in the recent interlaboratory comparison of hydrophone calibrations (Preston et al., 1988). The principle of the method has been detailed in (Dunn et al., 1977). Briefly, it involves a 2.5 to 3.2 mm diameter type 440 stainless steel ball glued to a nylon filament of approx. 0.5 mm diameter with a small amount of Silastic™. The ball is suspended vertically in the (unfocused) sound field. The filament length is typically 10 cm and the displacement of the ball is measured with a horizontal cathetometer. For small angle deflection, the spatial-peak temporal-average intensity can be computed as (Preston et al., 1988): Ispta = F_r c/πa2Y, where F_r is the radiation force, a is the radius of the sphere, c is the speed of sound in the medium (usually degassed and distilled water), and Y is the acoustic force radiation function (Dunn et al., 1977, Hosegawa and Yosioka, 1977).

Y depends on the ratio of the ball diameter to the wavelength in the medium, the medium density, and the elastic properties of the ball (Dunn et al., 1977). The technique has been utilized in the frequency range .25 MHz - 10MHz. Reproducibility of the technique is +/-2% and estimated uncertainty +/-10% (Preston et al., 1988).

Strain gauge technique Acoustic power measurements of medical ultrasonic probes using strain gauges connected in Wheatstone bridge configuration has been suggested in (Anson and Chivers, 1982). The temporal resolution of the strain gauge ultrasonic therapeutic exposimeter is determined by dimensions and flexibility of material of the cantilever arm on which the strain gauges are cemented and the absorbing target is mounted. Range over which the the acoustic power is measured is typically 10 mW - 10 W for semiconductor gauges (up to 25W for foil gauges). While the technique is in principle applicable in the frequency 1 - 20 MHz, there are limitations in the lower frequency end due to the size of the target and in the upper frequency end due to target design and attenuation. In addition, a power supply and electronic read out of the unbalanced bridge voltage are needed. Prior to performing measurements, a calibration has to be performed. The lowest detectable level or sensitivity of the strain gauge technique is about 50 mW (Anson and Chivers, 1982).

3. Thermal Devices

Thermal techniques utilize acoustic absorption in order to determine the intensity or the acoustic power. Again, depending on the approach, the thermal methods can either determine the total power of the acoustic beam or measure a local value of intensity within a beam profile (cross section). Calorimetric techniques will determine the total power, while thermocouple techniques will yield local values of intensity.

Thermocouple probes The immediate advantage of a thermocouple is that it can be fabricated with an extremely small measurement volume, and therefore is well suited to measure local values of intensity (Hawley et al., 1962). It should be pointed out that the thermocouple measures relative acoustic intensity and absolute intensity measurement would require prior calibration in a known field.

Uncoated or transient probes having resolution on the order of 10 microns can be made. Spatial resolution of coated probes is usually lower. However, the thermocouple devices do not respond to rapid changes of the ultrasonic field and their temporal resolution is typically on the order of 1s.

Also, the probe needs to cool for several seconds between the measurements. Furthermore, the technique requires relatively high, on the order of 0.5 W/cm^2 intensity to give a satisfactory output which, in addition, is frequency dependent. However, while thermocouple measurements are frequency dependent, they have been successfully carried out in the frequency range 0.26 MHz up to 2 GHz (Hawley et al., 1962). Measurement accuracy is dependent upon the calibration procedure. Repeatability is dependent on frequency and ranges from approx. +/- 10% to 3% over the frequency range 0.26MHz-2GHz (Hawley et al., 1962).

Calorimeters Calorimetric devices are based on measurement of the heat generated when ultrasound energy is totally absorbed. If properly designed, the calorimeter measures time-averaged acoustic power. The principal problem in calorimetric measurements is to separate the heat due to the inefficiency of the acoustic source (heating generated by the source) from that due to the absorption of the acoustic energy (Zieniuk and Chivers, 1976). The most recent attempt to design a precise calorimeter capable to measure acoustic power in the range 1mW to 10 W over the frequency range 1 MHz to 15 MHz is detailed in (Zapf and Harvey, 1981). The device is currently in use at National Institute of Standards and Technology (NIST) and requires that the beam from the source does not exceed 25 mm diameter 5 mm from the transducer face. Measurement takes from 1 to 10 minutes and the accuracy of technique is +/-10% of reading +/-0.2 mW.

4. Measurement Systems

As already mentioned, comprehensive acoustic field characterization includes recording pressure time waveform and determining the beam pattern distribution. It was also noted that such field measurements are most conveniently carried out using miniature, calibrated PVDF ultrasonic hydrophone probes together with a computerized measurement system. An example of such system is shown in Fig. 6 (Lewin and Schafer, 1988). The characterization of the field is carried out in the following way. The hydrophone probe is scanned through the field along the acoustic axis z of the source using the xyz micromanipulator. The output from the probe is monitored on the oscilloscope. Once the point of interest is identified (it is usually the point corresponding to maximum pressure amplitude of the wave) the corresponding pressure-time waveform is captured and transferred to the digital oscilloscope. Wave analysis is subsequently carried out using the software algorithms and a PC computer. Next, the field is scanned in the direction perpendicular to the axis of the acoustic source in order to determine the beam pattern distribution. To minimize measurement

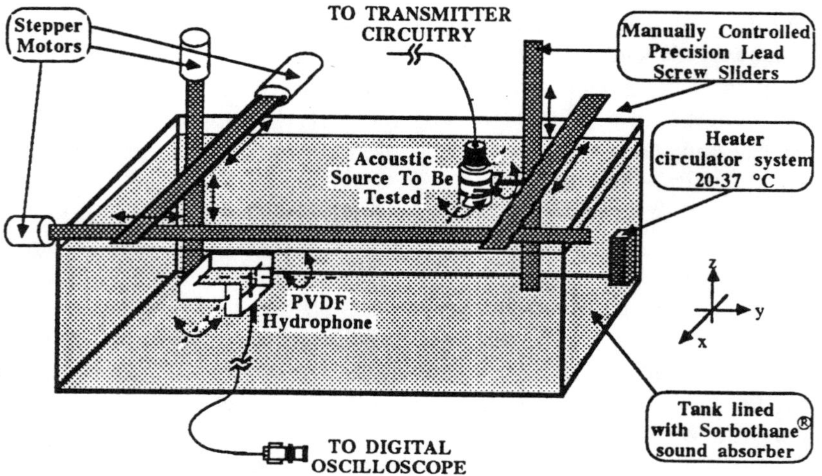

Fig. 6. A computerized system for characterization of ultrasound field. Reprinted with permission (Lewin and Schafer, 1988), ©1988, IEEE.

uncertainties this scan is usually carried out in the xy and yz directions (Harris, 1992). The measurements during scanning of the hydrophone are usually taken at the intervals corresponding to one half wavelength. The beam plot data are used to determine total acoustic power. A more detailed description of the system can be found in (Lewin and Schafer, 1988).

5. CONCLUDING REMARKS

Measurement systems used in ultrasound measurement practice together with ultrasonic field sensors have been described. The sensors are based on piezoelectric conversion of pressure into voltage and are used together with computerized measurement systems. The systems include micromanipulators for adequate scanning of the field, digital osciloscopes for capture of the pressure-time waveform, and a personal computer (PC) needed to carry out calculations of the relevant field parameters The attention was primarily focused on hydrophone probes and radiation force balance systems, because these devices are best documented and most widely available. While such devices as calorimeters and thermocouples may offer an alternative to hydrophone probes and radiation force balances, they often require customized design. Although hydrophone probes require periodic recalibration using absolute methods, at present, they offer the most convenient approach to modern, real time dosimetry requirements. If properly designed and calibrated, they yield both spatial and temporal field characteristics. In addition, scanning the field allows total acoustic power in to be determined. Hydrophone probes together with a sensitive radiation force balance constitute a powerful combination for quantitative ultrasonic exposimetry measurements. The major limitation of the current piezoelectric hydrophone technology is the finite size of the probes, often comparable with the wavelength of the field measured. Future advances in the ultrasound dosimetry will most likely introduce hydrophone probes based on fiber optic technology. Finally, a word on the overall uncertainty using hydrophone probes and associated electronics is appropriate. Typically, this uncertainty is about +/- 20% when pressure amplitude of the wave is measured in the frequency range 1-20 MHz. This figure includes contributions due to absolute calibration of the probes (approx. +/-12%), corrections due to nonlinear propagation phenomena (typically 4%), and electronic instrumentation (about 4%). When intensity values are calculated, this overall uncertainty doubles.

References:

AIUM/NEMA (1981) Standards Publication UL-1-1981, Washington D.C. National Electrical Manufacturers Association.

Anson, L.W., Chivers, R.C. (1982) A Rugged Strain Gauge Ultrasonic therapeutic Exposimeter, J. of Medical Engineering and Technology, v. 6, pp. 150-152.

Bobber RJ. (1970) Underwater Electroacoustic Measurements. GPO. Washington DC. 1970.

Baboux, JC., Djelouah, H. and Perdrix, M. (1988) Interferometric measurements of transient ultrasonic field: application to hydrophone calibration, IEEE Ultrasonics Symposium Proc., pp. 857-861.

Chan, H.L.W., Chiang, K.S. and Gardner, J.L. (1988), Polarimetric optical fiber sensor for ultrasonic power measurement, IEEE Ultrasonic Symp. Proc., pp. 599-602.

Chivers R.C. (1986), Time delay spectrometry for ultrasonic transducer characterization. J. Phys. E. Sci. Instr., 19, 834-843.

Daft, C. M. W., Siddiqi, TA, Fitting, D. W. , Meyer, RA and O'Brien, WD. Jr. (1990) In vivo fetal ultrasound exposimetry, IEEE Trans. UFFC-37(6), pp. 501-505, 1990.

DeReggi AS, Edelman S., Roth SC, et.al, Piezoelectric Polymer Receiving Arrays for Ultrasonic Applications, J. Acoust. Soc. Am., 61:S17, 1977.

Dunn, F. , Averbuch, A.J., O'Brien, W.D.Jr., (1977) A primary method for the determination of ultrasonic intensity with the elastic sphere radiometer, Acustica, v.38, pp. 58-61.

FDA (1973).Ultrasound dosimetry, Chapter 4, p. 153-201, In: Interaction of ultrasound and biological tissues, Reid and Sikov, eds., DHEW Publication (FDA) 73-8008, GPO, Washington D.C.

FDA (1985) 510(k) Guide for Measuring and Reporting Acoustic Output of diagnostic Ultrasound Devices, Wash. D.C.

Haran, M.E. (1979) Visualisation and measurement of ultrasonic wave fronts. IEEE Trans. Proc. 1979, 67, pp. 454-456.

Harris, G.R. (1988) Hydrophone measurements in diagnostic ultrasound fields, IEEE Transactions UFFC 35, p. 81-101, 1988.

Harris, G.R. (1982) Sensitivity considerations for PVDF Hydrophones using the spot poled membrane design,   IEEE Trans. Son. Ultrason., SU-29, pp. 370-377.

Harris, G.R. (1992) Ultrasound safety standards (somewhere in this volume)

Hasegawa, T. and Yosioka, K.(1969). Acoustic radiation force on a solid elastic sphere. J. Acoust Soc. Am. 46, pp. 1139-1143.

Hawley, S.A. , Breyer, J.E. and Dunn, F. (1962) Fabrication of Miniature Thermocouples for UHF Acoustic detectors, Review of Scientific Instruments, v. 33, pp. 1118-1119.

Hussey , R.G. and Bly, S.H.P.(1988) A portable ultrasonic power meter for routine calibration of ultrasonic therapy devices. Journal of Clinical Engineering v. 13 (2), 109-113.

IEC (1991). Measurements and characterization of ultrasonic fields using hydrophones in the frequency range 0.5-15 MHz. IEC Publication 1101

IEEE (1989) Guide for Medical Ultrasound Field Parameter Measurements, IEEE Std 790-1989.

Lewin, P.A., Jensen, F. (1979) The use of miniature ultrasonic transducers for measurements in penetrable media. Proc. Ultrasonics International '79, Butterworth Publishers, p. 279-286.

Lewin, P.A. (1981) Miniature Piezoelectric Polymer Ultrasonic Hydrophone Probes, Ultrasonics, 19, pp. 213-216.

Lewin, P.A. and Chivers, R.C. (1981) Two Miniature Ceramic Ultrasonic Probes, J. Phys. Sci. Instrum. E., v. 14, 1420-1424,

Lewin, P.A. (1983) Polymer hydrophones in biomedical applications, IEEE Ultrasonics Symp. Proc. , pp. 822-825.

Lewin, P.A. (1985) Linearity of the polymer probes, Proc. WFUMB 1985, Sydney, Australia, p. 35, Pergamon Press.

Lewin P.A. and Schafer M.E.(1986), Ultrasonic probes in measurement practice, MD&DI, p.40-45.

Lewin, P.A., Schafer, M.E.and Chivers, R.C.(1987) Factors affecting the choice of preamplification for ultrasonic hydrophone probes, Ultrasound in Med. Biol. v. 13, (3), pp. 141-148.

Lewin, P.A. and DeReggi, A.S. (1988) Short range transducer applications. In: Wang TT, Herbert JM, Glass AM, eds. The applications of ferroelectric polymers, New York, The Blackie Group Publishers, pp. 162-189.

Lewin, P.A and Goldberg, B.B. (1989) Ultrasound bioeffects for perinatologists. Gynecology and Obstetrics (J.J. Sciarra, ed.) Ch.71, pp. 1-14, J.B. Lippincott Co.

Lewin, P.A , Gilmore, J.M. and Schafer, M.E. (1989) PVDF sensors for quantitative acoustic shock wave measurements. In Ultrasonics International'89 Conf. Proc, Butterworth and Co Publishers, pp. 548-553.

Lewin, P.A. and Schafer, M.E. (1991) Shock wave sensors: I Requirements and design. J. Lithotripsy and Stone Disease. Vol. 3.(1), pp. 3-17.

Ludwig, G. and Brendel, K. (1988). Calibration of hydrophones based on reciprocity and time delay spectrometry. IEEE Trans. UFFC 35(2): pp. 168-174.

Meeks, S. , Ting, R. (1984) The evaluation of static and dynamic stress on the piezoelectric and dielectric properties of PVDF. J. Acoust. Soc. Am. v.74, pp. 1010-1012.

O'Brien, W.D. (1978), Ultrasonic Dosimetry: In :Ultrasound: its application in medicine and biology, P.1. F.J. Fry, ed., pp. 343-391, Elsevier.

Pedersen, P. C. and Lewin, P.A. (1988)Applications of the time delay spectrometry for calibration of ultrasonic transducers, IEEE Trans. UFFC, 35(2): 185-205.

Platte, A. (1985) Polivinylidene fluoride needle hydrophone for ultrasonic applications, Ultrasonics 23, pp. 113-118.

Preston, RC, Bacon, DR. Livett, AJ, et al. (1983) PVDF membrane hydrophones and their relevance to the measurement of acoustic output of medical ultrasonic equipment, J. Phys. Sci. Instr., 16, 786-796.

Preston, R.C., Bacon, DR, Harris, GR, Lewin, PA ,Mc Gregor JA , O'Brien, WD and Szabo, T.L. (1988) Interlaboratory comparison of hydrophone calibrations. IEEE Trans. UFFC 35(2), pp. 206-213.

Preston, R.C. (1988) The NPL Ultrasound Beam Calibrator, IEEE Trans. UFFC-35, p. 122-139, 1988.

Rooney, J. A. (1973) Determination of acoustic power outputs in the microwatt-milliwatt range, Ultrasound in Medicine and Biology, Vol. 1, pp. 13-16.

Schafer, M.E. and Lewin, P.A.(1988) A computerized acoustic output acquisition system, IEEE Trans. UFFC, v. 35, pp. 102-109.

Schafer, M.E. Calibration techniques, In: Ziskin, M.C. and Lewin, P.A. (eds., 1992) Ultrasound Dosimetry, CRC Press (to be published).

Sherar, M.D. and Foster, F.S. (1989) The design and fabrication of high frequency poly (vinylidenefluoride) transducers. Ultrasonic Imaging, 11, pp. 75-94.

Shombert DG, Smith SW, and Harris, GR (1982) Angular Response of Miniature Ultrasonic Hydrophones. Med. Physics 9, 484-492.

Tancrell, RH. , Wilson, DT and Rickets D. (1985), Properties of PVDF polymer for sonar, Proc IEEE Ultrasonic Symposium, pp. 624-629.

Zapf, T.L., Harvey, M.E.,Larsen, N.T. and Stoltenberg, R.E. (1983) Ultrasonic calorimeter for beam power measurements, NIST Technical Note 686.

Zieniuk, J. Chivers, RC, (1976) Measurement of ultrasonic exposure with radiation force and thermal methods, Ultrasonics 14, pp. 161-172.

Ziskin, M.C. and Lewin, P.A. (eds., 1992) Ultrasound Dosimetry, CRC Press (to be published).

ULTRASOUND SAFETY STANDARDS

Gerald R. Harris

Center for Devices and Radiological Health
US Food and Drug Administration
5600 Fishers Lane
Rockville, MD 20857

1. INTRODUCTION

Ultrasonic energy has been used in medical practice for approximately 40 years in a variety of therapeutic and diagnostic applications. For many of these applications, standards or guidelines have been or are being developed to help assure that the ultrasonic energy is applied to the patient as safely and efficaciously as possible. Furthermore, in the area of industrial ultrasound, exposure criteria have been adopted by a number of organizations to help limit inadvertent occupational exposure from equipment-generated airborne ultrasound.

In this paper some of the more important aspects of these medical and industrial ultrasound safety standards and guidelines are presented. With respect to medical uses of ultrasound, standards activities in three prominent areas are described; namely, therapeutic ultrasound, diagnostic ultrasound, and extracorporeal shock-wave lithotripsy (ESL). Since in therapeutic applications a beneficial effect from the ultrasound is desired, the objective of the therapy standards is to specify performance criteria so that instruments are capable of delivering a prescribed amount of ultrasonic energy to the patient. In diagnostic applications, on the other hand, the goal is to obtain clinically useful images or data without producing any deleterious changes in the exposed tissues. Here the standards call for relevant exposure information to be provided so that risk can be minimized. In ESL, damage is a desired result, in that dissolution of kidney stones or gallstones is the intent of the treatment. Here again, details of exposure are necessary to optimize stone fragmentation while minimizing damage to surrounding tissues.

It will be noted that in the standards and guidelines described below, the various devices are characterized in most cases in terms of basic exposure characteristics of the ultrasound field (e.g., pressure, power, intensity). This is because the concept of dosimetric quantities, that is, quantities related to absorbed energy and subsequent biological effects, has not been well-defined for ultrasonic radiation (Duck, 1987). However, the exposure information commonly reported is believed to be relevant in assessing risk due to the thermal and mechanical effects that can result from the interaction of ultrasound and tissue. Moreover, in the medical field some dosimetric-type units are beginning to be introduced, as described in Sec. 3.3 and discussed elsewhere in these Proceedings (O'Brien, 1992). Also contained in these Proceedings is a description of many of the terms and quantities relevant to ultrasound exposure that are used in this paper (Bly and Harris, 1992).

2. THERAPEUTIC ULTRASOUND

Both national and international standards have been developed for ultrasonic therapy products. Common to all of these standards is that the output exposure level (e.g., ultrasonic power, effective intensity) must be indicated clearly to the operator. However, differences exist in the details of this indication, as well as in other areas of performance and labeling specification.

The oldest active standard is the international recommendation prepared by the International Electrotechnical Commission (IEC) Technical Committee (TC) 29, Electro-acoustics (IEC, 1963). This standard currently is being revised by IEC TC87, Ultrasonics. Notable national standards have been developed by the United States (US) (FDA, 1978; Stewart et al., 1980; Ferguson, 1985) and Canada (Standard for ultrasound therapy devices, 1984). Important features of these three standards are summarized in Table I and are discussed below.

2.1 Output Indicator

The US and Canada call for both ultrasonic power and effective intensity to be indicated (W and I_{eff}, respectively, in Table I), whereas in the IEC standard the indication of both quantities is optional. Furthermore, the US and Canada call for temporal average (TA) values in the continuous wave (CW) mode, and temporal peak (TP) values in the amplitude-modulated (AM) mode. For the US and Canada, the error in the ultrasonic power indication must be $\leq \pm 20\%$ for all settings greater than 10% of the maximum power setting. For the IEC, the accuracy requirement is the same below 50% of the maximum setting, and is $\pm 15\%$ above the 50% level.

TABLE I. Comparison of Three Ultrasonic Therapy Standards

	IEC	US	Canada
Output Indicator	W or I_{eff}, or both. TA - CW & AM	W and I_{eff}. TA - CW Mode TP - AM Mode	Same as US
Timer Control	Not Specified	Yes	Yes
Effective Radiating Area	Based on "Baffle" Measurement	Based on Hydrophone Measurement - % of Peak	Same as US
Spatial Maximum Intensity	$I_{max} \leq 2I_{eff}$	BNR Specification	None
Effective Intensity Limit	None	None	$I_{eff} \leq 3$ W/cm^2
Temporal Specifications in AM Mode	Pulse Shape, Duty Factor, & PRR in Manual	Pulse Shape, PD, PRR, & TP/TA I_{eff} Ratio on Generator	Same as US

2.2 Timer Control

Both the US and Canada require that all therapy units contain an adjustable timer capable of stopping the treatment at the end of a preset time. The timer indication must be accurate to within 30 seconds for settings of less than five minutes, to within $\pm 10\%$ for settings between five and ten minutes, and to within one minute for settings greater than ten minutes. The IEC standard does not address the issue of a timer control.

2.3 Effective Radiating Area (ERA)

All three standards require that the ERA be labeled on the applicator. The IEC specifies that the ERA must be within $\pm 10\%$ of the labeled value, Canada's requirement is $\pm 20\%$, and the US simply states that the accuracy must be provided. Significant differences exist, however, in the techniques used to measure the ERA. The IEC specifies a "baffle" technique, in which the ERA is determined by measuring the power passing through successively larger circular apertures and noting the aperture size for which 90% of the total ultrasonic power is transmitted. Because this technique is impractical for the variety of transducer configurations available, the US and Canada adopted a definition for ERA based on a hydrophone scanning technique. In this approach, the ERA is defined as the area consisting of all points on a surface 5 mm from the applicator face at which the intensity is 5% or more of the spatial-peak intensity on that surface. While an improvement on the baffle technique, the scanning technique is still subject to problems (Bly et al., 1989). Therefore, the IEC's TC87 is currently examining a third method for determining the ERA based on a proposal by Hekkenberg (1988). If this new approach is acceptable, it is expected that the US and Canadian standards would be modified accordingly.

2.4 Spatial Maximum Intensity and Effective Intensity Limit

Because the ultrasonic intensity distribution across the beam is not uniform, there will be places in the beam where the local intensity is greater than the effective intensity value indicated on the generator. Conceivably, excessive heating of a small volume of treated tissue could occur at these locations. Movement of the applicator will improve this situation, but some limit on the local intensity, or some knowledge of the degree of nonuniformity of the beam, should be of value. To this end, the IEC has placed a limit on its "maximum intensity," (denoted I_{max} in Table I and defined as $4/\pi$ times the power passing through a baffle with a 1 cm diameter aperture), stating that it must not exceed twice the effective intensity. The US has taken a different approach, requiring that the manufacturer describe the spatial distribution of the ultrasonic field in the user's manual, and give on the applicator label the ratio of spatial-peak intensity to the effective intensity. By multiplying this ratio (called the Beam Nonuniformity Ratio, or BNR) by the effective intensity indicated on the generator, the maximum point intensity can be determined. The Canadian standard has no such requirements, but it does place an upper limit of 3 W/cm^2 on the effective intensity.

2.5 Temporal Specifications in AM Mode

With regard to temporal specifications such as pulse duration (PD), pulse repetition rate (PRR), and duty factor (DF = PD*PRR), the US and Canadian standards call for this information to be on the generator. Furthermore, these two standards require that the ratio of the temporal peak effective intensity to the temporal average effective intensity be labeled.

3. DIAGNOSTIC ULTRASOUND

3.1 International Activities

The IEC has been active for a number of years in developing standards related to measurement and characterization of medical ultrasonic fields and patient exposure from these fields. A summary of some of the efforts of TC87 with respect to diagnostic ultrasound safety is given in Table II (cf. Preston, 1989).

In addition to these measurement and labeling standards efforts, TC87 also is considering exposure and dosimetric criteria for medical ultrasound equipment. Initial efforts have concentrated on identifying biophysical end-points (e.g., temperature rise) and how these end-points can be determined from exposure measurements. Also being studied is an equipment safety classification scheme based on these end-points.

A second Technical Committee developing diagnostic ultrasound safety standards within the IEC is TC62 (Electrical Equipment in Medical Practice). Subcommittees SC62B (Imaging Equipment) and SC62D (Electromedical Equipment), with assistance from TC87, are planning to coordinate efforts that would include the areas of electrical and mechanical safety as well as acoustic output.

TABLE II. Some Recent IEC Technical Committee 87 Activities
Related to Diagnostic Ultrasound Safety Standards

Title	Status
The absolute calibration of hydrophones using the planar scanning technique in the frequency range 0.5 MHz to 15 MHz	IEC Publication 1101 (1991)
Measurement and characterisation of ultrasonic fields using hydrophones in the frequency range 0.5 MHz to 15 MHz	IEC Publication 1102 (1991)
Requirements for the declaration of the acoustic output of medical diagnostic equipment	Central Office Document 87(CO)11
Ultrasonic power measurement in liquids in the frequency range 0.5 MHz to 25 MHz	Central Office Document 87(CO)10
Standard method of measuring and labeling for ultrasonic Doppler fetal heart beat detector	Central Office Document 87(CO)15

3.2 National Activities

Policies regarding diagnostic ultrasound exposure have been developed in several countries (Nyborg, 1989). For example, Canadian guidelines state that if the free-field spatial-peak, temporal-average intensity (I_{SPTA}) can exceed 100 mW/cm^2 in fetal modes, then the device should have an exposure control capable of reducing the I_{SPTA} below this level. Also, if the I_{SPTA} can exceed 500 mW/cm^2, fetal indications for use should not be claimed for the device unless estimates of local temperature rise in fetal tissue are provided (Canada, 1989).

In Japan, the Japanese Industrial Standards (JIS) Committee of the Japanese Standards Association has produced several standards that specify maximum acoustic output limits for certain devices or modes of operation (Maeda and Ide, 1986). These limits are given in terms of the spatial-average, temporal-average intensity (I_{SATA}) at the transducer surface, which is to be measured by dividing the total output power by the area of the transducer. Table III summarizes the Japanese limits.

TABLE III. Japanese (JIS) Maximum Acoustic Output Limits

Mode or Device	I_{SATA} Limit (mW/cm^2)
A-Mode	100
Manual B-Mode	10
M-Mode	40
Doppler (fetal)	10
Real-Time Linear Array	10

In the US, the regulation of all medical devices, including diagnostic ultrasound, is under the authority of the Food and Drug Administration (FDA). Before introducing a new diagnostic ultrasound device into the US market, a manufacturer must demonstrate to the FDA that the device is substantially equivalent in terms of safety and effectiveness to (1) a device that was being marketed on or before May 28, 1976, the date on which the Medical Device Amendments to the US Food, Drug, and Cosmetic Act were passed, or (2) a device currently legally marketed (Villforth, 1988). In part, this determination of substantial equivalence is based on comparing the application-specific I_{SPTA} and spatial-peak, pulse-average intensity (I_{SPPA}) of new and pre-Amendments devices (FDA, 1985, 1987, 1991a). In its regulatory guidelines the FDA refers to measurement procedures and definitions contained in a standard developed by the American Institute of Ultrasound in Medicine (AIUM) and the National Electrical Manufacturers Association (NEMA) (AIUM/NEMA, 1983). Both AIUM and NEMA are in the process of updating this standard.

Specification of exposure quantities such as I_{SPTA} and I_{SPPA} are based on measurements made in water. However, because water lacks the attenuation properties of tissue, water-measured results are sometimes "derated" to an estimated *in situ* value. In its simplest form, the derating model assumes a homogeneous tissue path in which the power decreases exponentially at a rate of α dB/cm-MHz (Stratmeyer, 1989). The FDA bases its comparisons on derated intensity values, where a value of 0.3 is used for α. This regulatory approach has proved useful in assisting FDA reviewers to address questions of relative safety between devices. The use of derated values allows a fairer comparison of devices operating at different frequencies or focal lengths, although, as FDA has pointed out, the derated values should never be uncritically interpreted as actual *in situ* intensities (Stratmeyer, 1989).

Because of suggestions that FDA's pre-Amendments output limits have been (1) erroneously interpreted as safety levels, and (2) a detriment to the development of promising new applications or techniques requiring higher outputs, FDA is considering an addition to its regulatory procedures (Merritt and Yin, 1989; Merritt, 1989). Under a proposed new approach, the issue of equivalent safety could be addressed via the combination of an on-equipment output display and related user education program, rather than by performing a comparison of application-specific exposure levels.

The specifications for the output display have been drafted in a standard written cooperatively by FDA, AIUM, NEMA, and other clinical and scientific organizations (AIUM, 1991). The standard defines two types of biologically relevant indices for the display: a Mechanical Index (MI) designed to provide a measure of the likelihood for a mechanical event such as cavitation to occur, and a Thermal Index, either soft tissue (TIS), fetal bone (TIB), or cranial bone (TIC), designed to provide an indication of the maximum temperature elevation that could occur (Ziskin, 1990; Thomenius, 1990; Apfel and Holland, 1991, O'Brien, 1992).

Key elements of this standard are summarized in Table IV. For simplicity and space reasons, the standard requires that only one index need be displayed at a time. In real-time (B-mode) imaging, the MI is displayed. In all other modes, including combined modes, the operator may choose either the TIS or TIB to be displayed. The intent here is to present the index of greatest concern for the mode being used. The Thermal Index is unlikely to be high in B-mode. For other modes, heating is of greater concern; moreover, the MI in other modes should not be significantly greater than in the B-mode. The only situation in which the MI would be available for display in modes other than the B-mode is in Doppler-only devices. Also, if a diagnostic ultrasound system is intended solely for adult cephalic applications, then the Thermal Index display must include the TIC.

If the index value at the highest output setting is less than 1.0, then no display is necessary. However, if the maximum index value can exceed 1.0, then the index must be displayed when it equals or exceeds 0.4, to assist in keeping the exposure level as low as reasonably achievable (i.e., the ALARA principle) (Merritt, 1989; Ziskin, 1990). Also, the display increments shall be no coarser than 0.4, 0.6, 0.8, 1, 2, 3, 4, etc.

TABLE IV. Key Elements of US Output Display Standard

Element	Comment
Visual Display of Output	Mechanical Index in real-time imaging (B-mode); Thermal Index in all other modes
Threshold for Display	No display if max. index < 1; Otherwise, display when index ≥ 0.4
Increment for Display	≤ 0.2 (index < 1) ≤ 1.0 (index ≥ 1)
Default Settings for Display	Appropriate to application

To help minimize patient exposure, the standard calls for devices to incorporate application-specific output default settings. These default levels are set upon power-up, entry of new patient ID data, or change from a non-fetal to a fetal application.

The goal of the output display standard is to make device operators aware of the ultrasonic output of their instrument via the real-time display, so that the ALARA principle can be implemented easily and effectively, and prudent risk-benefit decisions can be made. As mentioned above, a user education program is essential in achieving this goal, and work has begun on developing the educational message, identifying the clinical target groups, and selecting the most effective means for conveying the message to these groups (AIUM, 1990).

4. EXTRACORPOREAL SHOCK-WAVE LITHOTRIPSY (ESL)

No national or international standards have been adopted for ESL devices at this time. However, certain activities are beginning in the IEC. In TC87's Working Group 7 (Surgical Instruments), plans are underway to draft measurement guidelines. Also, TC62 (Subcommittee SC62D) plans to write a safety standard that will include areas such as electrical and mechanical safety, and positioning repeatability.

In the US, the FDA has developed a draft guidance document on measuring and reporting shock-wave characteristics as part of its regulatory review responsibilities (FDA, 1991b). This guidance was written to assist manufacturers of ESL devices in testing and characterizing shock waves, and in submitting the results of these tests to FDA as part of investigational device exemption (IDE, the FDA clinical studies regulation) or premarket approval (PMA) applications. The document covers the measurement data that should be provided, including information on the shock pulse waveform and focusing characteristics, and on the instruments (e.g., hydrophones) and procedures used to make these measurements.

This last item regarding instruments is significant because shock waves are inherently destructive, and finding hydrophones that are both durable and accurate has been a challenge. The guideline recommends that, for faithful reproduction of the shock-wave pulse waveform, piezopolymer hydrophones of the spot-poled membrane type be used (Lewin and Schafer, 1992). The sensitivity vs. frequency, effective size, angular response, and amplifier performance (if applicable) must be provided for each type of hydrophone used.

Table V summarizes key features of the temporal and spatial information called for in the FDA guidance. Referring to the table under Temporal Information, $p(t)$ and $P(f)$ are the temporal shock-wave pulse and its frequency spectrum, respectively, measured at the focus. The peak positive and negative pressures (p_c and p_r), along with the rise time and pulse width (t_r and t_w), are all to be calculated from the $p(t)$ waveform. In addition, values of p_c must be provided for all shock-wave generator output settings available to the user.

Under Spatial Information, various plots of p_c and p_r must be made by scanning the hydrophone through and around the focal region. From these plots the -6 dB focal dimensions can be determined, and by integrating the pulse pressure-squared integral ($\int p^2(t)dt$) over the -6 dB focal surface, an estimate of the energy per pulse can be obtained.

Finally, the guidance asks for a description of the measurement technique used along with an error analysis, including effects of shock-to-shock variability on measurement accuracy. This last point is especially important for spark-gap generators, because the exact point at which the electrical discharge occurs is not always stable in space (Coleman and Saunders, 1989).

TABLE V. Summary of Measurement Data
Specifications in FDA Draft ESL Guidelines

Temporal Information	Spatial Information
• $p(t)$ at focus	• p_c - lateral & axial plots through focus; contour plots around focus
• $P(f)$ at focus	• p_r - lateral plot through focus
• p_c & p_r at focus	
• t_r & t_w at focus	• Focal dimensions (-6 dB) - lateral & axial
• p_c vs output	• Energy per pulse

5. INDUSTRIAL ULTRASOUND

Exposure criteria in the form of third-octave-band sound pressure level (SPL) limits vs. frequency have been advanced by several national and international organizations, as described in Herman and Powell (1981), WHO (1982), Acton (1983), INIRC/IRPA (1984), ACGIH (1991), and Canada (1991). Table VI contains some of these criteria. The frequency in the table is the mid-frequency of a one-third octave band. All SPL values are for exposures exceeding four hours per day. For exposures of less than four hours per day, permissible increases in the limits in Table VI have been proposed (WHO, 1982; INIRC/IRPA, 1984).

TABLE VI. Limits for Occupational Exposure to Airborne Ultrasound

(dB re 20 μPa)

Freq. (kHz)	IRPA (1983)	ACGIH (1991)	Canada (1991)	Japan (1971)	Sweden (1978)	USSR (1975)
16	-	80	75	90	-	85
20	75	105	75	110	105	110
25	110	110	110	110	110	110
31.5	110	115	110	110	115	110
40	110	115	110	110	115	110
50	110	115	110	110	115	110

Many of the values in Table VI are based on the works of Grigor'eva (1966), Parrack (1966), or Acton (1975, 1983). The IRPA limits, which were developed in cooperation with the Environmental Health Division of the World Health Organization, are based on the recommendations of Acton and Grigor'eva; the American Conference of Governmental Industrial Hygienists (ACGIH) and Sweden have adopted the work of Parrack.

Most of the values in Table VI are within ± 5 dB of each other, with the notable exception of those at 20 kHz. The main reason for this difference is that Acton lowered his original limit at 20 kHz from 110 dB to 75 dB, because part of the energy in the band centered on this frequency lies in the upper audible range for certain people who operate industrial ultrasound equipment (Acton, 1975). Although some have followed Acton's amended recommendation, its acceptance has not been universal.

In addition to these criteria, the US Air Force limits the SPL to 85 dB for frequencies from 12.5 kHz to 40 kHz, and Norway permits a level of 120 dB for frequencies above 22 kHz.

REFERENCES

ACGIH (1991) 1991-1992 threshold limit values for chemical substances and physical agents and biological exposure indices, American Conference of Governmental Industrial Hygienists, Cincinnati, OH, 76.

Acton, W.I. (1975) Exposure criteria for industrial ultrasound, Ann. Occup. Hyg. **18**:267-268.

Acton, W.I. (1983) Exposure to industrial ultrasound: hazards, appraisal and control, J. Soc. Occup. Med **33**:107-113.

AIUM (1990) Diagnostic ultrasound educational workshop, summary and recommendations and working group reports, American Institute of Ultrasound in Medicine, Rockville, MD.

AIUM (1991) Standard for real time display of thermal and mechanical indices of diagnostic ultrasound equipment, American Institute of Ultrasound in Medicine, Rockville, MD.

AIUM/NEMA (1983) Safety standard for diagnostic ultrasound equipment, J. Ultrasound Med. **2**(suppl.):S1-S50.

Apfel, R.E., Holland, C.K. (1991) Gauging the likelihood of cavitation from short-pulse, low-duty cycle diagnostic ultrasound, Ultrasound in Med. & Biol. **17**(2):179-185.

Bly, S.H.P., Harris, G.R. (1992) Ultrasound sources and human exposures, in Non-Ionizing Radiation: Proceedings of IRPA 2nd International Workshop, Vancouver, B.C., May 10-14.

Bly, S.H.P., Hussey, R.G., Kingsley, J.P., Dickson, A.W. (1989) Sensitivity of effective radiating area measurement for therapeutic ultrasound transducers to variations in hydrophone scanning technique, Health Physics **57**(4):637-643.

Canada (1989) Guidelines for the safe use of ultrasound: Part I - Medical and paramedical applications, Safety Code 23, Publ. 88-EHD-89, Canada Communications Group, Ottawa, K1A OS9.

Canada (1991) Guidelines for the safe use of ultrasound: Part II - Industrial and commercial applications, Safety Code 24, Publ. EHD-TR-158, Canada Communications Group, Ottawa, K1A OS9.

Coleman, A.J., Saunders, J.E. (1989) A survey of the acoustic output of commercial extracorporeal shock wave lithotripters, Ultrasound in Med. & Biol. **15**(3):213-227.

Duck, F.A. (1987) The measurement of exposure to ultrasound and its application to estimates of ultrasound 'dose', Phys. Med. Biol. **32**(3):303-325.

FDA (1978) Ultrasonic therapy products: Radiation safety performance standard 21 CFR 1050, 43 Federal Register, 7166-7172, Part IV.

FDA (1985, 1987, 1991a) 510(k) guide for measuring and reporting acoustic output of diagnostic ultrasound medical devices, December, 1985; Diagnostic ultrasound guidance update, January, 1987; 510(k) diagnostic ultrasound guidance update of 1991, November, 1991; FDA, Center for Devices and Radiological Health, Div. of Small Manufacturers Assistance, Rockville, MD 20857 (tel: 1-800-638-2041).

FDA (1991b) Draft of suggested information for reporting extracorporeal shock wave lithotripsy device shock wave measurements, FDA, Center for

Devices and Radiological Health, Div. of Small Manufacturers Assistance, Rockville, MD 20857 (tel: 1-800-638-2041).

Ferguson, B. H. (1985) A practitioner's guide to the ultrasonic therapy equipment standard, HHS Publ. FDA 85-8240, FDA, Center for Devices and Radiological Health, Div. of Small Manufacturers Assistance, Rockville, MD 20857 (tel: 1-800-638-2041).

Grigor'eva, V.M. ((1966) Effects of ultrasonic vibrations on personnel working with ultrasonic equipment, Sov. Phys. Acoust. 11:426-427.

Hekkenberg, R.T. (1988) On the accuracy of effective radiating areas for ultrasound therapy transducers, Leiden, The Netherlands: TNO, Medical Technology Unit; MTD/88.050.

Herman, B.A., Powell, D. (1981) Airborne ultrasound: Measurement and possible adverse effects, HHS Publ. (FDA)81-8163, NTIS no. PB81-240459, FDA, Center for Devices and Radiological Health, Rockville, MD 20857.

IEC (1963) Testing and calibration of ultrasonic therapeutic equipment, International Electrotechnical Commission, Publ. 150, Geneva.

INIRC/IRPA (1984) Interim guidelines on limits of human exposure to airborne ultrasound, International Non-Ionizing Radiation Committee (INIRC) of the International Radiation Protection Association (IRPA), Health Physics 46(4):969-974.

Lewin, P.A., Schafer, M.E. (1992) Ultrasound: measurement and instrumentation, in Non-Ionizing Radiation: Proceedings of IRPA 2nd International Workshop, Vancouver, B.C., May 10-14.

Maeda, K., Ide, M. (1986) The limitation of the ultrasonic intensity for diagnostic devices in the Japanese industrial standards, IEEE Trans. Ultrason. Ferroelec. Freq. Contr. 33(2):241-244.

Merritt, C.R.B. (1989) Ultrasound safety: What are the issues?, Radiology 173(2):304-306.

Merritt, C.R.B., Yin, L.L. (1989) AIUM/CDRH (FDA) regulatory discussion, Ultrasound in Med. & Biol. 15(suppl. 1):105-108.

Nyborg, W.L. (1989) Statements, recommendations, and guidelines for diagnostic equipment, Ultrasound in Med. & Biol. **15**(suppl. 1):85-90.

O'Brien, W.D., Jr. (1992) Ultrasound dosimetry and interaction mechanisms, in Non-Ionizing Radiation: Proceedings of IRPA 2nd International Workshop, Vancouver, B.C., May 10-14.

Parrack, H.O. (1966) Effects of airborne ultrasound on humans, Internat. Audiol. **5**:294-308.

Preston, R.C. (1989) Review of current IEC activities in acoustic output standardization of medical ultrasound equipment, Ultrasound in Med. & Biol. **15**(suppl. 1):101-103.

Standard for ultrasound therapy devices (1984) P.C. Code 1981-908, Can. Gazette Part II, 22 Apr 1981; amendment P.C. 1984-3737, Can. Gazette Part II, 12 Dec 1984; Can. Gov. Publ. Cntr., Dept. of Supply and Services, Ottawa K1A 0S9.

Stewart, H.F., Abzug, J.L., Harris, G.R. (1980) Considerations in ultrasound therapy and equipment performance, Physical Therapy **60**(4):424-428.

Stratmeyer, M.E. (1989) FDA model for regulatory purposes, Ultrasound in Med. & Biol. **15**(suppl. 1):35-36.

Thomenius, K.E. (1990) Thermal dosimetry models for diagnostic ultrasound, Ultrasonics Symposium Proceedings, Cat. No. 90CH2938-9, IEEE, Piscataway, NJ, 1399-1408.

Villforth, J.C. (1988) Medical devices - Non-ionizing radiation standards, in <u>Non-ionizing Radiations: Physical Characteristics, Biological Effects, and Health Hazard Assessment</u>, Repacholi, M.H., ed., IRPA Publ., London, 427-446.

WHO (1982) Ultrasound, Environmental Health Criteria 22, World Health Organization (WHO), Geneva, 136-140.

Ziskin, M.C. (1990) Update on the safety of ultrasound in obstetrics, Seminars in Roentgenology **XXV**(4):294-298.

PART IV

OPTICAL RADIATION
ULTRAVIOLET
INFRARED

OPTICAL RADIATION

Alastair McKinlay

National Radiological Protection Board
Chilton, Didcot, Oxon, OX11 0RQ, UK

INTRODUCTION

This paper is an introduction to optical radiation in the context of its role as a potential hazard. The paper summarises; terms, quantities and units relevant to describing the quality, quantity and temporal and spatial distribution of optical radiation; the physical properties of optical radiation pertinent to its hazard potential; the biological effects of optical radiation; standards for protection; the basic methods of producing optical radiation employed in commonly used sources and examples of their emission characteristics. Most of these topics are dealt with in more detail in subsequent chapters.

TERMS, QUANTITIES AND UNITS

Historically it was known that there were other radiations, that were invisible to the human eye, on both sides of the visible spectrum. These came to be known as 'obscure rays' in the 19th Century and are now known as ultraviolet radiation (UVR) and infrared radiation (IRR). It is now understood that UVR, light and IRR constitute optical radiation and that this is only a very small part of a continuum of electromagnetic radiations, the electromagnetic spectrum (EMS) which encompasses also high energy gamma and x-rays, radiofrequency radiation, microwave radiation and low frequency electric and magnetic fields.

The physical and biological interactions of these radiations with matter and with living systems vary markedly but they all obey the same physical laws and the inherent difference between gamma radiation at one end of the EMS and radiofrequency radiation at the other is the magnitude of the frequency or the wavelength. These two quantities are related by the expression

$$v = f \lambda \tag{1}$$

where $v =$ the speed of propagation of the electromagnetic radiation in a medium in metres per second

       $f =$ the frequency of the radiation in hertz (cycles per second)

       $\lambda =$ the wavelength of the radiation in the medium in metres

and either can be used to describe the position of a particular radiation in the EMS. Frequency is the invariant fundamental quantity whereas wavelength, which varies with the transmitting medium, is more widely used to describe optical radiation.

The interactions of optical radiations with and their penetration through tissue vary considerably with wavelength and the most widely used and most useful divisions into wavelength regions are those defined and promulgated by the International Commission on Illumination (CIE) (CIE, 1987) viz.

UVC: 100 to 280 nm	IRA:	760-780 to 1400 nm
UVB: 280 to 315 nm	IRB:	1.4 to 3.0 μm
UVA: 315 to 380-400 nm	IRC:	3.0 μm to 1 mm
light: 380-400 to 760-780 nm		

Whilst these divisions are based on the relative spectral efficacies of important biological effects, the effects can overlap wavelength regions.

The quantity and spatial and temporal distribution of optical radiation are described in terms of the so-called radiometric quantities and units, table 1. These quantities are physically absolute in the sense that they do not depend on the spectral response of a receptor or detector.

Lasers are unique in their ability to emit optical radiation at effectively one wavelength. All other optical sources emit the radiation as a spectral continuum typically containing peaks and troughs. Because of the great dependence of the type and magnitude of biological (including hazardous) effects on the wavelength of the radiation it is important to have information about the spectral emissions of such sources. For hazard analysis the most useful information is a set of data comprising spectral irradiance (W m^{-2} nm^{-1}) measurements at the target or, for resolved images on the retina, spectral radiance (W m^{-2} sr^{-1} nm^{-1}) data for

Table 1 Radiometric quantities and units

Quantity	Unit
Radiant energy	joule (J)
Radiant flux	watt (W)
Irradiance	$W\ m^{-2}$
Radiant intensity	$W\ sr^{-1}$
Radiance	$W\ m^{-2}\ sr^{-1}$

sr is the abbreviation for the unit of solid angle - the steradian

the source.   The biologically effective irradiance ($W\ m^{-2}_{effective}$) for any specific biological effect is obtained by multiplying the spectral irradiance at each wavelength by a factor (biological weighting or hazard function) that quantifies the relative efficacy of radiation of that wavelength for causing the effect.

Figure 1   Spectral hazard functions   for: ultraviolet radiation (UVR), blue light (BLUE), retinal burn (BURN) and aphake (APHAKE) hazards (INIRC 1991, and ACGIH, 1991)

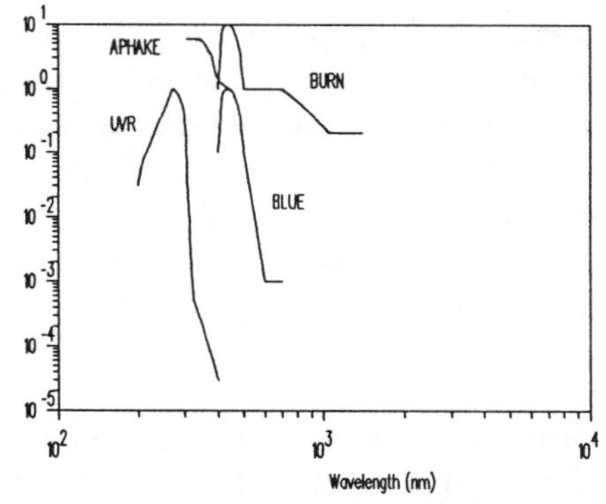

Such factors are obtained from action spectra (spectral efficacy curves) which are based on experimentally obtained values for the quantity of radiation of each wavelength which will cause the effect to a specified degree. In constructing action spectra in this way all the data are normalised to the datum at a chosen wavelength, usually at the most efficacious. By summing the calculated biologically effective irradiance over the exposure time the biologically weighted radiant exposure ($J\ m^{-2}_{effective}$) is obtained. This quantity is often referred to as dose.

With respect to the assessment of potential ocular hazards important examples of the application of this method are: the determination of (biologically) effective UVR using the UVR hazard function promulgated by the INIRC (INIRC, 1991) and; the determination of potential retinal blue-light and retinal burn hazards using the ACGIH retinal blue-light and retinal burn hazard functions (ACGIH, 1991), figure 1.

PRODUCTION OF OPTICAL RADIATION AND ITS PHYSICAL CHARACTERISTICS

Production of optical radiation

Non-laser sources

In almost all non-laser sources of optical radiation the radiation is produced either by heating a solid material to incandescence or by an electrical discharge in a gas or vapour.

Incandescence

When a material is heated a large number of energy transitions occur within the molecules of the material and optical photons are emitted. An idealized (most efficient) radiator of such radiation is termed a 'black-body'. The total radiant power and its spectral distribution depend only on the temperature of the black-body. The radiation characteristics of a black-body can be expressed by 3 basic formulae, viz.

Planck's radiation law

This describes the spectral (wavelength) distribution of the radiant flux per unit area (spectral radiant exitance) emitted by a black-body as

$$W_\lambda = \frac{C_1}{\lambda^5[\exp-(\frac{C_2}{\lambda T})-1]}$$

(2)

where W_λ is expressed in W cm^{-2} μm^{-1}.

λ (wavelength) is in μm. T is the temperature in kelvins (K) and C_1, and C_2 are Planck's first and second radiation constants and equal to 3.74×10^4 and 1.438×10^4 respectively.

The relative spectral distributions of emission of a black-body as a function of temperature as determined using Planck's radiation law are illustrated in figure 2 for different sources.

Figure 2   Relative spectral emissions of a black-body as a function of temperature

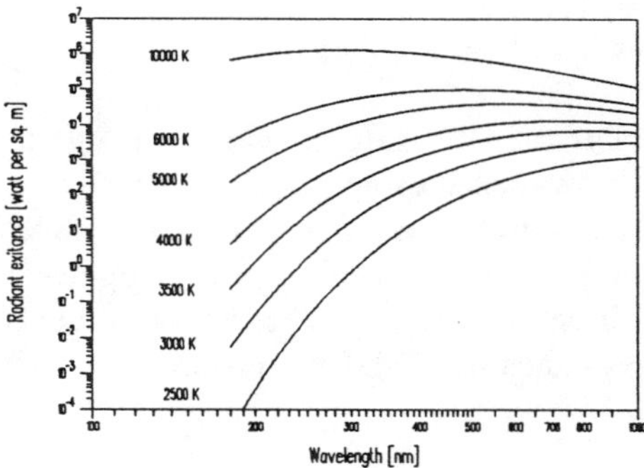

Stefan-Boltzmann law

By integrating (summing) over all wavelengths the area under an absolute plot of spectral radiant exitance against wavelength the total radiant flux per unit area (radiant exitance - W) can be determined. It is described by the equation:

$$W = \sigma T^4 \tag{3}$$

where σ = Stefan-Boltzmann constant (5.67×10^{-7} W cm^{-2} K^{-4}).

Wien's displacement law

Wien's displacement law expresses mathematically the relationship between the wavelength at which the maxima of the spectral distributions of the emission of black-bodies at different temperatures occur (λ_m) as a function of temperature (T); viz

$$\lambda_m = \frac{b}{T} \tag{4}$$

where b = 2897 μm K.

The retinal hazard potential of a high temperature incandescent source can be evaluated using spectral radiance data weighted with the retinal blue light and burn hazard functions and summed. By considering the source to be a Lambertian emitter, ie, it appears equally bright viewed from every direction, the spectral radiance L_λ can be derived from the spectral radiant exitance using the expression.

$$L_\lambda = \frac{W_\lambda}{\pi} \tag{5}$$

In practice no real material emits radiation with a black-body spectrum. However, tungsten at high temperatures (such as used for the filaments of incandescent lamps) and molten metals approximate to a black-body. The ratio of the actual spectral radiant exitance to the theoretical black-body spectral radiant exitance is termed the spectral emissivity. For tungsten at 3000 K (typical of a tungsten halogen lamp filament) the emissivity varies from an average of about 0.4 in the visible region to about 0.2 in the IRB region.

Electrical (gaseous) discharge

Optical radiation can be generated by electrical excitation of a low pressure gas or vapour. A current is passed through a gas or gases ionised to produce electrons and positive ions. A fluorescent lamp discharge is a typical example. The energetic electrons which produce the ionisation also excite the electrons of the gas atoms which subsequently de-excite resulting in the emission of characteristic radiations. Figure 3 illustrates a few of the multiplicity of possible transitions and related emissions of discharge lamps containing mercury vapour. The emission at 253.7 nm is used as a source of excitation of the phosphors of low pressure fluorescent lamps. By raising the pressure of the discharge to a few atmospheres the emission lines increasingly broaden eventually effectively forming a continuum. In some cases the 253.7 nm line emission will be self-absorbed by the vapour of the discharge.

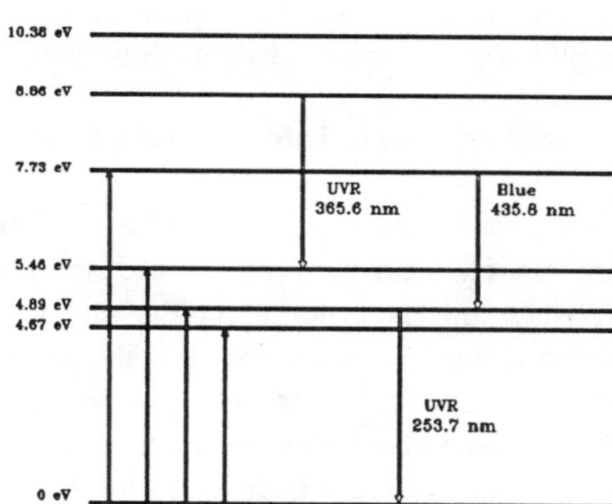

Figure 3   Simplified excitation and emission energy diagram for mercury vapour illustrating the energy transitions that result in the emission of 253.7 and 365.6 nm UVR and 435.8 nm (blue) light

Lasers

The production of optical radiation by a laser is essentially similar to the way that gaseous discharge non-laser sources operate in that the emissions are the result of atomic or molecular energy transitions in a material from low to high energy states (excitation) followed by high to low energy transitions (emission). However, whereas the emissions from non-laser sources result from spontaneous (random) transitions within the material those associated with lasers result from an avalanche effect of photons causing transitions (stimulation) from higher to lower energy states. This can be sustained only if more of the atoms or molecules of the material are in a higher energy state than in the lower energy state to which they de-excite. This distribution of occupation of energy states is termed population inversion. Amplification of the optical emission is achieved by ensuring that the beam is made to traverse the material repeatedly. Hence the number of emitted photons increases as a result of photon stimulation of other excited atoms or molecules to emit photons of the same energy (same wavelength) and in phase with each other.

All lasers have three basic components: an active material (active medium) whose atoms undergo the excitation and stimulated emission processes outlined above; a source of energy to achieve population inversion by excitation (pumping), and; a resonant optical cavity within which amplification takes place and from which the emitted radiation travels.

Lasers can be categorised in different ways but most often according to the type of laser active medium viz: solid state lasers, consisting of a host matrix of crystalline or vitreous material into which active atoms are introduced; gas (and vapour) lasers, consisting of 'pure' gas or vapour or a mixture of gases or vapours; semi-conductor lasers (laser diodes), consisting of 'n' and 'p' type material and; liquid lasers, consisting of a liquid containing an active medium such as an organic dye.

Physical characteristics of optical radiation

The physical characteristics of optical radiation govern the nature of its interactions with materials. These interactions are reflection, absorption and diffraction. Depending on the physical properties of the radiation the composition of the material and the geometry of the exposure one of these interaction processes is usually dominant. Consideration of these processes is fundamental to hazard evaluation of optical radiation and particularly with respect to its interaction with biological tissue. Damage can result only from the absorption of the radiation by the target tissue.

Reflection of optical radiation takes place at the interface between two media. There are two basic types of reflective processes, specular and diffuse. When optical radiation undergoes specular reflection at a surface the reflected ray lies in the plane of the incident ray and the angle of incidence (defined as the angle bounded by the incident ray and the normal to the surface) equals the angle of reflection (defined as the angle bounded by the reflected ray and the normal to the surface). Specular reflection occurs only when the physical irregularities on the reflective surface (roughness) are smaller than the wavelength of the radiation. For a surface where the irregularities are greater than the wavelength and where the irregularities are randomly oriented the radiation undergoes diffuse reflection. For a given surface, specular and diffuse reflectivities may change absolutely and relatively to each other depending on the wavelength of the radiation. A knowledge of the reflecting properties of materials is particularly important when assessing the likely effects of a laser beam striking a target material. Materials which, when dry are diffuse reflectors, may become specular reflectors when wet and consequently potentially more hazardous.

Refraction takes place at the interface between two media where the wavelength of the radiation in each medium is substantially different. In passing from one medium to a second the refracted ray lies in the plane of incidence and its direction of propagation is changed with respect to the incident ray. The ratio of the sine of the angle of incidence in the first medium to the sine of the angle of refraction in the second medium (defined as the angle bounded by the refracted ray and the normal to the surface) is a constant and is equal to the ratio of the refractive index n_2 of the second medium to that of the first n_1. In passing from a material of low refractive index to one of a higher refractive index, if the angle of incidence is increased to 90°, then the angle of refraction increases to a maximum value known as the critical angle. For a ray of light incident from air onto a glass surface ($n_2/n_1 = 1.5$) the critical angle ϕ_c is approximately 42°. Applying the principle of reversibility of light rays, ie, considering light rays travelling from the material of high refractive index n_2 to that of the lower one n_1, all incident rays will lie within a cone subtending an angle $2\phi_c$ and the corresponding refracted rays will lie within a cone subtending 2π. The refractive properties of glass are the basis for the focusing and reflecting characteristics of lenses and prisms respectively. The change in the refractive index that occurs at the interface between air and the cornea results in the main focusing action of the eye.

When optical radiation passes through a medium its propagation is affected in two important ways. Its velocity is different in the medium from that in free space and its intensity progressively decreases as it penetrates into the medium. A medium may display general absorption where optical radiation of different wavelengths are absorbed to the same extent, or selective absorption where radiation of different wavelengths are absorbed to different extents. No practical material acts as a general absorber except over limited ranges of wavelength. True absorption represents the removal of radiant energy from the incident beam; this being ultimately converted to heat in the absorbing material. However, radiant energy can also be removed by scattering processes. The intensity of the radiation at a depth d in an absorbing material may be represented by the equation

$$I_d = I_o \, e^{-\alpha d} \tag{6}$$

where I_o is the intensity of the radiation entering the absorbing material and α is the absorption coefficient of the material. In general α has two components, one due to the true absorption properties of the material α_a, the other due to its scattering properties α_s.

$$\alpha = \alpha_a + \alpha_s \tag{7}$$

Although in many cases either one of the components of α will be dominant, in some cases both true absorption and scattering may each play a significant role.

The interaction of optical radiation with small particles of refractive index different from that of the surrounding medium results in Rayleigh scattering. The size of the particles must be significantly smaller than the wavelength of the radiation. The intensity of the scattered radiation is proportional to the intensity of the incident radiation and, for a given particle size, is inversely proportional to the 4th power of the wavelength. As the wavelength of red light is approximately 1.8 times the wavelength of blue (violet) light, the scattering of blue light will be approximately 10 times that of red light provided that the scattering particle size is much less than the wavelength of blue light. This preferential scattering of blue light is demonstrated by the colour of the sky where practically all of the light that is seen in a clear sky is due to Rayleigh scattering by molecules of air in the large mass of air traversed. At sunset, the scattering preferentially removes blue light from the radiation reaching an observer and the sun appears red. The removal of the potentially harmful blue light in this way explains why even prolonged viewing of the sun at sunset does not result in retinal photochemical injury which would occur if the sun were similarly viewed when at large solar elevations.

When optical radiation passes through an aperture or past the edge of an object it spreads outward into the space that would not be irradiated had it passed without deviation; this is termed diffraction. Diffraction of light in this way results in the formation of a symmetrical pattern of light and dark (diffraction pattern). For an idealized aberration free system the angular diameter of the first dark ring of the diffraction pattern of a point image is

$$\beta = \frac{2.44\ \lambda}{D}\ rad \tag{8}$$

where λ is the wavelength of the radiation and D is the diameter of the circular aperture which limits the beam forming the primary image.

This contains 84% of the power in the image. Conventionally the resolving power of an optical instrument is determined by the diffraction patterns associated with two separate objects. The patterns are said to be resolved if the central maximum of one falls on the first dark ring of the other. The corresponding (minimum) angle of resolution is $\beta/2$ rad.

Applying this criterion to the human eye (assuming a pupil diameter of 3 mm) the minimum angle of resolution is 47 s of arc. This represents the theoretical (diffraction limited) maximum resolving power of the eye, however optical aberrations result in a correspondingly larger retinal image. Such considerations are important in determining thresholds for retinal injury and their interpretation in setting exposure limits for different viewing conditions.

BIOLOGICAL EFFECTS

Because of its limited penetration in biological tissues the direct hazardous effects of optical radiation are limited to the skin and to the eyes. Humans have been required to adapt to an optical radiation environment dominated by solar radiation. The eyes are normally protected from acute retinal injury, which can be caused by staring at the sun, by involuntary aversion responses. These responses protect the retina from visible radiation and from IRA (770-1400 nm) radiation both of which are focused onto the retina forming a resolved image of the sun. The cornea is normally physically protected from the acute effects of solar UVR by the recessed protective position of the eye in its socket and by the glancing incidence of the radiation. Exposed skin is at risk from both the acute and chronic effects of solar radiation. Even after adaptation, solar radiation may still be hazardous and it undoubtedly causes more adverse health effects than any human-made source of optical radiation. However, with scientific and industrial advances human-made sources of optical radiation have been developed with similar or even greater potential for injury.

The major physical interactions of optical radiation with biological tissues include specular and diffuse reflection, refraction, scattering and absorption.

Only energy absorption can result in the physical and chemical (photochemical) changes that result in a biological effect.

Effects on the eyes

The cornea absorbs nearly all optical radiation with wavelengths shorter than approximately 300 nm and longer than approximately 2 μm so that UVC, UVB, IRB and IRC are likely to be potentially more hazardous to this part of the eye than to the lens or retina.
The principal effect of the absorption of UVB and UVC by the cornea is photokeratitis and damage is generally limited to the epithelial cells of the cornea. However, there is also some evidence of possible long term effects of absorption of UVR by the cornea (Pitts et al, 1987).

The clarity of the lens is important for visual acuity and is achieved by its ordered structure. Damage to the lens whether as cell death or as disorder, tends to remain incorporated as a translucency or as an opacity. Studies with experimental animals have demonstrated the induction of cataracts by UVR (principally UVB) and by IRR and have established thresholds (Pitts, 1986). The penetration of UVA to the lens is much greater than of UVB. Animal data indicate that threshold lenticular damage is mainly limited to exposure to UVR in the range of wavelengths 295 to 325 nm. It is only over this wavelength region, and only for acute cataract in laboratory animals, that wavelength efficacy data have been established. However, it should not be implied that UVA is safe with respect to lens exposure. The ageing process in the lens is manifested in part by loss of elasticity and browning (brunescence); both of which may in part be caused by UVA.

Damage mechanisms in the retina are photochemical, thermal, thermo-acoustic and electrical. Which effect predominates depends on the duration of the exposure, the maximum irradiance at the retina and the wavelength of the radiation. Only photochemical and thermal are pertinent to non-laser sources. The size of the retinal image, the wavelength of the radiation and the duration of the exposure are the most important factors in determining whether injury follows a given energy absorption by the retina.

For radiation of wavelengths longer than 1 μm, absorption by the ocular media increases. For infrared radiations of wavelengths longer than 1.4 μm the lens and vitreous are particularly strongly absorbing and beyond 2 μm the cornea is effectively the sole absorber. The absorption of visible and particularly IRA radiation by both lens and pigmented iris is considered to play a role in the formation of lenticular opacities. The absorption of IRB and IRC by the cornea will also affect the thermal conduction cooling capacity of the interior structure of the eye, including the lens.

Effects on the skin

The absorption of UVR and particularly UVB by the skin may result in 'sunburn'. In its mildest form it consists of a reddening (erythema) of the skin that appears up to about 12 hours after exposure to UVR and gradually fades after a few days. In its most severe form, it results in inflammation, blistering, and peeling of the skin. The degree to which a person will experience sunburn depends critically on skin type. Quantitative studies of the spectral efficacy of ultraviolet induced erythema have been carried out for over 60 years. Statistical weighting of post-1945 data has led to the formulation of a reference erythema action curve (McKinlay and Diffey, 1987) which has been adopted widely for the purpose of spectral weighting and the evaluation of effective irradiances from sources.

The most serious long-term effects attributed to UVR exposure of the skin are the skin cancers. The common, but rarely fatal, forms of skin cancer are squamous and basal cell carcinomas, which are often collectively referred to as the non-melanoma skin cancers (NMSCs). Experimental studies clearly indicate that UVR is a causal agent of NMSC in mice and epidemiological studies clearly implicate solar ultraviolet as a causal agent in humans. Cutaneous malignant melanoma (MM) while much less frequently occurring than NMSC, is much more serious and accounts for the majority of deaths from skin cancer. There are no animal data for MM and the evidence for the indictment of UVR as a causal agent is based principally on epidemiological data. These indicate that short-term intermittent exposures to high levels of solar UVR, particularly at an early age, may be a contributing causal factor. The individual risk of MM is higher in people who have a large number of naevi (moles) and who sunburn readily and tan poorly with exposure to solar UVR.

There is evidence that cumulative exposure of the dermis of the skin to UVR (UVB and UVA) results in its premature ageing characterised by a dry, coarse, leathery and wrinkled appearance.

Other effects

Some localised skin and systemic changes in immunological reactions result from exposure to UVR radiation (Noonan et al, 1990). There is also evidence that exposure to UVR can activate and accelerate the growth of human viruses including human immunodeficiency virus (HIV) (Zmudzka and Beer, 1990). At present the significance of these observations with respect to the exposure of people to UVR is unclear.

RECOMMENDED MAXIMUM EXPOSURE LEVELS AND PROTECTION STANDARDS

Ultraviolet radiation

Protection standards and exposure guidelines with respect to occupational and public exposure to UVR are not comprehensive in that they have principally addressed the acute effects of exposure and less so the chronic effects. The reason for this is the lack of quantitative data on the chronic effects of UVR.

The most comprehensive exposure limits for UVR are those of the International Non-Ionizing Radiation Committee INIRC (INIRC, 1991) which are based on the threshold limit values (TLVs) for occupational exposure promulgated by the American Conference of Governmental Industrial Hygienists (ACGIH, 1991). They are intended to be applied to the working population but with some precaution they are recommended also for the general population. They represent conditions under which it is believed that nearly all people may be repeatedly exposed without adverse effect. However, they do not apply to UVR exposure of photosensitive individuals or of individuals concomitantly exposed to photosensitizing agents. They are summarised elsewhere in this volume.

The effective radiant exposures for broad-band sources are determined by temporal summation over an 8 h period of the spectrally weighted summation of the spectral irradiance from the source in the plane of the exposure using the UVR hazard function illustrated in figure 1.

Visible and infrared radiations

There are as yet no internationally agreed limits for visible and infrared radiation exposure from non-laser optical sources. However, ACGIH have made proposals for TLVs for visible and infrared radiations (ACGIH, 1991). An earlier form of these recommendations was included in a publication by the World Health Organisation (WHO) and the International Radiation Protection Association (IRPA) (WHO, 1982).

The current recommendations for visible and IRA radiations from sources which are resolved by the eye and imaged on the retina are based on data for typical retinal images of 700 to 1000 μm diameter or so. Photochemical and thermal hazards are considered separately through the 'blue light' B_λ and 'burn' R_λ hazard functions respectively, figure 1. The exposure limits are expressed in terms of radiance as a weighted spectral summation and are denoted mathematically as spectral summation formulae, described elsewhere in this volume.

With its limitation to viewing durations of less than 10 s, the use of the burn hazard function is directed to accidental viewing conditions not involving fixated staring, ie, for example, from an infrared heat lamp or other source, where there is no strong visual stimulus. More explicitly for these conditions ACGIH suggest the use of a spectral summation formula for near infrared radiance based on a 7 mm diameter pupil summed for all wavelengths between 770 and 1400 nm and a 100 W m^{-2} limit on irradiance.

For the special case of people who have had a lens removed (aphakes) there is an increased risk of retinal blue light hazard because of the resultant increased UVA and UVB transmission of the ocular media. Unless an ultraviolet absorbing 'intra-lens' has been implanted the blue light hazard function that should be used for the calculation of potential blue light retinal hazard is that shown as A_λ in figure 1 for aphakes, the so-called aphake photic hazard function. Under these circumstances the spectral summations are extended down to 300 nm.

Laser standards

The laser maximum permissible exposure levels are complex functions of parameters such as exposure duration, wavelength, pulse duration and repetition frequency, image size, etc. They are intended to be set below known hazard levels. Standards for broad-band radiation and standards for laser radiation are based on common data and include work with both coherent and non-coherent radiation. A system of classification according to the degree of hazard obviates the need for exposure measurements by most users (IEC, 1984).

SOURCES OF OPTICAL RADIATION, THEIR EMISSION CHARACTERISTICS AND HAZARD POTENTIAL

Lamps and lighting systems

Apart from lighting applications, lamps are found in many forms, in a wide range of environments and used by many trades and professions for a diverse range of applications that is ever increasing. They can be conveniently grouped under two broad headings; those that produce radiation by incandescence and those that produce it by a gas discharge.

Incandescent (heated filament) lamps

The incandescent lamp is the oldest type of lamp still in common use. Its emission results from the heating of a tungsten filament which generally is heated to between 2700 and 3000 K with powers up to 500 W and the peak spectral emission is in the infrared (IRA) region. The spectral emissions and radiances associated with 'conventional' tungsten filament lamps are such that in general they do not present either an UVR hazard to the skin or the eyes or a hazard to the retina.

In applications where much more power, or a physically much smaller source, is required tungsten (quartz) halogen lamps are often used. The combination of filament temperatures which are likely to be in the range 2900 to 3450 K and quartz bulbs results in a significantly higher level of emission of potentially harmful UVR compared with 'ordinary' tungsten filament lamps. Where such lamps are used in luminaires where exposure of people at close distances for long times is possible consideration should be given to the incorporation of appropriate UVR absorption filters in the luminaire. Some desk-top luminaires currently on sale have no such protective filters, table 2 (McKinlay et al, 1989).

Low pressure gas discharge lamps

In low pressure gas discharge lamps the filling gas is usually an inert gas. The commonest type of low pressure (non phosphor) discharge lamp is the 'neon' lamp. The low pressure mercury discharge lamp is often used for the purpose of germicide and disinfection. Such lamps are very efficient emitters of UVR. Approximately 50% of the electrical power is converted to UVR of which up to 95% is emitted at a wavelength of 253.7 nm. Germicidal lamps are hazardous if used carelessly.

Fluorescent lamps

The most common application of the low pressure discharge is in fluorescent lamps. The fluorescent lamp operates by means of a discharge between two electrodes through a mixture of mercury vapour and a rare gas, usually argon. Light is produced by conversion of 253.7 nm mercury emission radiation to longer wavelength radiations by means of a phosphor coating on the inside of the wall of the lamp. Lamps are available with many different phosphors and envelopes to produce a wide range of spectral emissions covering the visible, UVA and UVB regions. While the continuum emissions of fluorescent lamps are characteristic of the phosphors the narrow peak spectral emissions are dominated by the characteristic line emission spectrum of the low pressure mercury vapour discharge.

General lighting fluorescent lamps

These lamps are available in a range of physical sizes, powers and phosphors. The range of phosphors includes a large selection of 'near white' and 'special colour' lamps.

Data from a study of the amount of UVR emitted by white fluorescent lamps used in the UK for general lighting purposes indicated that people exposed for 1500 h per year would receive only 2-5 minimum erythemal doses (MEDs) per year (Whillock et al, 1988), table 2. These data differ somewhat from those obtained from measurements on general lighting lamps used in the USA where generally and specifically with some super high output lamps the levels of UVR emissions were higher (Cole et al, 1986). During the past few years the further development and improved design of general lighting fluorescent lamps has been evident in the production of compact fluorescent lamps. These lamps are essentially low wattage small diameter fluorescent tubes folded in a compact form.

They are most readily available commercially with cool white phosphors but other phosphors are also available. Their UVR emissions are similar to those of conventional fluorescent lamps (Whillock et al, 1991).

Low pressure sodium lamps

Like mercury, sodium metal can be vaporised and used together with a rare gas in a discharge and both low and high pressure discharges are used in practical lamp design. Sodium lamps are particularly used for street and other large area light applications where colour rendering is relatively unimportant. Low pressure sodium lamps have the highest luminous efficacies of commercial light sources. In general low pressure sodium lamps do not present an optical radiation hazard.

Table 2 A summary of measurement data for ultraviolet radiation from desk-top luminaires incorporating tungsten halogen lamps and from general lighting fluorescent lamps (McKinlay et al, 1989 and Whillock et al, 1988).

Type of lamp	Effective irradiance[a] (mW $m^{-2}_{effective}$)		Illuminance (lux)[a]
	ACGIH[b]	Erythema[c]	
Desk-top tungsten halogen 20 W	6.1	10.1	1400
Desk-top tungsten halogen 20 W	2.2	3.3	7600
Desk-top tungsten halogen 20 W	25.0	41.0	14000
Desk-top tungsten halogen 50 W	2.2	3.8	1600
'Cool white' general lighting fluorescent lamp 100 W	0.20	0.28	500

(a)   All tungsten halogen lamp measurements at 0.3 m from lamps

(b)   ACGIH occupational hazard weighted irradiance: the exposure limit is equivalent to 8 h exposure to 1 mW $m^{-2}_{effective}$

(c)   International Commission on Illumination reference action spectrum (McKinlay and Diffey, 1987)

'Special' applications fluorescent lamps

Apart from a number of special colour-rendering fluorescent lamps which are essentially variations of general lighting fluorescent lamps a number of special applications fluorescent lamps have been developed and are commercially available. An example is the so-called 'blacklight' lamp which uses a nickel/cobalt oxide (Woods' glass) envelope. The phosphor chosen for these lamps emits around 370 nm in the UVA. The glass is almost entirely opaque to light and the lamps are used for a number of commercial, scientific and industrial fluorescence purposes as well as for display and entertainment.

High intensity discharge (HID) lamps

The designation 'high intensity discharge' (HID) lamps includes the families of lamps often called high pressure mercury vapour, metal halide and high pressure sodium vapour lamps.

Mercury vapour lamps

High pressure mercury vapour lamps are widely used for industrial and commercial lighting, street lighting, display lighting, floodlighting and a large number of printing, curing and other industrial applications. The general construction of high pressure mercury lamps is a fused silica (quartz) discharge tube containing the mercury/argon vapour discharge mounted inside an outer envelope of soda-lime or borosilicate glass.

The outer glass envelope effectively absorbs UVR; consequently the quantity of potentially harmful UVR emitted by such lamps depends critically on the integrity of this envelope (FDA, 1988) although, even when intact, secondary filtration may also be required.

Metal halide lamps

The family of 'metal halide' lamps encompasses a number of different types of high intensity mercury lamps whose discharges all contain additives. The additives are most typically metal halides chosen to produce either a strongly coloured emission (usually a single halide), to produce a more broadly spectrally uniform emission (multi-halide) or to enhance the UVR (most often UVA) emission. Compared with 'ordinary' mercury high intensity discharge lamps the luminous efficacies of metal halide lamps are high.

High pressure sodium lamps

High pressure sodium lamps are similar in construction to but generally have a smaller diameter than high pressure mercury lamps. The inner tube is made from polycrystalline alumina and the outer envelope from borosilicate glass. The emissions are characteristic of the sodium vapour and as the vapour pressure is raised the emissions broaden across the visible spectrum.

Very high (super) pressure mercury, mercury xenon and xenon lamps

Where an optical source of very high radiance is required and of small size a very high pressure arc lamp may be used. These have a filling gas of mercury vapour, mercury vapour plus xenon gas or xenon gas. Metal halide types are also available. Two physical types are commonly used: the compact (or short) arc and the linear arc. Typical applications of compact arcs include projectors, searchlights and solar radiation simulators.

The spectral emission of xenon lamps, which at wavelengths shorter than infrared, closely matches that of a 'black-body' radiator at about 6000 K enables their use as solar radiation simulators. Their emission spectrum is continuous from the UVR through to the IRR regions. Emission peaks in the infrared principally between 800 and 1000 nm may be effectively removed by filtration. The luminance of compact xenon arcs may approach that of the sun (approximately 10^9 cd m^{-2}) and in some lamps with greater than 10 kW rating the luminance of the brightest spot may exceed 10^{10} cd m^{-2}.

In general lighting applications the outer glass bulb of mercury vapour lamps effectively absorbs ultraviolet radiation; therefore under normal operating and illumination conditions they do not present an ultraviolet radiation hazard. High intensity discharge lamps intended for incorporation in special equipment such as photopolymerisers and sunlamps may emit copious quantities of ultraviolet radiation. The hazards to the skin and eyes from such lamps are significant and equipment incorporating them should be effectively interlock to prevent unintentional exposure of people. All high intensity discharge lamps present potential retinal blue light hazards depending on their specific characteristics. Compact and linear arcs are particularly hazardous in this respect.

Gas and arc welding

Gas welding

Because of their comparatively low operating temperature the optical radiation hazards associated with gas welding processes are minimal. The use of standard welding filters necessary for comfort of viewing will prevent injury.

Arc welding

In comparison with gas welding processes, the emissions of optical radiations from arc welding are very high and many data on the optical radiation emissions associated with a variety of electric arc welding processes have been published (Sliney and Wolbarsht, 1980). Long term viewing of a welding arc is unlikely but not unknown. Under these conditions retinal photochemical injury is possible. Accidental exposure is limited by the aversion response and retinal injury should not occur within this time. Exposed skin may develop ultraviolet induced erythema. Clearly a combination of personal protection for the operator and the use of containment screens to prevent exposure of others is necessary.

Solar radiation

The sun approximates to a 'black-body' emitter with a temperature of about 5900 K. The mean solar irradiance just outside the earth's atmosphere is approximately 1370 W m^{-2} (the so-called 'solar constant') with an approximately $\pm$ 3% variation depending on the distance of the earth from the sun; the spectral distribution is shown in table 3. Values at the earth's surface are lower because of attenuation by the atmosphere whose effective thickness (with respect to the path travelled by solar radiation to an observer on the ground) varies with the angle of elevation of the sun. The air attenuation is lowest when the sun is high in the sky and at a maximum for the rising or setting sun. Solar irradiance and particularly those short wavelength components, ie, UVR and blue light, depends on solar elevation which in turn depends on latitude, time of day and time of year and on altitude, cloud cover and the degree of air pollution. Attenuation is most marked due to the absorption by ozone (particularly at wavelengths shorter than 300 nm) and by water vapour (in the infrared). As a consequence of the former solar radiation of wavelengths shorter than about 290 nm does not reach the surface of the earth. The presence of cloud plays a significant role in attenuating UVR. UVB and UVA irradiances are reduced due to scattering by water droplets and/or ice crystals in the clouds; the magnitude of such scattering processes is essentially independent of wavelength over the UVR range. Clouds provide their largest attenuation in winter when the fraction of sky likely to be covered by them is highest and when the paths travelled by UVR through them are greatest. In summer the values of UVR irradiance are likely to be much closer to those calculated for clear sky conditions.

Lasers

Lasers are unique among optical radiation sources in that they (generally) emit optical radiation at effectively a single wavelength (particular exceptions are the dye lasers) and (generally) in highly collimated beams (particular exceptions are the semiconductor lasers and laser arrays). Their emissions range over the entire optical radiation spectrum. Commonly used lasers emit power levels that range from a few microwatts to several kilowatts and pulsed radiation that can vary enormously in pulse duration, pulse energy and pulse repetition frequency. Detailed information about general applications of laser has been published (Sliney and Wolbarsht, 1980). A summary of commonly used lasers and examples of their applications is presented in table 4. Lasers present a range of potential hazards to the skin and the eyes.

Table 3 Spectral distribution of solar irradiance just outside the earth's atmosphere (Frederick et al, 1989)

Spectral region	Irradiance (W m^{-2})	Percent of total
UVC	6.4	0.5
UVB	21.1	1.5
UVA	85.7	6.3
light	532	38.9
IRR	722	52.8

Incandescent sources (other than lamps and welding arcs)

Many industrial processes involve the heating of materials to incandescence. Because of the relatively low temperatures of the majority of these sources the principal concern in relation to potential ocular hazards lies with the infrared radiation and to some extent with the visible radiation emitted.

In general the emissions from other incandescent materials will be up to a factor of about 3 lower than their black-body equivalents and will vary with temperature, wavelength, the material and the nature of its surface.

Table 4 A summary of commonly used lasers and examples of their applications

Laser	Spectral Emission(s)	Applications
excimer	193-351 nm	laser surgery, material processing
helium cadmium	325, 442 nm	alignment, surveying
dye lasers	350-1000 nm	instruments, dermatology
argon ion	350, 458-514.5 nm	holography, retinal surgery, displays
krypton ion	568, 647 nm	instruments, displays
helium neon	632.8 nm	alignment, surveying, holography
ruby	694.3 nm	ranging, dermatology
gallium arsenide	850-950 nm	optical fibre communications, ranging
neodymium glass/YAG	1.06 μm	material processing, radar/ranging, surgery
carbon dioxide	10.6 μm	material processing, radar/ranging, surgery

REFERENCES

ACGIH (1991) "Threshold limit values for chemical substances and physical agents in the workroom environment." (Cincinnati: The American Conference of Governmental Industrial Hygienists).

CIE (1987) International Commission on Illumination: International lighting vocabulary, CIE Publication No 17.4.

Cole, C. A., Forbes, P. D., Davies, R. E. and F. Urbach. (1986) "Effect of indoor lighting on normal skin." Report published by the New York Academy of Sciences, New York.

FDA (1988) "Performance standards for light-emitting products; paragraph 1040.30" in "Regulations for the administration and enforcement of the Radiation Control for Health and Safety Act of 1968." HHS Publication FDA 88-8035, (Rockville: Center for Devices and Radiological Health).

Frederick, J.E., Snell, H.E. and Haywood,E.K. (1989) "Solar ultraviolet radiation at the earth's surface". Photochem. and Photobiol., 50 (8), 443-450.

IEC (1984) "Radiation safety of laser products, equipment classification, requirements and users guide". IEC Publication 825, (Geneva: International Electrotechnical Commission).

INIRC (1991) "Guidelines on limits of exposure to ultraviolet radiation of wavelengths between 180 nm and 400 nm (incoherent optical radiation)" in "IRPA guidelines on protection against non-ionizing radiation" eds Duchene A S, Lakey J R A and Repacholi M H. (New York: Pergamon Press).

McKinlay, A. F. and Diffey, B. L. (1987) "A reference action spectrum for ultraviolet induced erythema in human skin". CIE Journal 6 (1), 17-22.

McKinlay, A. F., Whillock, M. J. and Meulemans, C. C. E. (1989) "Ultraviolet radiation and blue light emissions from spotlights incorporating tungsten halogen lamps", National Radiological Protection Board (NRPB) Report R228, (Chilton: NRPB).

Noonan, F. P. and De Fabo, E. C. (1990) Ultraviolet-B dose-response curves for local and systemic immunosuppression are identical. Photochem. Photobiol. 52:801-810.

Pitts, D. G. (1986) "A position paper on ultraviolet radiation in Hazards of Light", Myths and Realities, Eye and Skin ed. Cronly-Dillon J., Rosen E. S., Marshall J. (Oxford: Pergamon Press).

Pitts, D. G., Bergmanson, J. P. G., Chu, L. W-F., Waxler, M., Hitchins, V. M. (1987) "Ultrastructural analysis of corneal exposure to UV radiation". Acta. Ophthalmol. (Copenh.) 65:263-273.

Whillock, M. J., Clark, I., McKinlay, A. F., Todd, C. and Mundy, S. (1988). "Ultraviolet radiation levels associated with the use of fluorescent general lighting, UVA and UVB lamps in the workplace and home", National Radiological Protection Board (NRPB) Report R221, (Chilton: NRPB).

Whillock, M. J., McKinlay, A. F., Kemmlert, J. and Forsgren, R. G. (1990) "Ultraviolet Radiation Emissions from Miniature (Compact) Fluorescent Lamps". Lighting Research and Technology 23 (3), 125-128.

WHO (1982) "Environmental health criteria 23. Lasers and optical radiation". (Geneva: World Health Organization).

Zmudzka, B. Z.; Beer, J. Z. (1990) "Activation of human immunodeficiency by ultraviolet radiation". Photochem. Photobiol. 52:1153.

Recommended general reading on Optical Radiation Hazards and Protection Standards

INIRC (1991) "IRPA guidelines on protection against non-ionizing radiation" eds Duchene A S, Lakey J R A and Repacholi M H. (New York: Pergamon Press).

McKinlay, A. F., Harlen, F. and Whillock, M. J. (1988) "Hazards of optical radiation. A guide to sources, uses and safety" (Bristol and Philadelphia: Adam Hilger).

Sliney, D.H. and Wolbarsht, M.L. (1980) "Safety with lasers and other optical sources, a comprehensive handbook" (New York and London: Plenum Press).

MEASUREMENTS AND BIOEFFECTS OF INFRARED
AND VISIBLE LIGHT

David H. Sliney

Laser Microwave Division
US Army Environmental Hygiene Agency[1]
Aberdeen Proving Ground, MD 21010=5422 USA

Introduction

There are many occasions where one views bright light sources such as the sun, arc lamps and welding arcs. The adverse effects from viewing such sources has been studied for decades and during the last two decades guidelines for limiting exposure to protect the eye have been developed. The guidelines were fostered to a large extent by the growing use of lasers and the quickly recognized hazard posed by viewing laser sources.

Adverse Effects from Exposure to Intense Light Sources

In this chapter, ultraviolet radiation hazards will not be considered. With UVR hazards considered in another chapter, the remaining hazards result from visible radiation (i.e., light) and infrared (IR) radiation. At wavelengths greater than about 550 nm, thermal injury mechanisms dominate. However, in the blue-light spectral region, it is also possible for a bright light source to pose a photochemical hazard at wavelengths less than 550 nm. The principal retinal hazard resulting from viewing bright light sources is photoretinitis; e.g., *solar retinitis* with an accompanying scotoma (blind spot) which results from staring at the sun. Solar retinitis was once referred to as "eclipse blindness" and associated "retinal burn." Only in recent years has it become clear that photoretinitis results from a photochemical injury mechanism following exposure of the

---

1. The opinions or assertions herein are those of the author and should not be construed as reflecting official positions of the US Department of the Army or Department of Defense.

retina to shorter wavelengths in the visible spectrum, i.e., violet and blue light. Prior to conclusive animal experiments at that time (Ham, et al., 1976), it was thought to be a thermal injury mechanism. However, it has been shown conclusively that staring at an intense light source which is rich in short-wavelength light (hereafter referred to as "blue light") can cause retinal injury.

Blue-light retinal injury (photoretinitis) can result from viewing either an extremely bright light for a short time, or a less bright light for longer exposure periods. The product of the dose-rate and the exposure duration always must result in the same exposure dose (in joules-per-square centimeter at the retina) to produce a threshold injury. This characteristic of photochemical injury mechanisms is termed *reciprocity* and helps to distinguish these effects from thermal burns, where heat conduction requires a very intense exposure within seconds to cause a retinal coagulation; other-wise, surrounding tissue conducts the heat away from the retinal image. Injury thresholds for acute injury in experimental animals for both corneal and retinal effects have been corroborated for the human eye from accident data. Occupational safety limits for exposure to UVR and bright light are based upon this knowledge. As with any photochemical injury mechanism, one must consider the *action spectrum*, which describes the relative effectiveness of different wavelengths in causing a photobiological effect. The action spectrum for photoretinitis peaks near 440 nm; whereas, light mediated effects upon the neuroendocrine system apparently peak in the green (Brainard, et al., 1991; Sack *et al.*, 1991) and therefore suggests a possible mediation by the collection of both (or either) night (rods) and daylight photoreceptors (i.e., the cones). Figure 1 shows the blue light hazard action spectrum.

Quantities and Units

Two sets of light-measurement quantities and units are useful in defining light exposure of the retina: *radiometric* and *photometric*. Radiometric quantities such as *radiance*--used to describe the "brightness"

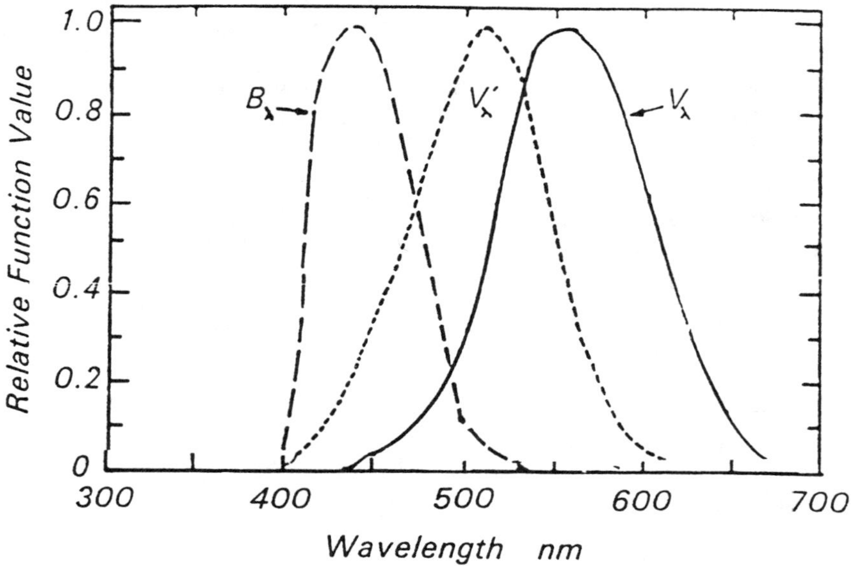

Figure 1. The CIE Spectral Sensitivity (Standard Observer) Curve V_λ for the Human Eye. For comparison, the ACGIH Blue Light Hazard Function B_λ is also provided.

of a source [in W/cm^2·sr] and *irradiance*--used to describe the irradiance level on a surface [in W/cm^2] are particularly useful for hazard analysis. Figure 2 illustrates the basic concepts of irradiance and radiant exposure and how they are related to radiant power and energy. More sophisticated concepts are *radiance* and *luminance*, which are particularly valuable because these quantities describe the source and do not vary with distance. The luminance or radiance of any extended source such as the lightbulb above are the same whether measured near the source or from across a large room. Photometric quantities such as luminance (brightness in cd/cm^2 as perceived by a human "standard observer") and illuminance in lux (the "light" falling on a surface) indicate light levels spectrally weighted by the standard photometric visibility curve which peaks at 550 nm for the human eye. To quantify a photochemical effect it is not sufficient to specify the number of photons-per-square-centimeter (photon flux) or the irradiance (W/cm^2) since the efficiency of the effect

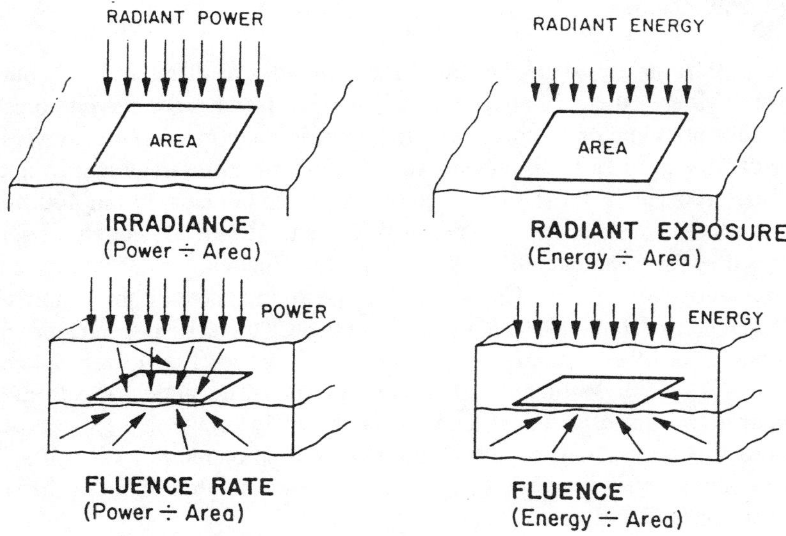

Figure 2. Irradiance and Radiant Exposure. Two other terms, fluence rate and fluence have the same units of W/cm^2 and J/cm^2, but have different geometrical meaning (they include scattered radiation in addition to incident radiation.

----

will be highly dependent on wavelength. Generally, shorter-wavelength, higher-energy photons are more efficient in producing a biological effect.

Photometric quantities are hybrid quantities which are defined by an action spectrum for vision--a photochemically initiated process. Photometric quantities may not have much value in describing retinal effects other than vision (Sliney, 1991). Scotopic quantities and units also exist (although they are not extensively used). Scotopic lumens define the light levels as weighted against man's rod sensitivity function. These units normally have no value in assessing hazards, but could be used to determine the visual transmission of colored filter eye protection.

A specialized unit of photopic retinal illuminance is the *troland (td)* which is the product of source luminance (cd/m^2) and pupillary diameter squared (mm^2). This unit is widely used in studies of retinal photochemistry and light-induced visual disturbances such as afterimages (flashblindness).

Unfortunately, since the spectral distributions of different light sources vary widely, there is no simple conversion factor between photometric (either photopic or scotopic) and radiometric quantities. This conversion may vary from 15 to 50 lumens/watt (1m/W) for an incandescent source to about 100 1m/W for the sun or a xenon arc, to perhaps 300 to 400 lm/W for a fluorescent source (Sliney and Wolbarsht, 1980). Although the action spectra for photokeratitis (Pitts, 1978; Zuclich, 1980); for acute cataractogenesis (Pitts, 1977); and for short-term, acute photoretinitis for the monkey (Ham, 1989) have been published, the action spectra of some other types of light damage appear to be quite imprecise or even unknown (Lanum, 1978; Williams and Baker, 1980). The action spectrum for neuroendocrine effects is still only approximately known. This suggests the need to specify the spectrum of the light source of interest as well as the irradiance levels if one is to compare experimental results from different experimental studies.

Calculating Retinal Irradiance

Retinal irradiance (exposure rate) is directly related to source radiance (brightness). It is <u>not</u> readily related to corneal irradiance (Sliney, 1984). Formula [1] gives the general relation for any eye's optical system, where E_r is the retinal irradiance (W/cm^2), L_s is the source radiance, W/cm^2·sr), f is the effective focal length of the eye (cm), d_e is the pupil diameter (cm), and τ is the transmittance of the ocular media. The more general relation is:

$$E_r = \pi L_s . \tau . d_e^2 / 4f^2 \tag{1}$$

This formula was derived by considering the equal angular subtense of the source and the retinal image at the eye's nodal point (see Sliney and Wolbarsht, 1980, for a complete derivation). The transmittance τ of the ocular media in the visible spectrum for younger humans (and most animals) is as high as 0.9 (i.e., 90%) (Geeraets and Berry, 1968). If one uses the effective focal length f of the adult human eye (Gulstrand eye), where f = 1.7 cm, and d_e (2 to 7 mm), then from equation [1], it is possible to calculate irradiances (dose rates) E_r at the human retina:

$$E_r = 0.27 L_s \cdot \tau \cdot d_e^2 \tag{2}$$

Exposure of the anterior structures of the human eye to infrared (as well as ultraviolet) radiation may also be of interest; and the relative position of the light source and the degree of lid closure can greatly affect the proper calculation of the exposure dose to the cornea and lens. [For ultraviolet and short-wavelength light exposures, the spectral distribution of the light source can also be important.]

Retinal Injury Hazards

In the past two decades a large number of studies have been reported on the adverse effects of prolonged light exposure of the vertebrate retina (especially the mammalian eye). Several reviews and collections of such studies have been published (Ham, 1989; Lanum, 1978; Marshall, 1983; Sperling, 1980; Williams and Baker, 1980; Waxler and Hitchens, 1987). The interpretation of this research and the assessment of its relevance to human exposure conditions requires accurate dosimetry. A comparison of light levels used in the various published studies shows that many of the dosimetric interpretations of the several investigators were flawed (Sliney, 1984). While the great majority of investigators in this field have made light measurements designed to characterize the exposure conditions, the calculated levels of retinal exposure that have been published often vary by more than a thousand-fold amongst studies employing the same light sources. From a knowledge of optical constants of the experimental eyes and from knowledge of the lamp sources used in the experiments, it is possible to show that the retinal irradiances should be very similar for the same light source, even when light levels in the corneal plane vary enormously (Sliney, 1984).

Spectral Weighting Calculations

Luminous exposure and radiant exposure are quantities used to describe a total exposure dose from a flashlamp or for a more lengthy exposure. For example, light-induced retinal injury that only occurs after prolonged exposure (i.e., greater than 100 s) is generally agreed to result from a photochemical injury mechanism, rather than thermal injury. Two key factors distinguish a photochemical process from a thermal process. *Thermal injury* is a *rate process* and is dependent upon the volumic absorption of energy across the spectrum. By contrast, any photochemical process will have a long-wavelength cutoff where photon energies are insufficient to cause the molecular change of interest. A photochemical reaction will also exhibit reciprocity between irradiance (exposure dose rate) and exposure duration. Repair mechanisms, recombination over long periods and photon saturation for extremely short periods will lead to reciprocity failure. For the lengthy exposures characteristic of light damage studies, it is difficult to know what effective exposure time to use for an exposure calculation. Irradiance E in W/cm^2 times exposure duration t is equal to the radiant exposure H in J/cm^2, i.e.,

$$H = E \cdot t \tag{3}$$

While both E and H may be defined over the entire optical spectrum, it is necessary to employ an action spectrum for photochemical effects. The V_λ , V_λ, and B_λ curves of Figure 1 are examples of action spectra which may be used to spectrally weigh the incident light. With modern computer spread-sheet programs, one can readily develop a method for spectrally weighting a lamp's spectrum by a large variety of photochemical action spectra. These computations can be tedious, but straightforward and take the form:

$$E_{eff} = \Sigma E_\lambda \cdot A_\lambda \cdot \Delta\lambda \tag{4}$$

where A_λ may be any action spectra of interest. One then can compare different sources to determine relative effectiveness of the same irradiance from several lamps for a given action spectrum.

Examples

As one example, a typical cool-white fluorescent lamp has an illumination of 185 ft-cd (200 lux) and the irradiance is 0.7 mW/cm^2. The effective blue-light irradiance is found to be 0.15 mW/cm^2. From previous measurements (Sliney and Wolbarsht, 1980), the blue-light radiance is 0.6 $mW/(cm^2 sr)$ and the radiance is 2.5 $mW/(cm^2 sr)$. When fluorescent lamps are used in a phototherapy light box, a plastic diffuser reduces this luminance.

A 1,000-W tungsten halogen bulb has a greater radiance, but the percentage of blue light is far less. A typical blue-light radiance is 0.95 $W/(cm^2 sr)$ compared to a total radiance of 58 $W/(cm.sr)$. The luminance is 2600 cd/cm^2--a factor of 3000 times brighter than the cool white fluorescent. But since the retinal image is small, eye movements spread the exposure over a much larger area.

The above example illustrates the importance of considering the size of a light source and the impact of eye movements in any calculation of retinal exposure dose. Only in medical situations is the eye restrained, e.g., during eye surgery or during certain ocular diagnostic procedures. The retinal irradiance calculation in this case could include a geometrical factor for averaging the retinal exposure over only the area exposed to the light source (e.g., an operating microscope). If a light source were to occupy 20% of the environmental field of view for a given task, and if the surface area and the remainder of the environment were black, then the average retinal irradiance would be reduced to at most 20% of the direct exposure calculation. Of course, an individual in this instance might be able to avoid any direct exposure.

Environmental Exposure of the Eye

When one considers natural light exposure of the human retina out-of-doors, one is struck by the fact that fluorescent lamps have a much higher blue-light radiance than the normal terrain or sky. Furthermore, the sun has a strong blue and UV component only near midday when people do not look at it. When the sun is at angles of about 5° to 20° above the horizon, one can stare at it for short periods without risk of injury (Sliney, 1985). This illustrates the enormous importance of considering both spectral and geometrical factors in light damage experiments and in assessing environmental risks and the merits of some sunglasses. The natural aversion to bright light and the use of diffusing fixtures and baffles normally prevents hazardous acute retinal exposure from the sun and man-made lamps. Figure 3 shows the retinal irradiances experienced when viewing a variety of bright light sources.

Only in the outdoors is the human eye exposed to significant levels of UVR. Exposures of the eye to welding arcs and other open arcs is very rare because of the use of eye protectors. The well known acute effect of excessive UVR exposure to the eye, photokeratoconjunctivitis, normally results from exposure to a limited band of wavelengths below about 315 nm, i.e., the UV-B and UV-C bands. The terrestrial solar spectrum in this spectral region varies enormously with time of day.

Time-of-Day

The exposure to UVR and light outdoors constantly changes during the day. Subjectively, we are largely unaware of the degree of these changes. For example, the exposure rate of UV-B (the shortest

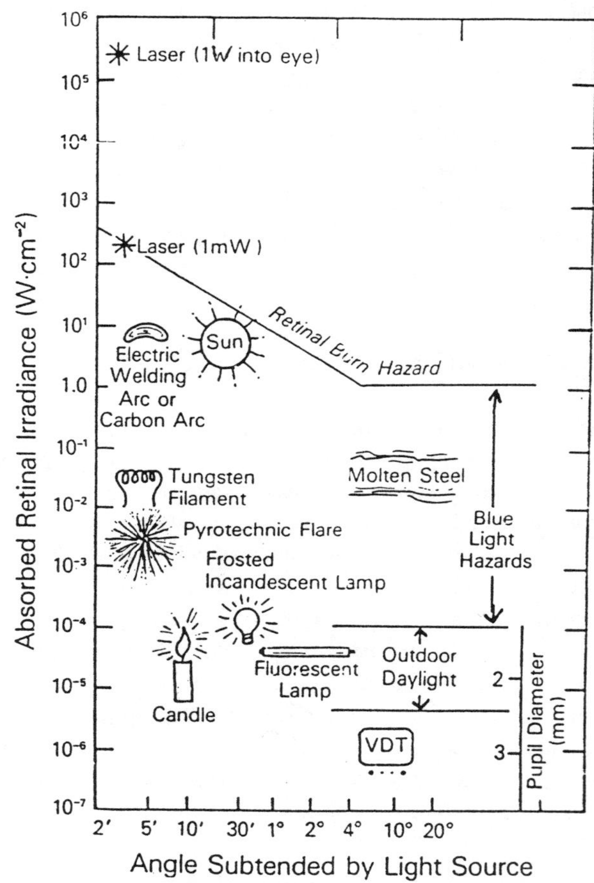

Figure 3. Relative Retinal Irradiances for Staring Directly at Various Light Sources. Two retinal exposure risks are depicted: retinal thermal injury which is image-size dependent, and photochemical injury which depends on the degree of blue light in the source's spectrum. The horizontal scale indicates typical image sizes. Most intense sources are so small that eye movements and heat flow will spread the incident energy over a larger retinal area. The range of photoretinitis (the "blue-light hazard") is shown to be extending from the normal outdoor light levels and above. The xenon arc lamp is clearly the most dangerous of non-laser hazards.

---

wavelengths in sunlight) varies remarkably with the time of day. When the sun is overhead at noon, the spectral irradiance at a wavelength of 300 nm is ten times greater that at either three hours before or after noon. An untanned person with fair skin would receive a mild sunburn in 25 minutes at noon, but would have to lie in the sun for at least two hours to receive the same dose after 3:00 (standard time). Such an enormous change does not occur in the visible part of the sun's spectrum, but some change still occurs. We are all aware of the red color of the sun at sunset; and an attempt to take a color photograph either very early or very late in the day will result in a picture that is yellowish or orange in hue. We are fortunate that the scattering of sunlight by air

molecules (due to Rayleigh scattering) favors UVR and blue light (hence the blue sky). For longer pathlengths through the atmosphere when the sun is low in the sky, much more UVR and sunlight is scattered and the sun which is white at noonday becomes yellow and then orange as less UVR and blue light are present in the direct rays. When the sun is overhead and "white," it would take only 90 seconds to stare at the sun and receive solar retinitis. A few hours later, with the sun much lower in the sky, it would take several minutes to reach a hazardous retinal dose, and it is virtually impossible at sunset. Thus, the geometry of exposure as well as the spectrum (hue) plays a major role in determining the hazards of viewing the sun.

Fortunately, we seldom look directly overhead when the sun is very hazardous to view, and the sun is not very hazardous to view when the sun is sufficiently low in the sky to fall within our normal field-of-view. Furthermore, when the sun is greater than about 10° above the horizon, we squint, thus shielding the retina from direct exposure (Sliney, 1983). Although the cornea is more sensitive to UVR injury than the skin, one seldom experiences a corneal burn when out in sunlight. Again, the geometry of exposure helps. When the sun is overhead and UVR exposure is most severe, the brow ridge and specular reflectance of the sun greatly reduce corneal exposure.

Exposure from Snow Reflections

The greatest retinal exposure to blue light and to UVR (either indoors or outdoors) is most likely to occur when one is outside standing or walking in a field of snow without eye protection. Under these conditions, one can readily be exposed to levels exceeding occupational exposure limits. As most skiers know, failure to wear ski goggles can result in erythropsia (red vision) and photokeratitis, sometimes loosely referred to as "snow-blindness." This can result from the unusually intense reflection of light and

UVR from the snow. Fresh snow reflects 85 % of the UVR compared to only 1 or 2 % from grass; thus, the less intense overhead UVR of the winter sun is incident upon the eye in far greater intensities than in summer sunlight (19). In addition, the excessive retinal illumination can result in perceptible "red vision" (erythropsia) upon entering the ski lodge. A reduction in night vision can also be detected that evening, if tested.

Exposure Limits

A number of national and international groups have recommended occupational or public exposure limits (ELs) for optical radiation [i.e., ultraviolet (UV), light and infrared (IR) radiant energy]. Although most such groups have recommended ELs for UV and laser radiation, only one group has recommended ELs for visible radiation (i.e., light). This one group is well known in the field of occupational health--the American Conference of Governmental Hygienists (ACGIH). The ACGIH refers to its ELs as "Threshold Limit Values," or TLVs and these are issued yearly, so there is an opportunity for a yearly revision. The current ACGIH TLV's for light (400 nm to 760 nm) have been largely unchanged for the last decade, although they have been on a tentative list for much of that time. They are based in large part on ocular injury data from animal studies and from data from human retinal injuries resulting from viewing the sun and welding arcs. The TLVs also have an underlying assumption that outdoor environmental exposures to visible radiant energy is normally not hazardous to the eye except in very unusual environments such as snow fields and deserts.

On the international scene there are currently no limits for ocular exposure to visible light except for the special case of laser radiation. The International Non-ionizing Radiation Committee (INIRC) of the International Radiation Protection Association (IRPA) published Guidelines on Limits of Exposure to Laser Radiation in 1985 and revised them in 1988 (IRPA, 1985, 1988). INIRC guidelines are developed through collaboration with the World Health Organization (WHO) by jointly publishing criteria documents which provide the scientific data base for the exposure limits (WHO, 1982). The ACGIH TLVs and INIRC EL's are generally in agreement with those of the International Electrotechnical Commission, the American National Standards Institute (1986), the Health Council of the Netherlands (1979), and others.

Eye Protection

As previously explained, the eye is remarkably well protected from environmental light unless one forces oneself to stare at the sun or unless one is exposed to UVR reflected from snow. Modern sunglasses generally filter out significant UVR, but they seldom filter out more than 85% of visible or infrared radiant energy (Sliney, 1983) and it would be dangerous to think that sunglasses would permit one to safely stare at the sun or a welding arc. In industry, infrared absorbing glass lenses are used around furnaces in steel mills and in the glass industry. Welding filters are the most common form of eye protection to protect against the one major industrial retinal hazard: the welding arc.

From the turn-of-the-century when Crooke's glass was introduced, the design of eye and face protection for the welder remained little changed for 75 years (Sliney, 1981; Sliney and Wolbarsht, 1980). When properly used, Crooke's glass fixed shade filters offer effective protection for the welder; however, they have significant ergonomic shortcomings. When one considers the total welding process, the welder must lower and raise the helmet or filter each time an arc is started and quenched. The welder must work blind just prior to striking the arc, which may compromise safety and making it difficult to start the weld at precisely the position desired. Furthermore, the helmet is frequently lowered and raised with a sharp snap of the neck and head which can lead to neck strain or more serious injuries. Faced with this uncomfortable and cumbersome procedure, some welders frequently initiate the arc with the helmet in the raised position--leading to 'welder's flash' ('arc-eye' or photokeratitis).

However, in recent years, a number of new, more ergonomic welding helmet designs have been developed and sold which feature auto-darkening welding filters. Under normal ambient lighting conditions, a welder wearing a helmet fitted with an auto-darkening filter can see well enough with the eye protection in place to perform tasks such as aligning and fixturing the parts to be welded, precisely positioning the welding equipment and striking the arc. Then, in the most typical helmet designs, light sensors detect the arc flash and direct an electronic drive unit to switch a liquid crystal filter from a light shade to a preselected dark shade. This relieves the welder from constantly raising his helmet between welds and also provides increased protection from ultraviolet radiation (UVR) from any adjacent welding operation. Mechanically operated dark-shade filters, generally activated by an air-line pressure switch, which switch into position immediately before the arc is struck are also made, but are generally more cumbersome than the electro-optic, liquid crystal autodarkening systems.

The aforementioned drawbacks of fixed shade filters are largely eliminated by auto-darkening filters which explains the widespread popularity of these newer systems. A properly designed auto-darkening filter enables the welder to produce more welds of higher quality while enjoying improved comfort and safety. Production managers therefore favor auto-darkening protective helmets, despite the added cost, because of the marked improvement in weld quality, quantity and consistency--due primarily to the welder's ability to view the work continuously and begin the weld at precisely the desired spot--and because of reduced training time for new welders who no longer must cope with blind periods during the weld cycle.

While it is relatively easy for the production manager to recognize the benefits of auto-darkening filters, one frequently encounters inquires as to the safety of these newer systems. There are currently no approved safety standards for these devices in North America. The physical characteristics such as closure time and shades are generally difficult to measure. Some have been concerned about whether after-images (flash-blindness) can result in impaired vision in the workplace? Do the new types of filters really offer equivalent or better protection than conventional fixed filters? The lack of objective standards means that the welder, the safety manager or industrial hygienist is frequently confronted with a decision based on the varying attributes claimed by the manu-facturers of the different units. There is currently a European effort to develop a standard which eliminates the need for concern about hazardous ocular exposure (Buhr and Sutter, 1989).

Despite the fact that most current auto-darkening filters will meet the recommended draft standard, not all systems are equivalent. Filter closure speeds and light and dark shade values vary, as does the weight of each unit. The temperature dependence of performance, variation of shade with electrical battery degradation, the 'resting state shade,' and other technical factors vary depending upon each manufacturer's design. Sometimes these variations may have an impact upon safety.

Current auto-darkening filters vary in switching speed from 1/10th second (0.1 s) to faster than 1/10,000th second (100 s). Buhr and Sutter (1989) have indicated a means of specifying the maximum switching time, but their formulation varies relative to the time-course of switching. Switching speed is crucial, since it gives the best clue to the really important (but unspecified) measure of how much light will enter the eye when the arc is struck as compared to wearing a fixed filter of the same working shade number. A surge of light lasting for several milliseconds is emitted when the arc is struck (Eriksen, 1985). If too much light enters the eye for each switching during the day, the accumulated light energy dose could potentially be hazardous ('blue-light hazard'). Most current configurations with switching speeds of the order of 10 ms or less will provide adequate protection from the "blue-light hazard (photoretinitis). However, the shortest switching time of the order of 100 μs (0.1 ms) has the advantage of reducing 'transient adaptation' effects.

Transient adaptation is the visual experience caused by sudden changes in one's light environment which may be accompanied by discomfort, glare and temporary loss of detailed vision. If one looks directly at a bright light bulb and then immediately tries to read some lines of very small print, anyone can understand the effects of transient adaptation. Since there is an initial surge of enhanced light during the first 20 milliseconds as the electric arc is struck the faster switching speeds are particularly important for auto-darkening welding filters. When repeated light flashes are experienced many times during the day, vague subjective symptoms such as eye-strain or headache may result. Hence the advantage of very fast switching speeds may not become apparent until a welder has used that type of filter for a day or two. For these reasons, the welder should try different filter systems to determine if a faster switching speed may be needed in his operation.

Conclusions

Aside from pulsed lasers, visible and near-infrared radiation exposure of the eye and skin is seldom a significant risk because of one's natural aversion to bright light and the thermal pain sensation. The sources identified in this chapter that require eye protection to preclude blue-light photoretinitis are: staring at any open-arc processes, welding, the sun, and any light source that is very uncomfortable to stare at. The natural aversion to viewing bright light sources is remarkably good at indicating when one can encounter a hazard. Of course a light source with a great deal of UVR or IR can be dangerous even if it is not sufficiently bright to create a strong visual aversion response. IR protection is needed to protect against industrial heat cataract in the glass industry and steel industry.

References

1. American Conference of Governmental Industrial Hygienists (ACGIH) (1991), TLV's, Threshold Limit Values and Biological Exposure Indices for 1990-1991, American Conference of Governmental Industrial Hygienists, Cincinnati, OH.

2. ACGIH (1991), Documentation for the Threshold Limit Values, 4th Edn., American Conference of Governmental Industrial Hygienists, Cincinnati, OH.

3. ANSI (1986), Safe Use of Lasers, Standard Z-136.1-1986, American National Standards Institute, Laser Institute of America, Orlando, FL.

4. Brainard, G.

5. Buhr, E., Sutter, E., Dynamic Filters for Protective Devices, In:"Dosimetry of Laser Radiation in Medicine and Biology," Mueller, G.J., and Sliney, D.H., (eds.), SPIE, vol. IS-5, SPIE, Bellingham, WA, 1989.

6. Eriksen, P., Time Resolved Optical Spectra from MIG Welding Arc Ignition, Ame. Ind. Hyg. Assn. J., 46:101-104, 1985.

7. Fender, D.H., 1964, Control of the Eye, Science Amer., Vol. 211, pp. 24-33.

8. Gerathewohl, S.J. and Strughold, H., 1953. Motoric Response of the Eyes When Exposed to Light Flashes of High Intensities and Short Durations, Journal of Aviation Medicine, Vol. 24, pp. 200-207.

9. Ham, W. T., Jr. (1983) The photopathology and nature of the blue-light and near-UV retinal lesion produced by lasers and other optical sources (M. L. Wolbarsht, ed.) "Laser Applications in Medicine and Biology, New York, Plenum Publishing Corp. (1989).

10. Ham, W.T., Jr., Ruffolo, J.J. Jr., Mueller, H.A., and Guerry, D., III, (1980) The nature of retinal radiation damage, dependence on wavelength, power level and exposure times, Vision Res. 20:1105-1111.

11. Health Council of the Netherlands (1979), Acceptable Levels for Micrometer Radiation, Rijswijk, Gezondheidsraad.

12. IRPA, International Non-Ionizing Radiation Committee (1985), Guidelines for Limits of Human Exposure to Laser Radiation, Health Physics, 49(5):341-359; & 54(5):573-575.

13. Lanum, J. A. (1978) The damaging effects of light on the retina, empirical findings, theoretical and practical implications. Surv. Ophthalmol. 22, 221-249.

14. Lerman, S. (1980) Radiant Energy and the Eye, MacMillan & Co., New York.

15. Mainster MA. Spectral transmission of intraocular lenses and retinal damage from intense light sources. Am J Ophthalmol. 1978; 85:167-170.

16. Marshall J. Light damage and the practice of ophthalmology. In: Rosen

E Arnott E and Haining W (eds.). Intraocular Lens Implantation. London: Moseby- Yearbook, Ltd., 1983.

17. Noell, W. K. (1980) Possible mechanisms of photoreceptor damage by light in mammalian eyes, Vision Res., 20, 1163-1171.

18. Pitts, D.G., and Cullen, (1981), Determination of infrared radiation for acute ocular cataractogenesis, Albrecht von Graefes Arch Klin Ophthalmol., 217:285-297.

19. Sack, R.L., Lewy, A. J., White D.M., Singer, C.M., Fireman M.J., Vandiver R. (1990) *Arch Genl Psychiatry*, 47:343-351.

20. Sliney DH. Eye protective techniques for bright light. Ophthalmology 1983; 90(8):937-944.

21. Sliney, D.H. "Dynamic Welding Filters," presented at the International Congress of Ergophthalmology, San Francisco, CA, 1981.

22. Sliney, D.H. and Wolbarsht, M.L., "Safety with Lasers and other Optical Sources," New York, Plenum Publishing Company, 1980.

23. Sliney DH. Physical factors in cataractogenesis--ambient ultraviolet radiation and temperature. Invest Ophthalmol Vis Sci, 1987;

24. Sliney DH: Quantifying retinal irradiance levels in light damage experiments Current Eye Research, 3(1):   , 1984.

25. Sperling HG (ed.). "Intense Light Hazards in Ophthalmic Diagnosis and Treatment, Proceedings of a Symposium." Vision Res. 1980; 20(12):1033-1203.

26. Sykes, S. M., Robison, W. G. Jr., Waxler, M., and Kuwabara, T (1981), Damage to the monkey retina by broad-spectrum fluorescent light. Invest. Ophthal. Vis. Sci. 21, 425-434.

27. Varma SD and Lerman S (eds.). Proceedings of the First International Symposium on Light & Oxygen Effects on the Eye. Oxford:IRL Press, 1984. [also published as Current Eye Res. 1984; 3(1).

28. Waxler M and Hitchens V (eds). Optical Radiation and Visual Health. Boca Raton: CRC Press, 1986.

29. Williams TB and Baker BN (eds.). The Effects of Constant Light on the Visual System. New York: Plenum Press, 1980.

30. World Health Organization [WHO], (1982), Environmental Health Criteria No. 23, Lasers and Optical Radiation, joint publication of the United Nations Environmental Program, the International Radiation Protection Association and the World Health Organization, Geneva.

31. Yarbus, A.L., 1967, Eye Movements During Fixation of Stationery Objects, Plenum Press, NY.

32. Zuclich JA. Ultraviolet-induced photochemical damage in ocular tissues. Health Phys, 56(5):671-682, 1989

ULTRAVIOLET STUDIES

David H. Sliney

Laser Microwave Division[1]
US Army Environmental Hygiene Agency
Aberdeen Proving Ground, MD 21010-5422 USA

Introduction

The ultraviolet radiation (UVR) spectrum is generally split into three photobiological spectral bands by the Commission International de l'Eclairage (CIE, 1988). These are the UV-A (315-400 nm); the UV-B (280-315 nm) and the UV-C (100-280 nm). UV-C does not penetrate through the ozone layer. Hence, exposure of man to UV-C is limited to open-arc processes, such as arc welding. The CIE bands are widely used to distinguish between different biological effects and hazards. Most UVR biological effects result from photochemical interactions and *reciprocity* between irradiance and duration exists; i.e., a total exposure dose in J/m^2 will produce the same effect whether the energy is delivered in 1 μs or 1 ks (16.7 minutes). Figure 1 summarizes the spectral ranges where biological effects occur.

The UVR wavelengths below 315-320 nm (i.e., UV-B and UV-C) are generally thought to be the most "actinic" (or, effective) wavelengths most responsible for adverse health effects such as skin cancer and environmental cataract from solar radiation. Acute effects such as photokeratitis ("welders' flash or snow-blindness") and erythema (sunburn) occur from excessive exposure not only to sunlight but also from some artificial light sources.

---

1. The opinions or assertions herein are those of the author and should not be construed as official policies of the Department of the Army or Department of Defense.

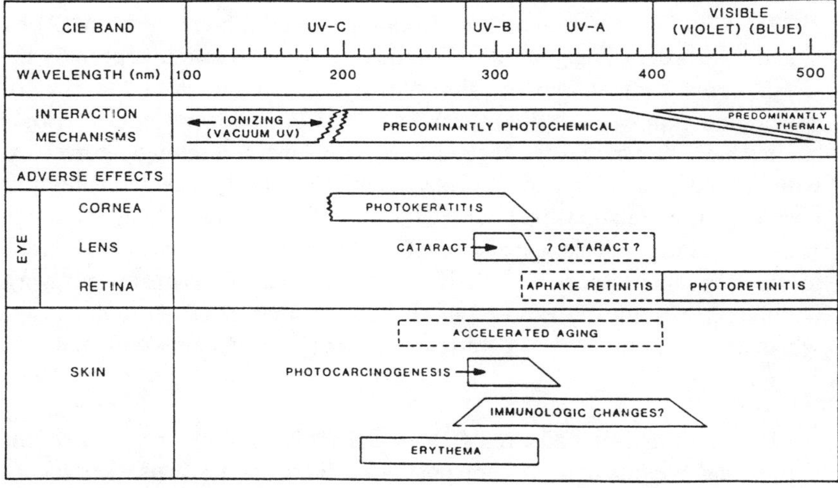

Figure 1. The biologic effects of UVR.

UVR Biological Effects

Although skin cancer and cataract are the primary examples of the adverse effects of long-term, chronic exposure to UVR, far more is scientifically known about the acute effects of erythema and photokeratitis. The action spectra (curves of relative effectiveness of each wavelength in eliciting a photobiological effects) are far better understood than action spectra for chronic effects. Knowledge of the action spectra alone is not sufficient to establish occupational exposure limits, however. The severity of each biologic effect, the transient nature of the effect and other factors play a role (INIRC, 1985).

As an example of how different wavelengths cause different degrees of an adverse effects, consider "sunburn," i.e., erythema. Erythema resulting from exposure to the longer UV-B wavelengths (295-315 nm) is more severe and persists longer than the slight skin reddening from exposure to the 254-nm emission from a low-pressure mercury lamp. The increased severity and time course of the erythema results from deeper penetration of these wavelengths into the epidermis. Maximum sensitivity of the skin occurs at 295 nm. The minimal erythemal dose (MED) reported in more recent studies for untanned, lightly pigmented skin range from 6-30 mJ • cm^{-2} (Hausser, 1928; Coblentz, *et al.*, 1931; Luckiesch, *et al.*, 1930; Everett, *et al.*, 1965; Freeman, *et al.*, 1966; Berger, *et al.*, 1968; Willis, *et al.*, 1972). Skin pigmentation and conditioning (tanning) increase the MED by at least one order of magnitude. Chronic exposure to sunlight especially the UV-B component accelerates the skin aging and increases the risk of developing skin cancer (Fitzpatrick *et al.*, 1974). Several epidemiologic studies have shown that the incidence of skin cancer is strongly correlated with latitude, altitude and sky cover.(WHO, 1972). Exact quantitative dose-response relationships have not yet been established, although fair-skinned individuals, especially of Celtic origin, are much more prone to develop skin cancer. Relatively few studies have examined work populations chronically exposed to artificial sources of UV-B to determine whether there is an increased skin cancer risk (Cunningham-Dunlop, *et al.*, 1977).

Actinic ultraviolet radiation (UV-B and C) is strongly absorbed by the cornea and conjunctiva. Overexposure of these tissues cause keratoconjunctivitis, commonly referred to as welder's flash, arc-eye, etc. Pitts has characterized the course of ordinary clinical photokeratitis (Pitts and Tredici, 1971; Pitts, 1974). The latent period varies inversely with the severity of exposure ranging from 1.5 to 24 hours, but usually occurs within 6-12 hours. Conjunctivitis follows and may be accompanied by erythema of the facial skin surrounding the eyelids. The individual has the sensation of a foreign body or sand in the eyes and may experience photophobia, lacrimation and blepharospasm to varying degrees. The acute symptoms last from 6-24 hours and discomfort usually disappears within 48 hours, but the individual is visually incapacitated during this period. Only rarely does exposure result in permanent ocular injury.

Threshold data for photokeratitis in humans have been established by Pitts and Tredici (1971) for 10 nm wavebands from 220 to 310 nm. The maximum sensitivity of the human eye was found to occur at 270 nm. The wavelength response or action spectrum is not as marked as in the case of erythema with the thresholds varying from 4-14 mJ.

Wavelengths above 295 nm can be transmitted through the cornea and are almost totally absorbed by the lens. Pitts et al have recently shown that cataracts can be produced in rabbits by wavelengths in the 295-320 nm band.(13) Thresholds for transient opacities ranged from 0.15 to 12.6 J/cm^2 depending on wavelength with a minimum threshold at 300 nm. Permanent opacities required greater radiant exposure. No lenticular effects were noted in the wavelength range 325 to 395 nm with radiant exposures of 28 to 162 J/cm^2. Whether chronic exposure to lower levels will produce lenticular opacities has not been determined; however, Taylor and colleagues at the Johns Hopkins University in Baltimore have shown that solar UV-B exposure increases the risk of cataract incidence among Maryland watermen (Taylor, 1989).

Exposure Limits

The occupational and public exposure limits (ELs) for UVR have been reasonably fixed within the 200-315 nm spectral region for more than 20 years (Sliney, 1972; Sliney, 1985; INIRC, 1986). Updates were made in 1986 by INIRC by extending the earlier hazard action spectrum into the UV-A and later by extending the lower boundary from 200 nm to 180 nm (ACGIH, 1991).

The development of occupational exposure limits (ELs) to UVR has entailed a study of the risks of both acute and chronic injury to both eye and skin. In the UV-B and UV-C regions, an action spectrum curve can be drawn which envelopes the threshold data for acute effects obtained from recent studies of minimal erythema and keratoconjunctivitis (Sliney, 1972). This EL (or Threshold Limit Value, TLV) curve does not differ significantly from the collective threshold data considering measurement errors and variations in individual response and is well below the UV-B cataractogenic thresholds. Figure 2 shows the envelope exposure limit (solid line) and thresholds for photokeratitis (triangles). Two new data points (o) to extend the action spectrum in the UV-C at 193 nm are also shown (Sliney *et al.*, 1991).

Repeated exposure of the eye to potentially hazardous levels of UV is not believed to increase the protective capability of the cornea as does skin tanning and thickening. It often comes as a surprise to many that following UVR exposure of the skin, it is the thickening of the stratum corneum provides most of the additional protection (conditioning) against further exposure, rather than does melanogenesis.

Sources of UVR Exposure

The most commonly encountered man-made sources which pose potential UVR hazards are: quartz-envelope mercury (germicidal) lamps,

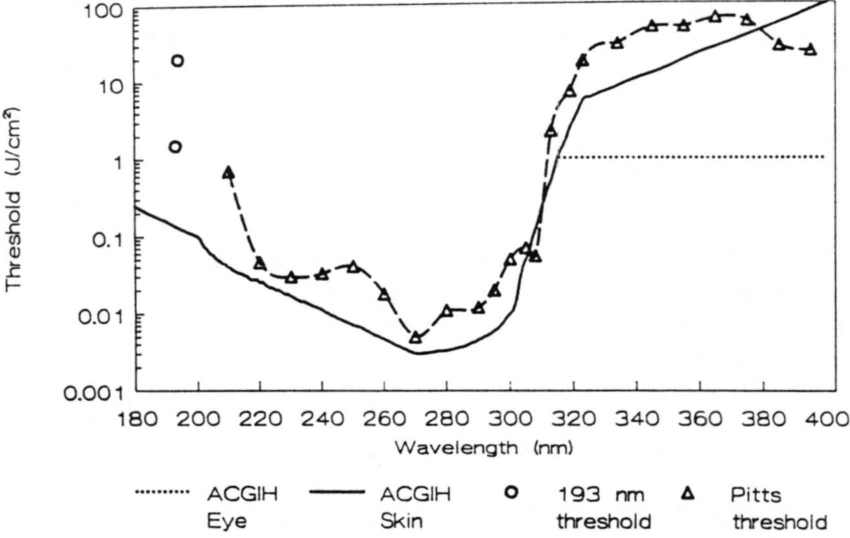

Figure 2. ACGIH TLV for UVR from 180 nm to 400 nm is shown for exposure of skin (solid line) and eye (dotted line). Also, for comparison, are shown photokeratitis thresholds (Δ). Two new photokeratitis threshold points (o) at 193 nm reported by Sliney *et al.* (1991) are shown.

---

xenon-arc lamps and welding arcs. Many lamps including the common fluorescent and incandescent lamps emit some UVR, but normally emit only trace amounts of UVR and are rarely of health concern. Nevertheless, under certain circumstances, UV-B from even some incandescent quartz-halogen lamps can produce an erythema.

Sunlight Exposure

During the past 20 years, concerns of UVR induced skin cancer and depletion of the ozone layer in the stratosphere led to a worldwide program of outdoor, daily UVR measurements and epidemiological

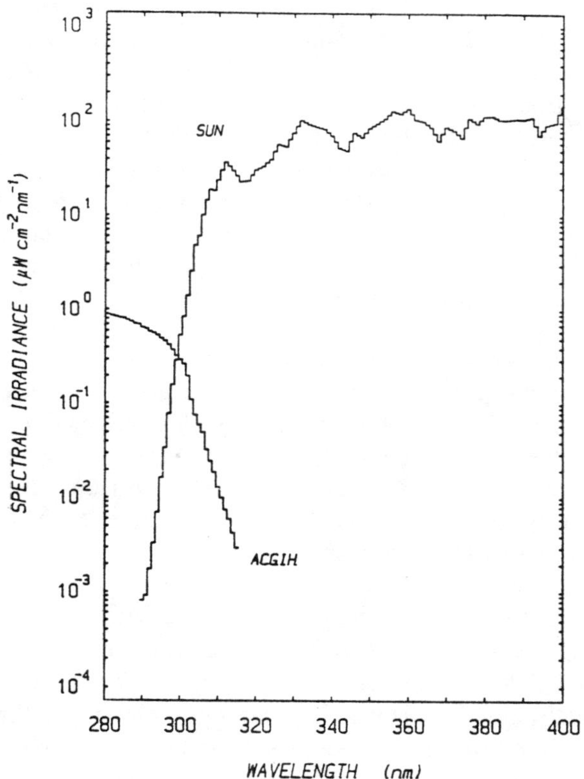

Figure 3. The ACGIH Action Spectrum for UV Hazard Assessment (very similar to erythemal action spectrum from 300 - 320 nm) and the Solar Spectrum. The ACGIH action spectrum is closely fit by some UVR radiometers. However, because of the small overlap of the terrestrial solar spectrum with the action spectrum, problems of stray light need to be dealt with by constant checks with a filter which blocks wavelengths less than 320 nm (adapted from Sliney, et al., 1990).

----

studies of skin cancer. Maps of annual UVR exposure were compiled for epidemiological studies of skin cancer (Scotto, 1980). Accurate spectral measurements of the actinic UV-B radiation in sunlight are very difficult because of the rapidly increasing terrestrial solar spectrum which increases by a factor of more than five orders of magnitude between 290 and 320 nm

(Figure 3). Nevertheless, extensive measurements of ambient UVR in this spectral band have been performed world-wide (Schulze, 1962; Schulze, 1969; Henderson, 1970; Sundararaman, 1975; Garrison, 1978; Hoikkala, 198?; Mercherikunnel, 1980; Doda, 1980; Kostkowski, 1982; Ambach, 1983; Livingston, 1983; Blumthaler, 1983; Blumthaler, 1985a, 1985b; Kolari, 1986; Rosenthal, 1985; Rosenthal, 1987?, Sliney, 1991). In most instances, longer wavelength UVR (UV-A) has also been measured and is available in many of the previously referenced sources. Measurement of terrestrial solar UV-A radiation is not as challenging as measuring UV-B, since there is not a strong variation of the spectrum with zenith angle and the spectral distribution is relatively flat.

Despite the extensive UVR measurements, their interpretation in relation to human exposure has been difficult because of two factors: a) the strongly changing UV-B spectral irradiance with solar position throughout the day, with the season and with latitude, and b) the geometry of exposure of the exposed individual. Unlike visible radiation (light) from the sun which is largely unscattered, the UVR reaching the earth's surface is strongly scattered by atmospheric molecules.

The total solar radiation arriving at the earth's surface is termed *global radiation*, and is comprised of two components, conveniently referred to as the *direct* and *diffuse* components. Measurements of terrestrial UVR are most frequently global radiation, i.e., the radiant energy falling upon a horizontal surface from all directions (both direct and scattered radiation). In addition, reflected UVR from the terrain (i.e., the albedo) is also of interest.

Atmospheric molecular scattering (referred to as Rayleigh scattering), is responsible for the fact that approximately half of UVR at 300 nm is in the diffuse component rather than in the direct rays of the sun. The ratio of diffuse to scattered radiation steadily increases from less than 0.2 at 400 nm to 0.6 at 300 nm (Garrison, 1978).

Essentially all of the measurement programs have been limited to the direct component and total diffuse components of sunlight. However, for studies of hazards to the eye and much of the skin of a standing individual, the actinic UVR incident upon the body from terrain reflectance and horizon sky is of far greater importance. Most environmental measurements of UVR were all of direct and/or global UVR. While such measurements would be of interest in calculating the UVR exposure dose to a prone individual, they are of very limited value in estimating exposure of the eye and shaded skin surfaces. Sliney et al. (1986) and Rosenthal, et al., (1987) reported on measurements of outdoor ambient UVR measurements which emphasized this reflected component to the eye.

The ambient levels of ultraviolet radiation (UVR) incident upon the eye and skin vary dramatically with latitude, time of day (solar zenith angle), and geometry of exposure. The first two factors have been extensively studied for meteorological purposes and to quantify UVR exposures for epidemiological studies of geographical incidence of skin cancer. The last factor--geometry of exposure--has often been neglected in studies of chronic exposure to sunlight and in related epidemiological studies of skin cancer and cataract with the result that erroneous conclusions were derived (Sliney, 1986). The UVR irradiance in the 295-315 nm spectral band has been measured in several geographical locations at several elevations with varying ground reflectance factors to provide a better quantitative estimate of UVR exposure in different climates (Sliney, 1986). This data is of value to better quantify the ambient UVR exposure to exposed vertical skin areas in order to permit more accurate epidemiological studies of the relative causal factor (i.e., the etiological factor) of UVR exposure in the development of skin cancer and changes in the anterior segment of the eye, such as cataract. The studies showed that the reflectance of the terrain plays the greatest role in varying UVR exposure to the eyes and to shaded areas of the skin, and that cloud cover plays a lesser role than most would expect.

The UVR incident upon shaded areas of the skin and the eye from ground reflections and from the sky near the horizon is of primary importance to any epidemiological study of UVR induced corneal changes, cataractogenesis, pterygium, or potential retinal changes. Apart from some relative dose studies by Diffey, et al.(1982) and Rosenthal et al. (1990), this data has been unavailable. Quite obviously, the ocular exposure varies significantly from skin exposure, since photokeratitis is seldom experienced during sunbathing, yet the threshold for UV photokeratitis is less than that for erythema of the skin. Indeed, the variation in ground reflectance would explain some of the apparent contradictions of epidemiological studies of cataract where cataract appeared to be at a lower incidence in high mountain areas with a large global UVR exposure than at sea-level sites where UVR was reduced because of the longer atmospheric path (Sliney, 1986).

UVR Variation with Zenith Angle

Because of the exponential attenuation of the ozone layer, slight changes in the slant path length of the sun's rays reaching earth will result in significant seasonal and daily variations in the terrestrial solar spectrum. Although visible light falling upon the ground may vary by only 20% from noon to 3:00 p.m. (local solar time) in the summer, the UVR spectrum undergoes a dramatic change during this period. At a wavelength of 300 nm the spectral irradiance decreases by 90% [from approximately 1.0 to 0.1 $\mu W/(cm^2$-nm)]! The result is that an untanned sunbather exposed for 23 minutes at noon may develop an erythema (sunburn), but one exposed for two hours from 08:00 to 10:00 would not.

Although we visually perceive some changes in ambient daylight, the preferential Rayleigh scattering of shorter wavelength radiation (i.e., UVR and blue-violet visible wavelengths), we are perceptually unaware of the dramatic variations in our UVR environment. Although most visible sunlight falling on the earth's surface is direct and causes distinct shadows when intercepted by a tree or parasol, this is not true for UVR. The majority of 300-nm UVR incident upon the body is indirect, scattered radiation. Hence, one can actually be sunburned while sitting in the shade of a beach umbrella even with little ground reflection. The principal point to be made here is that the UVR exposure of the eye, the underarms, and underneath the chin is from scattered radiation. Furthermore, direct solar radiation normally does not strike the cornea during the hours of peak sunlight when UVR irradiances are highest, from approximately 9:00 a.m. to 3:00 p.m. local solar time (i.e., 10:00 to 4:00 daylight savings time) (Sliney, 1986).

Measured Ground Reflectance in the UVR

Table 1 lists some typical ground surface reflectance values measured in various locales. These values were measured using a cosine-corrected hemispherical UV-B detector head of IL 730 radiometer (Sliney, et al., 1990). Reflectance is the ratio of "down"/zenith measurement. Compared to the high reflectance of fresh snow (85%), sand, concrete and asphalt reflect only 10/th that of snow, and grass and humus reflect only 1/100th. Hence, the ground reflectance in the UV plays a major, if not dominant, role in determining the ocular exposure to UVR. It is clearly for this reason that photokeratitis from outdoor daylight is so rare, and occurs only when one is out in the snow at midday.

The typical angular variations of E-eff across the sky differ with zenith angle for a sunny day and an overcast day. On a hazy day the irradiance at zenith is greatest, and actually exceeds the irradiance from the sun's position behind the clouds. Furthermore, the irradiance values near the horizon (70° to 90° zenith angles) do not vary dramatically with clouds: the cornea is

TABLE 1. Representative Terrain Reflectance Factors for Horizontal Surfaces Measured with a UV-B UVR Radiometer and Midday Sunlight (290-315 nm).

Material	Percent Reflectance
Lawn Grass, Summer, Maryland, California and Utah	2.0-3.7
Lawn Grass, Winter, Maryland	3.0-5.0
Wild Grasslands, Vail Mountain, Colorado	0.8-1.6
Lawn Grass, Vail, Colorado	1.0-1.6
Flower Garden, Pansies	1.6
Soil, clay/humus	4.0-6.0
Sidewalk, Light Concrete	10-12
Sidewalk, Aged Concrete	7.0-8.2
Asphalt Roadway, Freshly Laid (black)	4.1-5.0
Asphalt Roadway, Two-years Old (grey)	5.0-8.9
Housepaint, White, Metal Oxide	22
Boat Dock, Weathered Wood	6.4
Aluminum, Dull, Weathered	13
Boat Deck, Wood, Urethane Coating	6.6
Boat Deck, White Fiberglass	9.1
Boat Canvas, Weathered, Plasticized	6.1
Chesapeake Bay, Open Water	3.3
Chesapeake Bay, Specular Component of Reflection at Z=45 o	13
Atlantic Ocean, New Jersey Coastline	8.0
Sea Surf, White Foam	25-30
Atlantic Beach Sand, Wet, Barely Submerged	7.1
Atlantic Beach Sand, Dry, Light	15-18
Snow, Fresh (Two Days Old)	88(50)

exposed to an effective irradiance of the order of 0.1 $\mu W/cm^2$ during daylight hours where the sun is at elevation angles above 20° (i.e., $Z = 0°$ to 70°) when an individual is in an open field. However, when hills, trees, or buildings loom in one's field of view, this level is greatly reduced--often by an order of magnitude. When the sky is overcast, the lids open more widely; therefore, a slightly overcast sky can lead to a greater ocular exposure dose to a worker in an open field than when the sun is brightest with no clouds present!

When the sun was "out" (i.e., in an open sky), clouds near the horizon opposite the sun apparently reflected more UVR than would otherwise be

TABLE 2. UVR Exposure Factors Useful in Estimating Ocular Exposure.

---

1.   Using the global UVR measurements for a given location, one can multiply those values by the following exposure values to estimate ocular exposure (Sliney, 1986).

2. Because of the wide variation in atmospheric characteristics and ground reflectances, these factors should be considered approximate. The use of ocular UVR dosimeters should provide more accurate values for future epidemiological studies of neoplasias appearing at shaded areas of the skin and the conjunctiva.

Condition	Exposure Factor	
	No Hat	Brimmed Hat
A. Environmental		
Snowfield	0.8	0.45
Desert (sand only)	0.40	0.16
Desert (sage, cacti, etc)	0.25	0.12
Grasslands	0.16	0.05
Boating	0.25	0.10
Built-Up Urban Areas, No trees	0.20	0.08
Forest	0.001	0.001
B. Use Factors		
Wearing sunglasses		0.01 - 0.1
Exposure limited to ± 2 hours of local noon		0.9
Exposure in morning/afternoon; indoors ± 2 hours of noon		0.1
Always looking away from the sun		0.7
Always looking toward the sun with squint*		0.2

* Very difficult to estimate. Needs further research.

NOTE: To use this table, multiply use factor by environmental factor. This product is then multiplied by global irradiance or exposure for final estimate of orbital irradiance or orbital exposure dose.

present from the blue sky. This is not really surprising when one studies photographs of the sky taken through a narrow-band filter at 320 nm (Livingston, 1980). Such photographs reveal that the sky looks almost uniformly bright even when clouds are present; the clouds disappear into a uniformly hazy sky. When the sun is near the horizon and can be looked at without great discomfort (i.e., at $Z = 75°-90°$), the UVR effective irradiance is again of the order of 0.1 $\mu W/cm^2$, and the greatest UV exposure was obtained by directing the detector to zenith.

Cancer of the cornea is extremely rare, hence UV-B exposure to the eye is of interest. Since ocular exposure to most of the cornea and lens is strongly affected by the degree of lid opening, it is worthwhile to point out that the lids are wide open only when the sky is overcast or when the head is turned away from the sun over a grassy plain. Hence, the values in Table 2 are particularly important. The eye is actually exposed to an irradiance E-eff ranging from 0.03 to 0.5 $\mu W/cm^2$ when the lids are open and the sky is overcast or the sun is not in the field-of-view during midday hours when a hat is not worn. These values agree with the known threshold exposure doses for human photokeratitis when integrated over a daily exposure. Remember that most UV-B exposure occurs in the middle of the day, and the hours near sunrise and sunset do not contribute to the total exposure, unless one faces into the sun above the horizon.

Methods of UVR Measurement

Measurements of both directional and global UVR during daytime hours have been performed in various environments: snowfields, desert, mountain, and grassy terrains. Since laboratory studies show that severe sunburn, and skin carcinogenesis followed exposure to the 295-320 nm wavelength band, most measurement programs have concentrated on this spectral region (UV-B).

A variety of measurement instruments and techniques have been developed to measure solar UVR (Berger, 1976; Rundel, 1983a; deLuisi, 1983; Diffey, 1986; Kolari, 1986; Sliney, 1986; Sliney, 1990). For fixed installations where the objective is to measure the solar spectrum, it is normally essential to employ a double-monochromator to sufficiently reduce stray light in the UV-B spectral region (Garrison, 1978; Sliney, 1980; Kostkowski, 1982), whereas, for field measurements, a portable instrument having a spectral response that simulates the action spectrum of erythema (Berger, 1976) or a safety envelope (Sliney, 1980) has been attractive. Individual dosimeters of polysulfone film have also been employed for dosimetric studies (Challoner *et al.*, 1976; Diffey, 1986; Diffey *et al.*, 1987; Leach *et al.*, 1978; Holman *et al.*, 1983; Larko & Diffey, 1983; Schothorst *et al.*, 1987; Slaper, 1987; Rosenthal *et al.*, 1990).

Portable UVR Survey Instruments

Using simple UVR survey instruments, one can measure the effective irradiance, E-eff, with a cosine-receptor detector with different angular fields-of-view (FOV); e.g., hemispherical, or with a baffle to produce a given (e.g., 40°) cone, and in this regard, the simpler instruments have a distinct advantage over a bulky spectroradiometer.

An International Light Model 730 UV Radiometer, which had a spectral response closely following the ACGIH action spectrum for UVR health hazard evaluations was employed by Sliney (1991) and coworkers to measure UVR "effective irradiance" over different terrains. Figure 5 shows the ACGIH action spectrum and a typical spectral response of this instrument.

Whenever a simple instrument or dosimeter is used to measure UVR, it must be calibrated against the solar spectrum, and a spectroradiometer system with a double monochromator (to obtain low-stray-light spectral irradiance measurements) should be used to obtain the reference spectrum of the sun at different zenith angles. Figure 6 shows the spectral irradiance measured by Sliney (1986) for a zenith angle of 15°. Measurements of near-UV (UV-A) wavelengths between 320 nm and 400 nm normally show little variation of measured values with zenith angle when compared to the 295-320 nm band. Since UV-A is approximately three orders of magnitude less effective in producing most known biological effects of UVR, measurements in this spectral band are of less consequence.

To measure ground reflectance, one should use a detector with a hemispherical response. One first measures global UV by directing the detector toward zenith; (detector has a hemispherical FOV) and then directed downward along a vertical axis (i.e., Z = 180°), then vertically upward toward zenith (i.e., $Z = 0°$). The ratio of the two readings is then recorded as the diffuse reflectance. All measurements should be performed with the cosine-corrected hemispherical UV-B detector head of a radiometer. Reflectance is the ratio of ground "down"/zenith measurement.

Personal UVR Dosimeters

Personal dosimetry for both eye and skin exposure offers the best estimation of actual exposure of the eye and skin to terrestrial (or indoor) UVR exposure, since geometrical factors of exposure, body movements are incorporated into properly worn individual dosimeters. However, calibration of such dosimeters has been a problem, and their best value is to provide *relative* exposures for different parts of the body. Personal dosimeters of polysulfone film have also been developed and used in a number of dosimetric studies (Challoner *et al.*, 1976; Diffey, 1986; Diffey, *et al.*, 1987; Leach *et al.*, 1978; Holman *et al.*, 1983; Larko & Diffey, 1983; Schothorst *et al.*, 1987; Slaper, 1987; Rosenthal *et al.*, 1990).

The problem of accurate calibration results not only from the photochemical process and measurement errors, but in the difficulty of achieving a spectral response that closely mimics the erythema action spectrum. It has been shown that the thinnest sheets of polysulfone film offer the closest match with the erythemal action spectrum or with the ACGIH hazard action spectrum.

Industrial and Laboratory Exposure

The most common accidental overexposure to UVR in the industrial setting occurs from welding arcs. Although sometimes painful, the resulting transient photokeratitis ("welder's flash") and erythema ("sunburn") seldom lasts for more than two days, and seldom, if ever, lead to permanent injury. Photokeratitis seldom lasts more than 24-48 hours after the exposure, and sunburn may last a little longer, depending upon the wavelength. Wavelengths near 310-315 nm will produce the most severe grade of erythema that will last for the longest period (Urbach and Gange 1986).

Extensive measurements of welding arcs (Sliney & Wolbarsht, 1980) make it possible to predict the effective (ACGIH-weighted) UVR irradiance at a given distance from different types of welding arcs used on different metals with different shielding gases and arc currents. The most intense UVR irradiances occur around plasma-arc spraying operations and near argon-shielded tungsten arc welding of aluminum. The filtration factors of welding filters recommended for use with different arcs were shown to be far more than minimally necessary (Sliney & Wolbarsht, 1980).

Accidental exposure seldom occurs in the laboratory setting, but can result from inadequate precautions around any open-arc plasma, UV lamps, excimer lasers, short-arc lamps, and germicidal (quartz-mercury) lamps. In any case, the exposure in both the laboratory and industrial setting is normally a small fraction of the accumulated exposure from sunlight for most workers.

Conclusions

Environmental exposure is by far the greatest lifetime exposure of the skin and the eye to UVR. However, significant acute effects ("welder's flash" and "sunburn" result from failure to wear appropriate protective clothing and eyewear in the arc welding/cutting environment. The use of global UVR measurements to assess outdoor UVR exposure can lead to serious errors. Geometrical factors of human exposure must be taken into account. Many exposure sites will be irradiated differently depending upon sun position and the position of the body (prone or standing). Ground reflectance plays a major role in skin exposure and is of the greatest importance in determining UVR exposure of the eye and shaded skin surfaces (e.g., beneath the chin). Representative grasses and soils are only 1-2 % reflective in the UV-B spectrum, whereas snow is 50 - 85 % reflective in this spectral region. With the sun near zenith, the reflectance from open water is about 3.3 - 8 %, and from sand and concrete, about 10 %. Hence, while grassy terrain reflections account for very little skin and eye exposure, and less exposure to UV-B radiation than the horizon sky, sand and concrete are by far the greatest contributors to ocular exposure. In many settings, tree-lines limit horizon-sky exposure significantly. For example, measurements of UV-B in the Rocky Mountains (Colorado, USA), showed that mountain-sides covered much of an individual's field-of-view and limited exposure. Only in the plains, atop a mountain ridgeline, in desert regions, or out on the sea does one typically have a significant exposure from an horizon sky. Although overhead UVR measurements at high mountain elevations were significantly above those found in sea-level environments, the ocular exposure would often be less due to low ground reflectance. Most people in mountainous regions live in a valley or aside a mountain--not on the peak where the horizon is in the center of view. Residents of non-mountainous arid regions where grass is not present apparently experience the greatest UV-B doses. Thus one is left with the conclusion that local UVR measurements are necessary along with a study of the geometry of exposure for any careful epidemiological study.

The greatest exposures to shaded areas of the skin and to the eye result from ground reflections from snow, followed by sand, concrete pavements, wooden walkways, boat decks, and water, and not from the horizon sky or from green grass or foliage. In fact, the reflected levels of UVR from light sand should be sufficient to cause a threshold photokeratitis within exposure periods of six to eight hours.

The UV-B measurements of reflected UVR permit one to estimate the relative exposure of the most exposed skin surfaces (nose, tops of the ears and forehead) to the lesser exposed areas such as the bottom of the chin. By comparing the incidence of skin cancer in these two areas, one can estimate more accurately the "threshold" radiant exposure for human skin cancer. A further insight could be gleaned from the study of skin cancer incidence of persons spending most of their life outdoors on sand/concrete/asphalt surfaces, vs grassy surfaces.

REFERENCES

1. Ambach, W. & Rehwald, W. (1983) Measurements of the annual variation of the erythema dose of global radiation. <u>Radiat Environ Biophys</u>, <u>21</u>, 295-303.

2. ACGIH (American Conference of Governmental Industrial Hygienists) (1991) 1991-1992 Threshold Limit Values for Chemical Substances and Physical Agents and Biological Exposure Indices, Cincinnati, ACGIH.

3. Berger, D., F. Urbach and R.E. Davies (1967) The action spectrum of erythema induced by ultraviolet radiation, preliminary report, XIII Congressus Internationalis Dermatologiae) Munich, pp. 1112-1117. W. Jadassohn and C.G. Schirren, Eds. Springer-Verlag, New York.

4. Berger, D., (1976) Sunburning ultraviolet meter: design and performance. <u>Photochem. Photobiol.</u>, <u>24</u>, 587-593.

5. Berger, D. & Urbach, F. (1981) A climatology of sun burning UV-radiation. <u>Photochem. Photobiol.</u>, <u>35(2)</u>, 187-192.

6. Blumthaler M., Canaval, H. & Ambach, W., (1983) Messung der erythemwirksamen Dosis der solaren UV-B-Strahlung. <u>Med. Met.</u>, <u>2</u>, No. 2, 14-17.

7. Blumthaler, M., Ambach, W. & Daxecker. F. (1985a) Berechnung von bestrahlungdosen bei Anderung der solaren UV-B-strahlung in hinblick auf eine keratitis photoelectrica. <u>Klin. Mbl. Augenheilk</u>, <u>186</u>, 275-278.

8. Blumthaler, M., Ambach, W. & Canaval, H. (1985b) Seasonal variation

of solar UV-radiation at a high mountain station. Photochem. Photobiol., 42, No. 2, 147-152.

9. Challoner, A.V.J., Corless, D., David, A., Deane, G.H.W., Diffey, B.L., Gupta, S.P. & Magnus, I.A. (1976) Personnel monitoring of exposure to ultraviolet radiation. Clin. Int. Dermatol., 1, 175-179.

10. Coblentz, W.R., R. Stair and J.M. Hogue (1931) The spectral erythemic relation of the skin to ultraviolet radiation. Proc. Nat. Acad. Sci. U.S. 17:401-403.

11. Commission Internationale de L'Eclairage (International Commission on Illumination) (1988) International Lighting Vocabulary, 3rd ed. Pub. CIE No. 17 (E-1.1). Vienna.

12. Cunningham-Dunlop, S., B.H. Kleinstein and F. Urbach: A Current Literature Report of the Carcinogenic Properties of Ionizing and Non)Ionizing Radiation: I. Optical Radiation. Contract 210-76-0145, Nat. Inst. for Occup. Safety and Health. DHEW Pub. No. (NIOSH) 78-122. Cincinnati, Ohio (December 1977).

13. Deluisa, J. & Harris, M. (1983) A determination of the absolute radiant energy of a Robertson-Berger meter sunburn unit. Atmospheric Environment, 17, 751-758.

14. Diffey, B. (1987) A comparison of dosimeters used for solar ultraviolet radiometry. Photochem. Photobiol.,46, 55-60.

15. Diffey, B.L. (1987) Analysis of the risk of skin cancer from sunlight and solaria in subjects living in northern Europe. Photodermatology, 4, 118-126.

16. Diffey, B. L., Kerwin, M. & Davis, A. (1977) The anatomical distribution of sunlight. Br. J. Dermatol., 97, 407-410.

Diffey, B. L. (1979) The calculation of the spectral distribution of natural ultraviolet radiation under clear day conditions. Phys. Med. Biol., 22, 309-316.

17. Doda, D. & Green A. (1980) Surface reflectance measurements in the UV from an airborne platform. Appl. Optics, Part1., 19, No. 13, 2140-2145.

18. Everett, M.A., R.L. Olsen and R.M. Sayer (1965) Ultraviolet Erythema. Arch. Dermatol. 92:713-719.

19. Fitzpatrick et al., (1974) Eds. Sunlight and Man, University of Tokyo Press, Tokyo, Japan.

20. Foukal, P. & Lean, J. (1990) An empirical model of total solar irradiance variations between 1874 and 1988. Science, 247, 556-558.

21. Freeman, R.G., D.W. Owens, J.M. Knox and H.T. Hudson (1966) Relative energy requirements for an erythemal response of skin to monochromatic wavelengths of ultraviolet present in the solar spectrum. J. Invest. Dermatol. 47:586-592.

22. Frederick, J.E. (1990) Trends in atmospheric ozone and ultraviolet radiation: mechanisms and observations for the northern hemisphere, Photochem. Photobiol., 51(6): 757-763.

23. Gardiner, B., Kirsch, P. (1991) European Intercomparison of Ultraviolet Spectrometers, Panorama, Greece, 3-12 July 1991, Report to the Commission of the European Communities, STEP Project 76, Science and Technology for Environmental Protection, CEE.

24. Garrison, L., Murray, L., Doda, D. & Green, A. (1978) Diffuse--direct ultraviolet ratios with a compact double monochromator. Appl Opt., 17(5), 827-836.

25. Giese, A. (1976) Living with Our Sun,s Ultraviolet Rays, New York, Plenum Press.

26. Green, A, Sawada, T, Shettle, E. (1974) The middle ultraviolet reaching the ground, Photochem. Photobiol. 19, 251-159.

27. Hausser, K.W. (1928) Influence of wavelength in radiation biology. Strahlentherapie 28:25-44.

28. Henderson, S. (1970) Daylight and Its Spectrum, New York, American Elsevier Publishing Co.

29. Hietanen, M. (1990) Spectroradiometric and Photometric Studies on Ocular Exposure to Ultraviolet and Visible Radiation from Light Sources, Ph.D. Thesis, Department of Physics, University of Helsinki.

30. Holman, D.D.J., Gibson, T.M., Stephenson, M. & Armstrong, B.K. (1983) Ultraviolet irradiation of human body sites in relation to occupation and outdoor activity: field study using personal UVR dosimeters. Clin. Exp. Dermatol., 8, 279-285.

31. INIRC (International Non-Ionizing Radiation Committee) (1985) Guidelines for exposure to ultraviolet radiation, Health Phys.

32. Klein, W. & Goldberg, B. (1974) Solar Radiation Measurements / 1968-1973, Washington, D.C., Smithsonian Institute.

33. Kolari, P., Hoikkala, M. & Lauharanta, J. (1986a) Assessment of the dose response of polysulphone film badges for the measurement of UV-radiation. Photodermatology, 3, 228-232.

34. Kolari, P., Lauharanta, J. & Hoikkala, M. (1986b) Midsummer solar UV-radiation in Finland compared with the UV-radiation from phototherapeutic devices measured by different techniques. Photodermatology, 3, 340-345.

35. Kostkowski, H., Saunders, R., Ward, J., Popenoe, C. & Green, A. (1982) Measurements of Solar Terrestrial Spectral Irradiance in the Ozone Cut-Off Region, Self-Study Manual on Optical Radiation Measurements, Part III, - Applications Chapter 1, NBS Technical Note 910-5, National Bureau of Standards, Washington, DC.

36. Larko, O. & Diffey, B.L. (1983) Natural UV-B radiation received by

people with outdoor, indoor and mixed occupations and UV-B treatment of psoriasis. <u>Clin. Exp. Derm.</u>, <u>8</u>, 279-285.

37. Leach, J.F., McLeod, V.E., Pingstone, A.R., Davis A., & Deane, G.H. W. (1978) Measurement of the ultraviolet doses received by office workers. <u>Clin. Exp. Dermatol.</u>, <u>3</u>, 77-79.

38. McCulloch, E. (1970) Qualitative and quantitative features of the clear day terrestrial solar ultraviolet radiation environment, <u>Phys. Med. Biol.</u>, <u>15</u>,723-734.

39. Lindley, D (1988) CFCs cause part of the global ozone decline, <u>Nature (Lond)</u>, <u>323</u>, 293.

40. Livingston, W. (1983) The landscape as viewed in the 320 nm ultraviolet. <u>J Opt Soc Am</u>, <u>73(12)</u>, 1653-1658.

41. Luckiesh, M.L., L. Holladay and A.H. Taylor (1930) Reaction of Untanned Human Skin to Ultraviolet Radiation. <u>J. Opt. Soc. Am.</u> 20:423-432.

42. Mercherikunnel, A. & Richmond, J. (1980) <u>Spectral Distribution of Solar Radiation</u>. <u>NASA Technical Memorandum 82021</u>, Greenbelt, Goddard Space Flight Center.

43. Parrish, J., Anderson, R., Urbach, F. & Pitts, D. (1978) <u>UV-A, Biological Effects of Ultraviolet Radiation with Emphasis on Human Responses to Longwave Ultraviolet</u>, New York, Plenum Press.

44. Pitts, D.G. (1974) The human ultraviolet action spectrum. <u>Am. J. Optom. Physiol. Optics</u>, 51(12):946-960.

45. Pitts, D.G. and T.J. Tredici (1971) The effects of ultraviolet on the eye. <u>Am. Ind. Hyg. Assoc. J.</u> 32(4):235-246.

46. Pitts, D.G., A.P. Cullen and P.D. Hacker (1977) Ocular effects of ultraviolet radiation from 295 to 365 nm, <u>Invest. Ophthal. Vis. Sci.</u>, 16(10):932-939.

47. Rosenthal, F. S., Lew, R.A., Rouleau, L.J. & Thomson, M. (1990) Ultraviolet exposure to children from sunlight: a study using personal dosimetry. <u>Photodermatol. Photoimmunol. Photomed.</u>, <u>7</u>, 77-81.

48. Rosenthal, F., Safran, M. & Taylor, H. (1985) The ocular dose of ultraviolet radiation from sunlight exposure. <u>Photochem. Photobiol.</u>, <u>42</u>, 163-171.

49. Rosenthal, F., Phoon, C., Bakalian, A. & Taylor, H. (1988) The ocular dose of ultraviolet radiation to outdoor workers. <u>Invest Ophthalmol Sci.</u>, <u>29</u>, 649-656.

50. Rundel, R. (1983) Promotional effects of ultraviolet radiation on human basal and squamous cell carcinoma. <u>Photochem. Photobiol.</u>, <u>38</u>, No. 5, 569-575.

51. Rundel, R. (1983) Action spectra and estimation of biologically effective UV radiation. <u>Physiol. Plant</u>, <u>58</u>, 360-366.

52. Rundel, R. & Nachtwey, D. (1983) Projections of increased non-melanoma skin cancer incidence due to ozone depletion. <u>Photochemistry and Photobiology</u>, <u>38</u>, No. 5, 557-591.

53. Schippnick, P.F. & Green, A.E.S. (1982) Analytical characterization of spectral actinic flux and spectral irradiance in the middle ultraviolet. <u>Photochem. Photobiol.</u>, **35** 89-101.

54. Schothorst, A.A., Slaper, H., Schouten, R. & Suurmond, D. (1985) UVB doses in maintenance psoriasis phototherapy versus solar UVB exposure. <u>Photodermatology</u>, <u>21</u>, 213-220.

55. Schotohorst, A., Slaper, H., Telgt. D., Alhadi, B. & Suurmond, D. (1987) Amounts of ultraviolet B (UVB) received from sunlight or artificial UV sources by various population groups in the Netherlands. In: Passchier, W.F. & Bosnjakovic, B.F.M, eds, <u>Human Exposure to Ultraviolet Radiation: Risks and Regulations</u>, Amsterdam, Excerpta Medica, pp. 269-273.

56. Schulze, R. (1962) Zum strahlungsklima der erde (On the radiation climate of the earth). <u>Strahlentherapie</u>, <u>119</u>, 321-348.

57. Schulze, R. & Grafe, K. (1969) Consideration of sky ultraviolet radiation in the measurement of solar ultraviolet radiation. In: Urbach, F., ed, <u>The Biological Effects of Ultraviolet Radiation (Proceedings of the First International Conference)</u>, Oxford, Pergamon Press, pp. 359-373.

58. Scotto, J., Fears, T. & Gori, G. (1980) <u>Measurements of Ultraviolet Radiations in the United States and Comparisons with Skin Cancer Data</u>, U. S. Department of Health, Education and Welfare Publication No. (NIH)80-2154.

59. Slaper, H. (1987) <u>Skin Cancer and UV Exposure: Investigations on the Estimation of Risks</u>, PhD Thesis, Utrecht, pp. 40-42.

60. Sliney, D.H. (1972) The merits of an envelope action spectrum for ultraviolet radiation exposure criteria. <u>Am. Ind. Hyg. Assoc. J.</u>, 33:644-653.

61. Sliney, D.H. (1983) Eye protective techniques for bright light. <u>Ophthalmology</u>, <u>90(8)</u>, 937-944.

62. Sliney, D.H. (1986) Physical Factors in Cataractogenesis: Ambient Ultraviolet Radiation and Temperature. <u>Investigative Ophthalmology & Visual Science</u>, <u>27(5)</u>, 781-790.

63. Sliney, D.H. (1987) Unintentional exposure to ultraviolet radiation: risk reduction and exposure limits. In: Passchier, W.F. & Bosnjakovic, B.F.M, eds, <u>Human Exposure to Ultraviolet Radiation: Risks and Regulations</u>, Amsterdam, Excerpta Medica, pp. 425-437.

64. Sliney, D.H. & Wolbarsht, M.L. (1980) <u>Safety with Lasers and Other Optical Sources</u>, New York, Plenum Publishing Corp.

65. Sliney, D.H., Wood R.L., Jr., Moscato, P.M., Marshall, W. J., & Eriksen, P. (1990) Ultraviolet exposure in the outdoor environment: measurement of ambient ultraviolet exposure levels at large zenith angles, pp. 169-180, in (Grandolfo, M,, Rindi, S., Sliney, D. Eds.), <u>Light, Lasers, and Synchrotron Radiation</u>, New York, Plenum Press.

66. Sundararaman. N., St. John, D. & Venkateswaran, S. (1975) <u>Solar Ultraviolet Radiation Received at the Earth's Surface Under Clear and Cloudless Conditions</u>, Springfield, Va., National Technical Information Service.

67. Taylor, H.R. (1989) Ultraviolet radiation and the eye: an epidemiologic study, <u>Tr. Am. Ophthal. Soc.</u>, 87:802-853.

68. Urbach, F. & Gange, R. (1986) <u>The Biological Effects of UV-A Radiation</u>, New York, Praeger Publishers.

69. Urbach, F., Berger, D., Robertson, D. & Davies, R. (1978) Field measurements of biologically effective UV radiation and its relation to skin cancer in man.

70. Willis, I., A. Kligman and J. Epstein (1972) Effects of long ultraviolet rays on human skin: photoprotective or photoaugmentative. <u>J. Invest. Dermatol</u>. 59:416-420.

71. Zuclich, J.A. and J.S. Connolly (1976) Ocular damage induced by near ultraviolet laser radiation. <u>Invest. Ophthal</u>. 15(9):760-764.

LASER : CHARACTERISTICS AND EMISSIONS

L. COURT

D. COURANT

Centre de Recherches
du Service de Santé des Armées
et
Commissariat à l'Energie Atomique
Direction des Sciences du Vivant
Grenoble - Fontenay aux Roses

The 1960's saw the first developements of LASER, an acronym for *LIGHT AMPLIFICATION BY STIMULATED EMISSION OF RADIATION*, research in different fields of applications, and wide and rapid diffusion of lasers in all aspects of human activity. The laser technology appears in industry, in research laboratory, lasers displays, alignment procedures, consumer devices, scientific applications, in modern communication, in microelectronics fabrication, in medicine and surgery. With laser technology it is possible to measure, align, cut, join, heat-treat, mark, scribe, drill holes, shape, polish and burnish and produce modifications to chemical processes, with great precision, minimal waste of materials, a very small-heat affected zone, a minimal noise often without any real contact. Laser penetration in medicine, surgery and chemotherapy has advanced rapidly in the past ten years with the most commonly used lasers as Carbon Dioxide, Nd-Yag with contact and non contact, visible (Argon, He-Ne, Krypton, Ruby ...), UV excimer lasers. The industries considered to be most benefitted by high power laser developments are, automotive, ship-building, offshore, aerospace, nuclear and chemical processing. Simultaneously the rate of development of military applications has been phenomenal. Laser technology has the potential of influencing and generating whole new applications in both consumer and service industries. Unfortunatly the improper use or design of any laser system can produce undesirable effects and it is important to distinguish:

- direct exposure to laser beam
- indirect exposure to reflected specular and diffuse radiation
- associated hazards such as high voltage electricity, toxic fumes,

chemical risks, compressed gazes, mechanical traps and fire.

The specific properties of laser beams explain their special and significant hazards. The majority of low power lasers types are without significant risks, but many industrial applications, in the field of medicine, communications, military and industrial applications are very hazardous. Experimental investigations of laser effects upon all biological systems or tissues and specially the eye and the skin, studies of interaction mechanism, laser beam-biological tissues are important. The key to protections is the setting of standards for safe laser exposure of the human body, by the development of protection guidelines and simultaneously the production of effective laser protective devices.

I.   COMPONENTS OF LASERS

The laser system has three basic components:

1. A Lasing Medium

To have a suitable energy levels, it must have at least one excited state (metastable state) where electrons can be trapped and not immediately and spontaneously dropped to lower state. When the medium is exposed to the appropriate pumping energy, the excited electrons are trapped in the metastable states long enough for a population inversion to occur: there are more electrons in the excited state than in the lower state to which these electrons decay when stimulated emission occurs. Generally most laser action involves four or more energy levels, but such action is possible with only two energy levels.

2. A Pumping System (ie an energy source)

These systems pump energy into the laser material and increase the number of electrons trapped in the metastable energy level. The available pumping systems are optical pumping, electron collision pumping and chemical reaction.

> ● optical pumping uses a strong source of light such as a xenon flashtube or another laser ruby laser, or more generally solid state and liquid lasers

• an electric current through the laser material or accelerating electrons from an electron gun to impact on the laser material (He-Ne laser) are electron collision pumping (semiconductor laser)

• the energy released in the walking and breaking of chemical bonds (H Fluoride lasers) is a chemical pumping (liquid and gas lasers)

3. A resonant optical cavity, formed by placing a mirror at each end of the laser material so that a beam of light may be reflected from on mirror to the other, passing back and fourth through the laser material. The beam passes through the material many times and is amplified each time. One of the mirrors is only partially reflecting and permits part of the beam to be transmitted out of the cavity.

II   LASER OPERATION: STIMULATED EMISSION

Four stages appear:

1. the pumping system supply the energy to the laser material

2. it produce a population inversion

3. the transition from the metastable energy level to the lower energy levels is emitting photons of precisely the same wavelength, phase and direction. With the population inversion the spontaneous decay of a few electrons from the metastable to a lower energy level starts a chain reaction.

4. the chain reaction continues with the reflection of photons by the end-mirror back into the material laser. A portion of photons will be reflected back into the cavity and the rest will emerge as a laser beam.

Solid state lasers employ glass or crystalline material, gas lasers, pure gas or mixture of gases, semiconductor lasers, transistor materials, liquid lasers, active material in a liquid solution or suspension.

III   MODE OF OPERATION

The rate at which energy is delivered distingish the different modes of lasers.

1. Some lasers are able to operate continuously. This mode of operation is called CONTINUOUS WAVE (CW): the peak power is equal to the average power output and the irradiance is constant with time.

2. Many lasers are operating in the PULSE MODE.

2.1. The long pulse mode or just pulsed mode are the mode for lasers with a pulse duration of a few ten microsecond to a few millisecond.

2.2 When a shutter between the mirrors turns on and off the beam rapidly, the mode of operation is normally called Q SWITCHED MODE. The duration of pulse (Q SPOILED or GIANT PULSE) is between a few tens of nanoseconds to a few microseconds. The energy is delivered in a much shorter time period and the Q switched lasers are able to deliver very high peak powers of several megawatts or even gigawatts.

2.3 With the synchronisation of the phases of different frequency modes or in "locked together" phases synchronisation ie in mode LOCKED operation, the different modes will interfere with one another to generate a beat effect. The laser output is observed as regularly spaced pulsations with trains of pulses each having a duration of a few picoseconds to a few nanoseconds.

The PULSE REPETITION FREQUENCY (PRF) is the number of pulses which a laser produces in a given time interval.

3. The Q factor for a particular mode of the laser cavity is given by an expression, ratio of the energy stored in the mode and the energy dissipated in mode per unit of time.

In a laser, oscillation occurs as soon as the stimulated emission gain is enough high to overcome the various losses suffered by the electromagnetic wave in the cavity. With a large population inversion, if the Q of the cavity is suddenly increased by an instantaneous and drastic reduction in these cavity losses, the gain will substantially exceed the losses and all the energy accumulated and stored will be discharged in one giant pulse. There are numerous Q switching techniques and these methods may involve mechanical (rotation of one of the cavity mirrors), electro-optical, magneto-optical, or acoustical effects or a saturation of atomic levels in materials introduced into the laser cavity (properties of materials called SATURABLE ADSORBERS which display a change in opacity directly related to the incident radiation intensity such as cryptocyanine in acetone for ruby).

4. CAVITY DUMPING: An intra cavity modulation is possible for laser systems in which the atomic system possesses a relatively long lifetime (YAG) as well as some short lifetime systems (Ar, He Ne). An acousto optic modulator is used, in a folded cavity created by three high reflectivity mirrors ; it is inserted at a point in the cavity where the beam attains its smallest spot. The laser is continuously pumped abovethreshold and the output compling, periodically varied with time, by activating the modulator which causes the acoustic beam to scatter the light beam passing through it. The system can provide repetition rates varying from 1 Hz to several MHz in variable pulses between 10^{-9} s and CW.

IV   MODE CONFIGURATIONS

In the laser cavity, the electromagnetic field configurations can be separate into transverse and longitudinal modes. In the longitudinal mode, the optical resonator is formed by placing mirrors (either plane or curved) perpendicularly to the axis along which the laser light will propagate. In the transverse modes, laser resonators are derived by using a scalar formation of HUYGEN'S PRINCIPE; the field at one mirror is assumed to be caused by the field distribution at the other mirror.

A wave having an electric field component given will also have a transverse magnetique field which is related to it through Maxwell's equations; the modes are designated by notation TEM $_{mn}$ TRANSVERSE ELECTROMAGNETIC WAVE. The beam geometries appear to have wave patterns across the beam which are identified by TEM_{mn} mode numbers. A laser operation in the TEM_{mn} mode can be considered as two lasers side by side, each with one-half the total power.

V   PHYSICAL PROPERTIES OF LASER RADIATION

The most important properties of laser light are:

- high radiance
- directionaly and very little divergence
- coherence manifested in both monochromaticity (temporal coherence) and uniphase wave front (spatial coherence)

1. HIGH RADIANCE AND BRIGHTNESS

The emission of laser is typically at very high levels of spectral radiance (SPECTRAL BRIGHTNESS). The sun emits about 7,000 W/ (cm^2.sr.μm) at its surface; lasers are capable of producing more than 10^{12} W (cm^2.sr.μm). Many small lasers exceeds the sun's radiance. Brightness is defined as the power per unit area per unit solid angle and lasers are the brightest radiation sources compared to conventional light sources (arc lamps, welding plasma etc....).

With an exposure duration of O.15 s, a 1 mW visible laser can come very close to producing a retinal injury.

2. DIRECTIONALITY AND DIVERGENCE

An obvious distinction between a laser and a conventional light source is that lasers emit radiation in a relatively narrow, often almost parallel beam. A laser operating in a single transverse mode can produce a beam whose divergence is limited only by the fundamental process of diffraction. The divergence, sometimes referred to as beam spread is the increase in the diameter of the laser beam with distance from the exit aperture. The divergence is taken as the full angle, expressed in radians, of the beam diameter measured between those points which include laser energy equal to 1/e of the maximum value. A Ne-He laser typically has a divergence of 0.5 to 1.5 milliradians. Divergence practically is the length of the arc subtended by a beam at great distance (a 1m diameter beam at one kilometer is 1 milliradian).

3. COHERENCE

The coherence is a very important property of LASER; coherence is manifested in both monochromaticity (TEMPORAL COHERENCE) and uniphase wavefront (SPATIAL COHERENCE).

3.1. TEMPORAL COHERENCE

The temporal coherence is the property of source emitting a wave train which retains a fixed phase over given coherence time; the temporal coherence is directly related to the monochromaticity, ie the bandwith of the radiation. The coherence length is the distance light travels during the coherence time with the speed of light, c. In other words, the coherence length is the maximum path length difference under which fringes are still observed; it is great for lasers, small for conventional light sources.

- the greater the monochromaticity of the beam i.e. the smaller the frequency spread (line width) is corresponding to the larger the coherence time of light beam

- lasers operating in a single longitudinal mode have extremely narrow line widths.

3.2   SPATIAL COHERENCE

An extended source has spatial coherency if two separate points can emit waves which retain a fixed phase relationship over a given time.   In conventional source, light is emitted randomly and the total energy is, on the average, radiated in all directions.

In contrast, the coherent laser light is generated over a sizable volume with the proper phase and focused by a lens, all the individual contributions by atoms in the lasing medium are in the correct phase and are adding up. The energy of the laser can be concentrated into a very small spot; the nearly plane wavefront from laser results in very high directionality better than from conventional light sources.

Spatial coherence determines the maximum enery of power per unit area which can be attained under optimum focussing conditions.

3.3.   LASER BEAM

SOURCES WITH POOR MODE STRUCTURE

The beams of such sources are usually described as collimated beams of diameter D_B, typically in the millimeter range and a divergence ϕ typically in the mrad range.

The ratio of the irradiance in the RETINA compared to that of the CORNEA is:

$$E_r = \left(\frac{D_B}{\varphi \cdot f_L}\right)^2 \cdot E_C = 10^5\, E_C$$

where $\quad D_B = 7$ mm

$\phi = 10^{-3}$ rad

f_L = focal length of eye, 25 mm

SOURCES WITH FULL COHERENCE GAUSSIAN (TEM $_{00}$) BEAM

The irradiance of a laser beam measured radially across the center is given by :

$$E_r = E_o \cdot e^{-k} \qquad \text{with} \quad k = \{2r/W\}^2$$

where E_r represents local irradiance at any point from the center of the beam E_o, at the center of beam irradiance. The beam diameter 2 W is defined by that radius r at which the irradiance has fallen to $E_w = E_o/e^2$.

4 WAVELENGTH

Laser emits electromagnetic radiation having wavelengths in the "OPTICAL REGION" (100 nm to 1 mm).

A laser beam is monochromatic but can emit sometimes radiation with several wavelengths. Typically, He-Ne laser emits red light at 632.8 nm and IRA at 1150 nm and IRB at 3390 nm. The spectral band are defined by the INTERNATIONAL COMMISSION ON ILLUMINATION (CIE).

These limits are useful in discussing the biological effects of optical radiation:

- UVR (ULTRA VIOLET RADIATION)
  between 100 nm and 400 nm subdivided into:
  UVC from 100 nm to 280 nm
  UVB from 280 nm to 315 nm
  UVC from 315 nm to 400 nm

- LIGHT OR VISIBLE RADIATION
  between 400 nm and 760 nm

- IR (INFRA-RED RADIATION)
  subdivided into:
      IRA from 760 nm to 1400 nm
      IRB from 1400 nm to 3000 nm
      IRC from 3000 nm to 1 mm (10^6 nm)

VI   <u>QUANTITIES AND UNITS</u>

The concepts of dose rate and dose are used. The optical radiation are expressed using the following quantities :

- SOURCE

The output of the light source is quantified in terms of ENERGY

RADIANT ENERGY	RADIANT POWER
Q	O(W)

with units

J joule or C calorie for pulsed laser	W WATT for continuous laser

- THE IRRADIATED AREA

The concentration of energy or power incident on irradiated surface is:

RADIANT EXPOSURE	IRRADIANCE
H	E

with units

$J\ m^{-2}$	$W\ m^{-2}$
$J\ cm^{-2}$	$W\ cm^{-2}$
or cal m^{-2}	(or mW, μW)
cm^{-2}	
(or mJ, μJ)	
(millijoule, microjoule)	(milliwatt, microwatt)

297

From extended source

	THE INTEGRATED RADIANCE Lp	RADIANCE L

with units

J/ (m^2.sr)	W/ (m^2.sr)
J/ (cm^2.sr)	W/(cm^2.sr)
or (mJ, μJ)	or (mW, μW)

VII   NATURE OF ILLUMINATION

The notion of direct, reflected, diffuse illumination must be introduced. The effects differ for point-source (direct viewing of collimated beam) and extended source or diffuse reflexions of laser.

The exposure limits, ELs, for extended source apply to sources which subtend a visual angle measured at the eye greater than "alpha minimum" (ALPHA min). The angular substence of a source (ie the apparent visual angle) is the angle at the eye which is subtended by the source and should not be confused with the beam divergence. Sources such as laser arrays, diodes and diffuse reflecting surfaces are considered extended sources of their angular subtense is equal to or greater than ALPHA min. All others lasers, ie collimated lasers which produce a small (nearly diffraction-limited) retinal image and point sources are intrabeam viewing cases (angular substense is less than ALPHA min). Any sources whose centers are separated by an angle less than ALPHA min are treated as extended source.

VIII   LIMITS APERTURE

The notion of aperture is function of the nature of irradiated area and the interaction mechanisms. Radiant exposure or irradiance may be averaged over a circular aperture of 1 mm diameter with a receptor having a cosine response except for comparison with the ELs for the eye in the retinal hazard spectral range between 400 and 1400 nm, where measurements should be made with detector having a 7 mm limiting aperture (pupil) with a receptor field of view of ALPHA min and except for comparison with all ELs for vawelengths between O.1-1 mm where the limiting aperture is 11 mm with a receptor having a cosine response.

WAVELENGTH in nanometers	SKIN	EYE in mm
400 - 760		7 mm RETINA
760 - 10000	1 mm	1 mm CORNEA
> 10OOO	11 mm	11 mm CORNEA

IX   TIME

The distribution of energy over time is an important paremeter and several quantities are to be precised :

- the duration of exposure pulsed or continuous illumination. All laser are CW of the duration of exposure, or pulse is greater than 250 ms.

- for pulsed illumination (single or multiple pulses), the pulse widht, the pulse repetition frequency, PRF, the pulse repetition period $T = 1/PRF$ or the duty cycle t/T.

- the time between two exposure for prolonged and repetitive exposure with the notion of total exposure time (chronical or long term exposure).

X   BIOLOGICAL AND HEALTH EFFECTS OF LASER RADIATION

1. OPTICAL PROPERTIES OF TISSUES

Laser radiation may be either absorbed, scattered or reflected from biological tissues.

1.1. UV RADIATION

In UVR, the absorbed energy may give use to photochemical reactions, but the depth of penetration of UVR into the human body is very restrictive. Penetration is somewhat greater at longer wavelengths and some penetration can take place into the layer below the epidermis at wavelength above 300 nm. Protein molecules are highly absorbing in the UV-C and dominate all absorption up to wavelengths of approximatively 300 nm.

Concerning the eye, at wavelengths less than 290 nm the cornea will be able to absorb incident UVR completely. Lens and tissues of the anterior part of the eye may be exposed to UVR if the wavelengths are greater than 290 nm.

1.2. VISIBLE AND IR RADIATION

Light and IR radiation of wavelengths less than 1400 nm penetrate skin and ocular tissues since water is relatively transparent at shorter wavelengths. The absorption by melanin (principle absorber at wavelengths less than 1000 nm) can lead to thermal injury of the iris and the retina in the eye and of the epidermis. Water is highly absorbing for wavelengths greater than 1400 nm; so, IR penetration depths are superficial.

1.3 INJURY MECHANISMS

Several mechanisms of destruction of biological tissues by laser beam exist: thermal, coustic and photochemical thermo process. The long duration exposure >1 ms cause destruction of tissue by thermal means (denaturation of proteins). Photo chemical phenomen may damage with from lengthy exposure ocular and skin tissues. For short pulses, equal to 10^{-9} (1 ns), non linear mechanisms appear (with Raman and Brillouin scattering, electron-avalanche breakdown, multi-photon absorption, ultra sonic resonance and acoustic shock waves).

1.4. SKIN EFFECTS

The skin effects of laser beam are considered to be less dramatic than those to the eye; eye damage is generally irreparable and irreversible. The exposure of the skin to high levels of optical radiation can cause depigmentation, burns and possible damage to underlying organs. The effects of UV lasers upon the skin are the same as for conventional UV sources = erythema from acute exposure and accelerated skin ageing and skin cancer from chronic exposure.

1.5 OCULAR EFFECTS

The effects differ with the character of the source and are different for direct ocular exposure (intrabeam viewing) and for viewing diffuse reflection of a laser beam or an extended source laser. The intraocular energy is limited by the diameter of pupil and the appearance of blink reflex after exposure (between 100 and 180 ms). The eye is generally the most critical organ in terms of vulnerability to laser radiation: for any given exposure in the visible region the retinal irradiance is always higher than that of the cornea or skin. The corneal, lens and eye media are transparent in the visible region.

IR-R greater than 1,400 nm can cause thermal injury to the cornea and conjonctiva. In the UV-C and UV-B regions, photokeratitis may be produced. In UV-A region photo keratitis may be produced only by chronic, high lever exposure. UV induced cataract production is an real problem.

A retinal injury which occurs in the central retina (macula) the most sensitive area of the retina is serious; injury to the paramacular or peripheral retinal region may have only a minimal effect upon vision.

XI   <u>SAFETY STANDARD AND CLASSIFICATION OF LASER DEVICE HAZARDS</u>

The INTERNATIONAL RADIATION PROTECTION ASSOCIATION with a working group on non-ionizing radiation, INTERNATIONAL NON IONIZING RADIATION COMMITEE (INIRC) examine the general concept of protection against adverse health effects of exposure to laser radiations. The major contribution of ANSI has been to define Maximum Permissible Exposure, MPE's, for the full range of laser parameters and to propose an hazard classification schema for lasers. The various classes are given in tables and are expressed in terms of an Accessible Emission Limit, AEL, ie the maximum amount of radiation to which a person could be exposed from a laser in this class.

1. DEFINTITION OF THRESHOLD

From dose response data obtained for different exposure conditions a probit analysis is performed. The EDn is the effective dose required to produce an observable response on n per cent of population confidence intervals about the dose response curve are calculated. From the curve data, the ED 50 is generally refered to as the damage threshold.

2. MAXIMUM PERMISSIBLE EXPOSURE (MPE)

From biological data and after statistical analysis the MPE for humans are derived. The MPE is obtained by dividing the ED 50 for the worst case condition using more sensitive significant criteria by a variable factor.

3. CLASSIFICATION OF LASER DEVICE HAZARDS
The hazard classification are defined by the output parameters and accessible levels of radiation.
The classes are :

1. laser system that are non hazardous (non-risk) ie any laser or laser system containing such a laser, that cannot emit laser radiation levels in excess of the AEL for class 1 for the classification duration (the longest daily exposure duration expected).

2. laser system (visible only) that are normally not hazardous by virtue of normal aversion responses (low risk).

3. laser systems where intrabeam viewing of the direct beam and specular reflections may be hazardous (moderate risk) sometimes divided into two subcategories a and b, where class 3a represents a low risk (equivalent to class 2) and is hazardous only if the beam if recollected by an optical instrument.

4. All lasers systems where even diffuse reflections may be hazardous or where the beam produces serious hazards (eye damage and skin hazard).

XII   <u>CONCLUDING REMARKS AND LASER HAZARDS RESEARCH</u>

The IRPA/INIRC, the WHO (WORLD HEALTH ORGANIZATION), the CEE and the LWO (LABOUR WORLD ORGANIZATION), the IEC (INTERNATIONAL ELECTROTECHNICAL COMMISSION) have published standards and guidelines on electrical, non-radiation and radiations hazards of laser (safety of laser products, equipment classification, requirements and users guide). Unfortunatly all biological risks of lasers are not known. However many military and industrial systems developers and generally all biomedical laser developers criticize currently laser safety standards for conservative and complex aspects. To review this position and to discuss this database is not always necessary, but effectively our knowledge must be precise and increased in several ways:

1. TEMPORAL DEPENDANCE OF RETINAL THRESHOLD
The states of subnanosecond laser retinal injury thresholds with precise

photometry are rare. The problem is to establish if the threshold for retinal injury in the picosecond regime is dependant upon peak irradiance rather than radiant exposure in other words if the temporal dependance exists. The fundamental problem is to know the nature of interaction mechanism with biological tissues for very high and very fast light pulses.

2. REPETITIVE PULSED EXPOSURE LIMITS

For international exposure to a point source train of laser pulses the $N^{-1/4}$ rule is generally in according to the theorical model and experimental data; but for near infrared sources with durations more long than O,5 s (gallium-arsenide, gallium-arsenide-phosphide and neodyme YAG laser) the present occupational exposure limits are not acceptable. It is conceivable that an visual science instrument could be developed where individuals would be fixed into a position where the same retinal area would be illuminated for long periods.

3. WAVE LENGTH DEPENDANCE OF RETINAL INJURY

It is necessary to study the wavelength dependance of retinal injury not only for pulsed lasers but for CW lasers as well for wavelengths between 600 and 1064 nm. It is the problem of "safe laser" and for the blue-violet wavelengths. Particularly insufficient data exist to allow adequate determination of the ocular hazard associated to semi conductor laser. These wavelengths may be more dangerous than predicted, even for thermal injury at pulsed duration less than phase 1 s.

4. RETINAL IMAGE SIZE DEPENDANCE OF INJURY THRESHOLDS OR SPOT-SIZE DEPENDANCE OF LASER RETINAL DOSIMETRY

New investigative techniques for the determination of the damage threshold show that present exposure limits for extended source viewing are not safe for retinal images larger than 300 um in diameter. Furthermore, while the spot-size dependance of the threshold damage level has been considered up to how as varying with time, recent data show that it is independant of the exposure duration. Therefore, it is suggested that more accurate exposure limits could be derived from intrabeam viewing criteria with the use of a multiplicative factor when apparent visual angle alpha is larger than the limiting angle alpha-min of 3.5 mrad. The exposure limit for large spot-sizes can be expressed in terms of radiant exposure or irradiance as follows H or E (EL es) = H or E (ELps) (alpha/alpha-min)$^{1/2}$. This formulation allows one to specify safety margins for large image diameters equivalent to those existing for intrabeam viewing conditions. It is the condition for a safe development of future medical applications of laser technology such as retinography, retinal spectroscopy or retinal holography involving very large images or the advent of applying non-linear optical effects to provide broad-band eye protection. The protection of eye is obtain with self focusing and induced scattering techniques by creating a larger retinal image for a distant laser sources. Generally it is necessary to perform a variety of experiments varying the size and the shape of exposed area, the wavelenghts, the NRF and the duration of pulses trains.

1. LONG TERM OR CHRONIC LASER EXPOSURES

Damage mechanisms and hazards have been carefully studied for acute exposure, but the effects of long-term chronic exposure to laser radiation are not yef fully understood. Functional as well as biological changes in eye tissues appear on the trained theses monkeys after long term chronic exposure in visible or near IR radiation at level lower than MPE. Other data bases are necessary to establish adequat standards.

2. DELAYED EFFECTS

The absence of acute effects does not mean that some damage has not occured at the cellular or molecular level. Delayed effects are always possible from level long IR or visible or UVR exposure. The retinal degradation or the lens lesion may be also occur from low level chronic exposure and the present MPE exposure limits do not permit to consider the absence of delayed risks from laser exposure to the skin and eye. Generally our knowledge of chronic visible IR and UVR dose effect relationships in man are very insufficient.

LASER INJURY MECHANISM

The exact nature of the laser damage mechanism is not very well understood:

- the spectral dependance does not appear to be predicted by only a thermal or thermoacoustic retinal damage mechanism
- blue-violet wavelengths may be more dangerous than predicted even for thermal injury and pulsed durations less than 1 s
- the non linear effect of ultrashort pulses are not really known and the corresponding data base on retinal injuries from very short pulsed IR or visible radiation, very rare.

REFERENCES

1.   IRPA guidelines on protection against non-ionizing radiation. Guidelines on limits of exposure to laser radiation between 180 nm and 1 mm. Chapter 4, 53-71 Ed. by A.S. DUCHENE, J.R.A. LAKEY, M.H. REPACHOLI, PERGAMON PRESS, 1991.

2.   IEC 825, Radiation Safety of Laser Products, Equipment Classification, Requirements and User's Guide, International Electrotechnical Committee, Geneva (1984)

3.   ANSI Z-136, 1-1986, ANSI Standard for the Safe Use of Lasers, American National Standards Institute, New York (1986)

4.   ACGIH, A guide for Control of Laser Hazards, American Conference of Governmental Industrial Hygienists, Cincinnati (1976)

5.   AFNOR, NFC 43-801 Sécurité des Rayonnements des Appareils à Laser. Classification des Matériels. Prescription et Guide de l'Utilisateur, UTE, Paris (1985)

6.   DIN, DIN 56912, Sicherheitstechnische Anforderungen für Bühnenlaser und Bühnenlaseranlagen (Safety Requirements for Lasers Used on Stages), Beuth Verlag GmbH, Berlin (1985)

7.   LASER APPLICATIONS IN MEDICINE AND BIOLOGY, Ed. by M.L. WOLBARS HT, Plenum Press (1971)

8.   LASERS IN INDUSTRY, Ed. by S.S. CHARSCHAN, Western Electric Series (1972)

9.   "SAFETY WITH LASERS AND OTHER OPTICAL SOURCES", SLINEY D.H. and WOLBARSHT M.L., Plenum Publishing, New York,(1980)

10. LASERS IN BIOLOGY AND MEDICINE, HILLENKAMP, F., SACCHI, C.A. and ARRECHI, T. Eds. Plenum Press, New York,(1980)

11. LASER BIOLOGICAL EFFECTS AND EXPOSURE LIMITS, COURT L., DUCHENE A. and D. COURANT, Eds. CEA/DPS FONTENAY AUX ROSES, (1988)

12. DOSIMETRY OF LASER RADIATION IN MEDICINE AND BIOLOGYG.I. MULLER and D.H. SLINEY, V 155, SPIE OPTICAL ENGINEERING Press, BELLINGHAM, WASHINGTON USA (1989)

13. LIGHT, LASER AND SYNCHROTON. RADIATION : A HEALTH RISK ASSESSMENT, M. GRANDOLFO, A RINDI and D.H. SLINEY, Eds NATO ASI SERIES B : PHYSICS Val 242, (1991)

NON-COHERENT OPTICAL SOURCES

Anthony P. Cullen and B. Ralph Chou

Optical Radiation Laboratory
School of Optometry
University of Waterloo
Waterloo, Ontario, Canada

Since prehistorical times human beings have relied upon incandescent flames to supplement the sun in the provision of heat, light and indirectly protection. It was not until the nineteenth century that the first electric discharge and incandescent filament lamps were introduced by Humphrey Davy and Thomas Edison respectively. The needs of industry and the general public have fostered a rapid development of all types of optical source with an emphasis on increasing intensity, efficiency and longevity. In addition attention to specific tasks and processes have resulted in an expansion of available sources.

Natural Sources

Natural sources of incoherent optical radiation include combustion (fire), incandescence (sun), electrical discharge (lightning), and luminescence (chemical and animal). Of these solar radiation is by far the most important from the point of view of providing illumination, infrared and ultraviolet radiation. Diurnal, annual and other variations in the levels of each optical waveband produce direct and indirect positive and negative effects on human health. Solar irradiance incident upon the Earth's atmosphere is 1367 Wm^{-2} (the solar constant) and approximates the spectral distribution of a blackbody radiator of $5,900°K$, however this is modified by atmospheric absorption (ozone, oxygen, carbon dioxide and water vapour) and scatter (Raleigh and Mie). Solar irradiance and spectral distribition at the Earth's surface (Figure 1) are also influenced by time of day, season, weather, atmospheric turbidity, variations in the ozone layer, and altitude. Under ideal conditions global solar irradiance may exceed half the solar constant and illuminance may exceed 100,000 lux.

Lightning, a natural electrical discharge in the atmosphere, produces ionized plasma at temperatures of the order $30,000°K$ but only for a few milliseconds. Optical radiation, which represents less than 1% of the emitted electromagnetic radiation, includes ultraviolet, visible and infrared

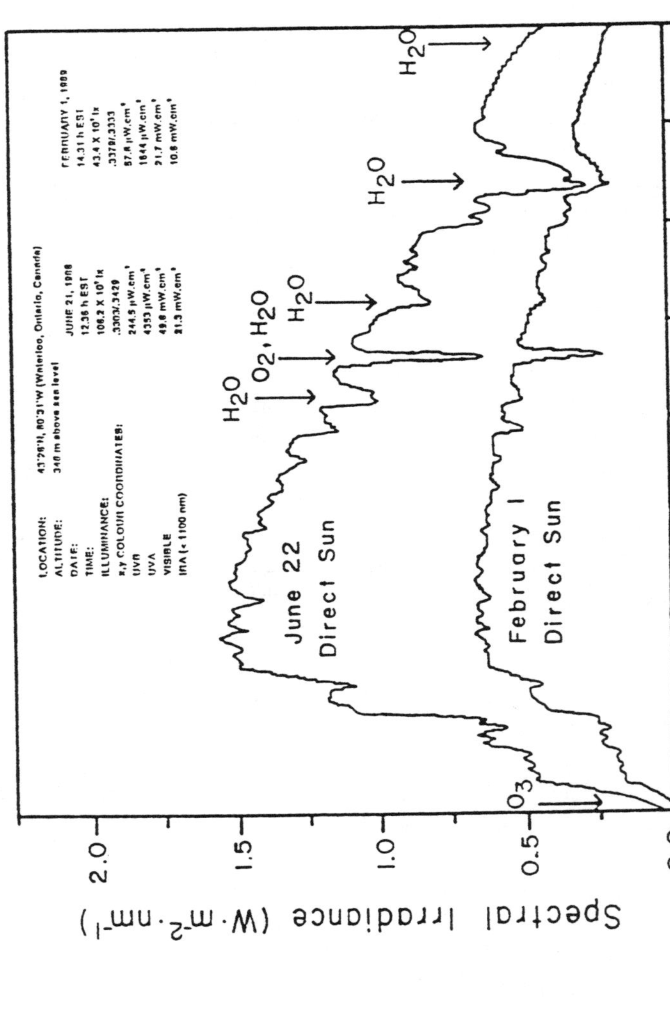

Fig. 1. Solar spectral irradiance as measured at Waterloo, Ontario, Can the winter and summer. Principal atmospheric absorption bands are indicated (from Cullen AP, 1991).

spectral bands with intense atomic emission lines for nitrogen and hydrogen. The radiance and luminance of naturally luminescent sources are so low that they may be considered insignificant.

Incandescent Filament Sources

For over a century modifications and improvements have been made to incandescent lamps including the material and design of the filaments, the introduction of gas filled lamps, innovations to the material, shape, frosting and tinting of the envelope(shell), and the incorporation of reflectors and Fresnel optics. Incandescent lamps range in size from the miniature ("grain of wheat"), operating at less than a watt, to beacons of ten thousand watts. As the primary sources of artificial lighting worldwide, these lamps are omnipresent in our lives be it at work, leisure, home or travelling.

During the nineteenth century the filaments were of carbon, iron, osmium or titanium however almost all lamps today use tungsten which may be alloyed with rhenium or thorium. Tungsten has been selected because of its strength, low cost, relatively high melting point and low vapour pressure. Since a filament source emits a spectrum which approximates that of a blackbody (more accurately greybody) at the same temperature it is clear that the hotter the lamp operates the "bluer" the light will be. Tungsten lamps operate between 2800°K and 3200°K so their spectral distribution peaks at the red end of the spectrum (Figure 2) and therefore they are safer from the ultraviolet and blue light hazard perspectives. In addition the lead glass envelopes of low wattage lamps absorb ultraviolet. High wattage lamps have silica glass or quartz envelopes. The filling of the envelope with inert gases (argon and/or nitrogen) increases the lifetime of the filament.

Tungsten-halogen lamps have become popular because of their whiter light and longer life. In order for the tungsten halogen cycle to occur the lamp must operate at a high temperature resulting in the need for a quartz envelope and heat absorbing filters (Figure 3). These lamps when used without additional filters have the potential to produce hazardous levels of ultraviolet.

Infrared heating and curing incandescent lamps and elements produce little or no visible(red) light since they operate in the 2000°K to 2500°K temperature band.

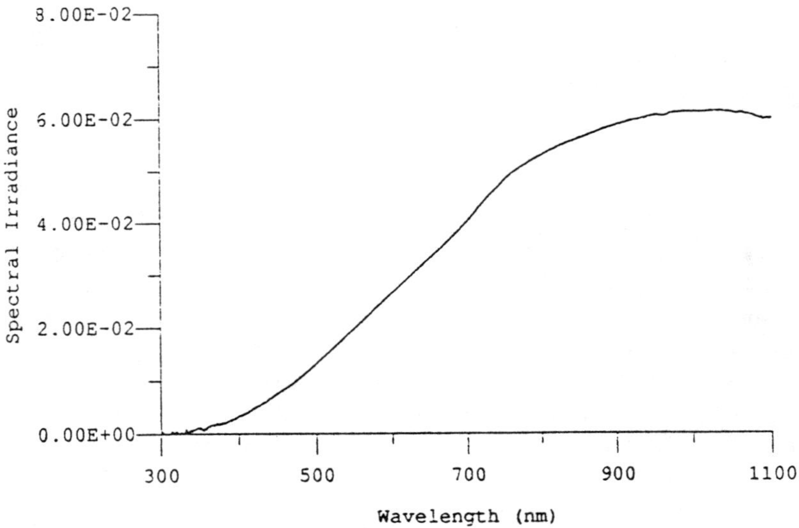

Figure 2. Spectral irradiance (Wm^{-2}nm^{-1}) produced by a 100W tungsten bulb at 50 cm.

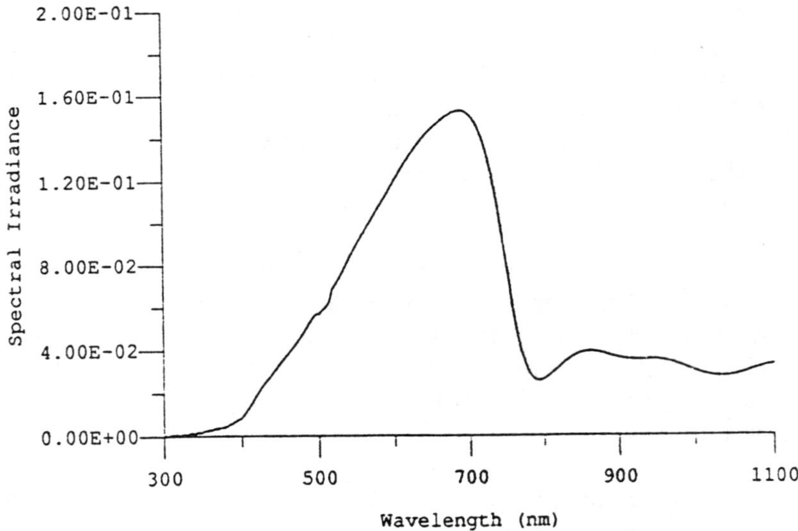

Figure 3. Specral irradiance (Wm^{-2}nm^{-1}) measured at 50 cm from a 25W tungsten halogen lamp with reflector and heat absorbing filter.

Low Pressure Discharge Lamps

The passage of an electrical discharge through a gas at low pressure results in the emission of spectral lines characteristic for the gas involved (Table 1). They require a high starting voltage in order to overcome the resistance of the gas until a relatively high operating temperature is achieved. A variety of these lamps are used for wavelength calibration of spectrometers and spectroradiometers. Advertising tube signs are low pressure discharge lamps filled with argon(blue), neon(red) and krypton(yellow). Low pressure mercury lamps are used for germicidal purposes in a variety of work environments; although similar lamps and low pressure sodium lamps are sometimes used for illumination they are less than ideal. The line spectra distort colour perception and cause other visual problems. In addition low pressure mercury lamps with quartz envelopes will produce both ultraviolet C and ultraviolet B.

Table 1

Colour	Element	Wavelength
Ultraviolet	Mercury	365.02
Violet	Mercury	404.66
Blue	Mercury	435.84
Blue	Cadmium	479.99
Blue	Hydrogen	486.13
Green	Mercury	546.07
Yellow	Mercury	576.9
Yellow	Mercury	579.1
Yellow	Helium	587.56
Yellow	Sodium	589.0
Yellow	Sodium	589.6
Red	Cadmium	643.85
Red	Hydrogen	656.28
Deep red	Helium	706.52
Infrared	Cesium	852.11
Infrared	Mercury	1014.00

Fluorescent Lamps

Fluorescent tubes are essentially low pressure mercury discharge lamps with a phosphor coating the inside of the glass envelope. The high emission of energy in the ultraviolet C (at 253.7nm) permits a wide range of phosphors to be used in order to design the spectral output of the tubes. Lamps are available with emission in the ultraviolet B (sunlamps), ultraviolet A ("blacklights") and the visible spectrum. The colour of common fluorescent tubes used for illumination is varied by variation in the phosphor or mixture of phosphors. Cool white (Figure 4) and warm white are more blue and more pink respectively. Lamps which claim to be equivalent to daylight may approximate solar spectral distribution but this is contaminated by the overlay of the mercury lines in the visible spectrum (Figure 5).

The term hot and cold may be applied to the type of cathode used in the lamp, usually these days it will be hot(tungsten filament) because of the advantages of a rapid start and greater efficiency. The efficiency and lifetime of fluorescent tubes may also be influenced by the type of ballast used.

Medium pressure mercury lamps

These lamps operate at a pressure of 1 to 10 atm (100 to 1000 kPa) and emit wavelengths in the UV, visible and infrared. The input power is much higher than that of the low pressure mercury lamp. The radiant output is rich in lines in the UVC, and UVB, and there is strong emission at 334 and 365 nm; all show pressure broadening. These sources are commonly found in industrial photochemical synthesis and photocuring processes, photocopiers, and sunlamps.

Metal halide lamps

The addition of small amounts of metal halides into the discharge tube of a medium pressure mercury lamp can enhance any given part of the optical spectrum. Metal halide lamps provide better colour and lumen efficiency for lighting applications. Antimony halide emits strongly in the UVC, and is used in germicidal lamps.

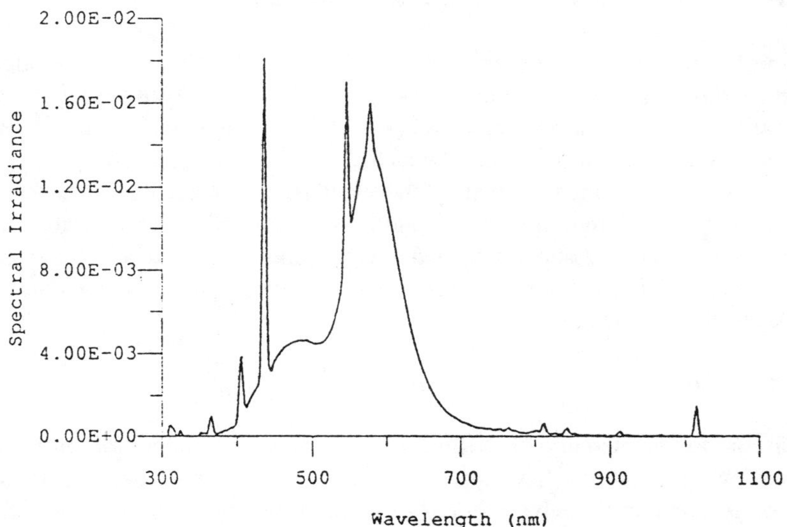

Figure 4. Spectral irradiance (Wm^{-2}nm^{-1}) at desk height produced by twin "cool white" fluorescent tubes. The mercury lines at 435.8 nm and 514.1 nm are prominent.

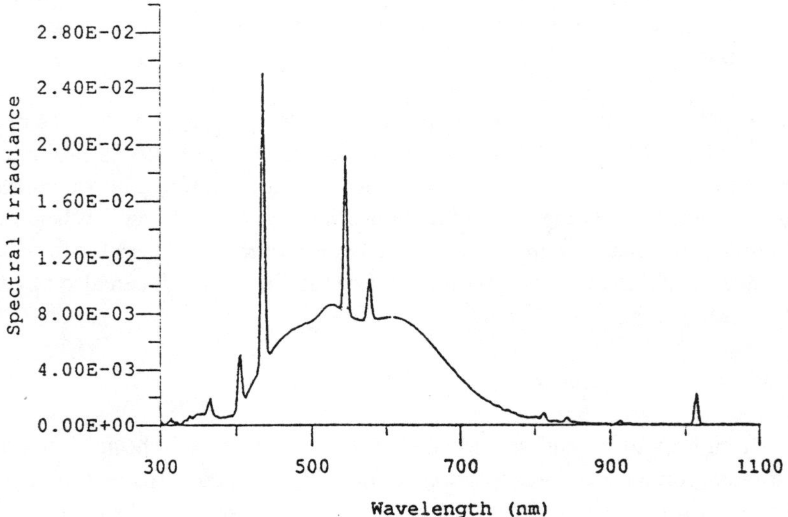

Figure 5. Specral irradiance (Wm^{-2}nm^{-1}) produced in the laboratory by "Vita-lites"®. This shows a closer approximation to the solar spectrum but has less red and the mercury spikes remain.

High pressure mercury and mercury-xenon lamps

These lamps operate at pressures above 10 atm (1000 kPa). Xenon is added to the discharge medium to provide a more stable arc. There are two forms available. The compact arc type has a bulb-shaped envelope and closely spaced electrodes which produces an arc several mm long. It operates at low voltage and high current. The capillary arc lamp has a discharge chamber up to 2 mm in diameter and 60 mm long with a thick wall. The spectrum shows pressure broadened mercury lines superimposed on a strong continuum. These lamps are used in photolithographic processes and in scientific instruments.

Xenon lamps

In contrast to mercury-filled discharge lamps, xenon lamps produce UV continuum radiation. Two types of lamp are available. The compact arc type is operated at high pressure (up to 60 atm or 6000 kPa), using a DC power supply at high current. The radiation emitted by this lamp type resembles that of a 6000 K blackbody, with some superimposed lines. The continuum ranges from the UV to the IR, with peak emission in the near IR. The compact xenon arc is commonly used as a solar simulator and as a high output light source for projectors, lighthouses and searchlights. It is also used as a continuum light source in a variety of scientific instruments.

The xenon arc may also be operated by discharging a high current pulse through the lamp. At low energy levels, the lamp produces continuum radiation similar to that of the continuous discharge lamp. The xenon flash tube is used in photographic flashlamps and strobe lighting. When very high voltage and current are used, the lamp becomes a UV-rich continuum source for photocuring and flash photolysis applications. It can also be used for laser pumping.

Sodium vapour lamps

Sodium vapour lamps are frequently used for street lighting and other illumination applications because of their high lumen efficiency and long service life. The low pressure type produces highly monochromatic light, predominantly the doublet at 589.6 and 589.0 nm (Figure 6). The high pressure lamp produces a weak continuum and pressure broadened line emission. Although a more efficient light source, the low pressure sodium lamp is less favoured for street lighting because of the resulting limited colour perception. The high pressure lamp is favoured primarily for esthetic, rather than economic reasons, as an illumination source.

Carbon arcs

The arc is formed when two carbon rods are touched together, then separated by a short gap while carrying an electric current. Carbon vapour is driven off from the rods by the arc, which excites the vapour to emit optical radiation. The resulting continuum radiation is rich in UV and short wavelength light. The output can be enhanced by using cored carbon rods filled with oxides of cerium or thorium. These sources are used for illumination.

Infrared and light emitting diodes

These P-N junction devices emit optical radiation when biased in the forward direction. Low input voltage and current produces a relatively high radiance from a small emitting area (Figure 7). Infrared LEDs are used in fibre optic communication systems. LEDs emitting visible light are used in visual display panels and as compact light sources.

Electroluminescent displays

Electroluminescence is the non-thermal conversion of electrical energy into light in a liquid or solid substance. One process is electron-hole recombination in a P-N junction device. The display panel is flat and has a wide parallax-free viewing angle. Power consumption is low. Blue, green, yellow or white light displays are available. Common applications include visual readout displays, some complex logic-circuit elements, and electroluminescent-photoconductive image intensifiers.

Welding flames and arcs

There is a wide variety of welding processes used in industry. Torch brazing and gas welding operations (including oxyacetylene welding) produce flames which are sources of IR-rich continuum radiation. Electric welding arcs are operated at a variety of voltage and current levels. The resulting arcs produce a continuum from the UV through the visible and IR regions of the spectrum, with numerous superimposed lines. The spectral composition of the emission is determined by the current level used, the type of metal being welded, presence and type of forming gas, and type of flux used (Sliney and Wolbarsht, 1980).

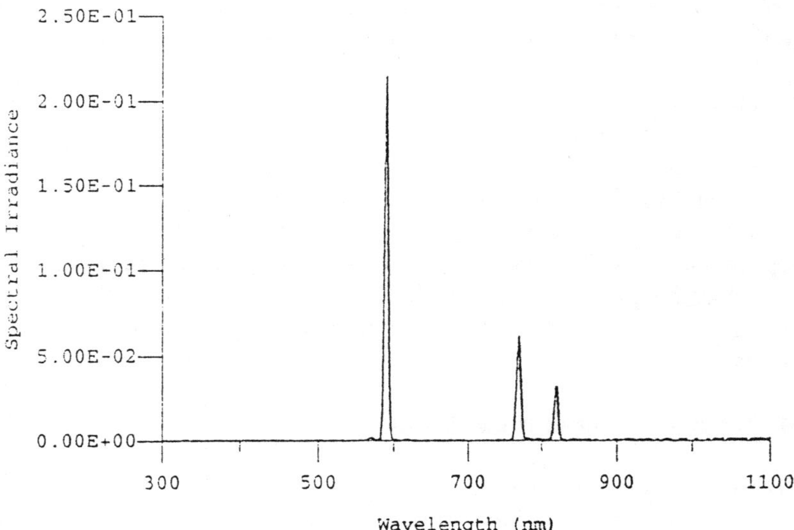

Figure 6. Low Pressure Sodium Lamp. The 589.6 nm and 589.0 nm sodium lines are superimposed. The lines at 759.4 nm and 822.7 nm represent oxygen.

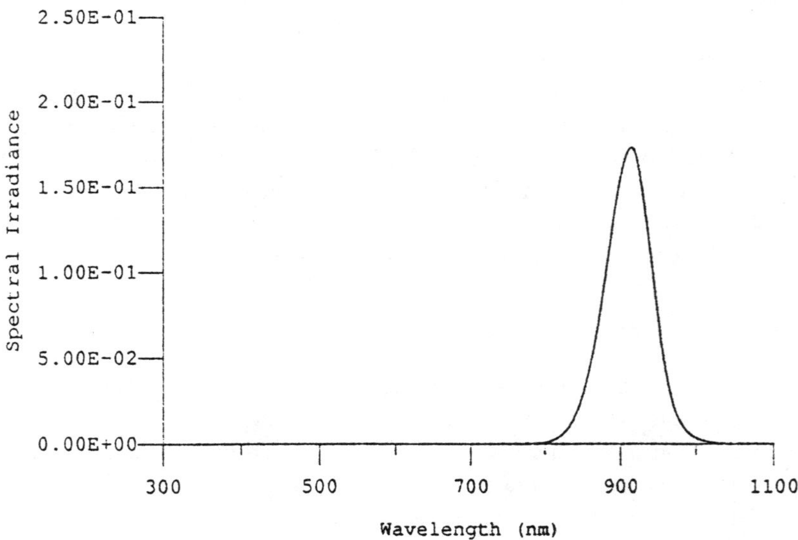

Figure 7. Infrared Emitting Diode. Spectral irradiance of a 4V pulsed LED at 5 cm revealing a much broader bandwidth than a laser diode.

Electric welding arcs are commonly associated with "welder's flash" or photokeratoconjunctivitis. This accounts for approximately 10% of all lost time industrial eye injuries, but has no known long-term consequences.

Electrical arcs and sparks

Electrical arcs and sparks from power generation and transmission line facilities produce an intense continuum superimposed with lines due to the copper and other constituents of the electrical conductor. The UVB emitted in arcs at up to 340 A at 16 kV is insufficient to cause a welder's flash injury at typical working distances of 2 m when a worker trips a secondary distribution system isolating switch (Chou and Cullen, 1991) (Figure 8).

Glassblowing flames

Glassblowers heat glass in a gas flame until it becomes plastic. The area of glass in contact with the flame glows with a bright green-white light. This is a continuum emission which is relatively rich in IR and long-wavelength light.

Video Display Terminals

In common with many electrical and electronic conveniences of modern life, the VDT is capable of producing a broad spectrum of electromagnetic radiation. The levels of UV emitted by VDTs have been a specific concern of some workers; our measurements (Table 2) indicate that, in an average office, a VDT will increase UVA irradiance of the user by only 20% and the provision of UV protection is not necessary. Indeed the levels of ultraviolet emitted by VDT's and office fluorescent tubes are six orders of magnitude less than encountered from sunlight.

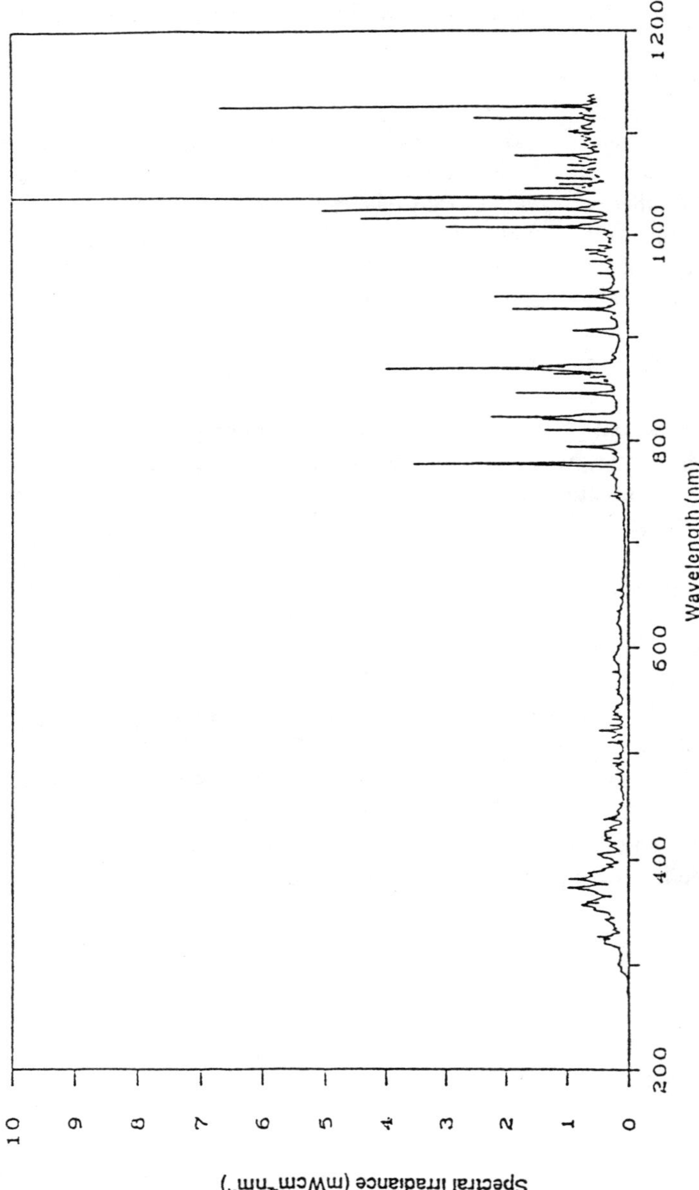

Figure 8. Spectral irradiance produced by an arcing in-line switch carrying 340 A at 27 kV, arc duration < 500 ms. (Chou and Cullen, 1991).

Table 2

Typical Ultraviolet Irradiance Levels

Source	UVB (290 320 nm)	UVA (320-400 nm)
VDT	< 1 pWcm^{-2}	6×10^{-7} mWcm^{-2}
Ambient in office (Fluorescent lighting)	0.1×10^{-5} mWcm^{-2}	$3 \times 10\text{-}6$ mWcm^{-2}
Direct sunlight (Midday June, Waterloo, Ontario)	0.2 mWcm^{-2}	4.4 mWcm^{-2}
Recommended standard	0.1 μWcm^{-2}	1 nWcm^{-2}

References

Chou, B.R., Cullen, A.P.(1991) Evaluation of ocular hazards due toelectric arc flash at an in-line switch. *Health Physics*, 61(4): 473-479.

Cullen, A.P.(1991) The Environment. Chptr 72, In Clinical Contact Lens Practice. Bennett, E.S. and Weissman, B.A., Eds. Lippincott, London and New York.

Diffey, B.L.(1989) Radiation Measurement in Photobiology, Academic Press, London and San Diego.

Kaufman, J.E., Christensen, J.F.(1984) Eds. IES Lighting Handbook, Reference Volume, Illuminating Engineering Society of North America, New York.

Sliney, D., Wolbarscht, M.(1980) Safety with Lasers and Other Optical Sources. Plenum Press. New York and London.

Thorington, L.(1985) Spectral, Irradiance and Temporal Aspects of Natural and Artificial Light. In the Medical and Biological Effects of Light, Wurtman R.J., Baum M.J. and Potts J.T., Eds. New York Academy of Sciences, New York.

MEDICAL APPLICATIONS OF OPTICAL RADIATION

L.D. Szabo

Department of Non-ionizing Radiation
National Research Institute for
Radiobiology and Radiation Hygiene
Budapest, Hungary

INTRODUCTION

The medical application of optical radiation is mainly phototherapy. "Phototherapy is the use of ultraviolet or visible radiation as a means of treatment". In 525 BC Herodotus related the strength of the skull to sunlight exposure. There appears to be no record of the experimental technique employed!

There is an almost universal belief in the health-giving properties of the sun, and exposure to UV lamps or sunlight (heliotherapy) has been prescribed for a variety of disorders. Treatment is not always carried out under controlled, scientific conditions but is more often governed by individual experience and prejudice. (Moseley, 1988)

In 1904 N. R. Finsen, a Danish physician was awarded the Nobel Prize for the medical application of optical radiation. Niels Ryber Finsen (1860-1904) graduated from the University of Copenhagen in 1890. As a medical student he started to study the effect of light on certain dermatological diseases. Suffering from a chronic skin disease himself, he believed in the healing power of the sunshine. He founded the Copenhagen Institute for Phototherapy in 1896 (Finsen, 1899). There he studied the effects of the sunlight of short wavelength and found that it is capable of killing certain bacterial cultures on the skin surface. He also stated that this was not a thermal effect. He achieved outstanding results in curing lupus vulgaris, a skin disease caused by Bact. tuberculosis. He described the physiological effect of blue, violet and ultraviolet lights, which became important when the X- and gamma-rays of much higher photon energy were discovered. Finsen developed some valuable instruments (e.g. the Finsen lamp: an arc-lamp of high intensity). The Finsen lamp, which emits mainly violet and UV radiation, was used for the so-called Finsen therapy (Finsen, 1899).

GENERAL CONSIDERATIONS

The ultraviolet (UV) radiation is of special interest (among the various optical radiations) because of its relatively high photon energy as compared to the other types of optical radiations. The photon energy 12.4 eV (at 100 nm) - 3.1 eV (at 400 nm) could lead to greater biological response. However the low penetration will restrict most of the direct biological responses to the superficial tissues (skin and eye). The UV radiation is the part of the electromagnetic spectrum lying between the softest ionizing (soft X-ray) on the one side and visible radiation on the other. Because of differences in physical properties and in biological responses, the UV-region has been subdivided:

- Less than 180 nm: no biological effect (the photons are absorbed by air)
- 180 nm - 300 nm : far UV region
- 300 nm - 400 nm : near UV region.

An other way of dividing the UV part of electromagnetic spectrum is arbitrary:

- UV-A: 400 nm - 315 nm: so called "black light"
- UV-B: 315 nm - 280 nm: the skin erythemal region
- UV-C: less than 280 nm: the germicidal region.

The wavelengths covered by optical radiation range from 100 nm to 1 mm. This wavelength region includes the ultraviolet part (from 100 nm to 400 nm), the visible part (from 400 nm to 800 nm) and the infrared part (from 800 nm to 1 mm) of the electromagnetic spectrum (Bosnjakovic,1988).

Until the last century, the main source of optical radiation was the sun. Today, beside the sun-light, the optical radiation is produced by several man-made radiation sources.

Man-made sources of UV radiation:

- Incandescent sources
    tungsten halogen lamps

- Gas discharges
    mercury lamps
    xenon lamps

    hydrogen lamps
    deuterium lamps
    flash tubes

- Fluorescent lamps
    lighting tubes
    sunlamps (UV-B)
    fluorescent tubes (UV-A)

- Lasers: excimer, nitrogen, He-Cd, etc.

MEASUREMENT OF OPTICAL RADIATIONS

The basic information concerning the measurements of optical radiations has been described recently. Sliney (1988) summed up the radiometric quantities used in optical radiation safety. The radiometric quantities (radiant energy, radiant power, radiant exposure, irradiance, radiant intensity, spectral irradiance and spectral radiance) were expressed in SI-units. He stated the instrumentation for hazard evaluation. Some UV safety meters are available which follow a UV hazard function (the cost of such an instrument is more than $20,000 US).

MEDICAL APPLICATIONS OF OPTICAL RADIATION

The optical radiations are often administered by physiotherapists in the treatment of a variety of human disorders. On the basis of European literature we have compiled a list of the Latin diagnosis where optical radiation (UV, heliotherapy, infrared, resp.) is being used (Lampert, 1952; Bozsoky and Iranyi, 1976; Nikolova, 1983; Callies, 1986). Refer to the following pages for the complete listings.

MEDICAL APPLICATIONS

Dermatology

Acne vulgaris
Acne rosacea
Acne conglobata
Seborrhea
Alopecia
Vitiligo
Granulosis rubra nasi
Eczema
Herpes zoster
Herpes simplex
Herpes duplex
Neurodermatosis
Scabies
Psoriasis
Scleroderma (collagenosis)
Tuberculosis cutis luposa

Internal Medicine

Tracheitis
Bronchitis
Bronchiolitis
Asthma bronchiale
Bronchiectasia
Pneumonia
Pneumosclerosis
Abscessus pulmonis
Emphysema pulmonum
Pleuritis
Pyelonephritis
Cystitis
Hypothyreosis (mixoedema)
Diabetes mellitus
Anaemia idiopathica
 hypochromica
Gastritis atonica
Gastritis acuta
Gastritis chronica
Rheuma

Pediatrics

Rachitis
Asthma bronchiale
Bronchitis
Pneumonia
Bronchoadenitis
Catarrh
Pleuritis
Diathesis axudativa
Incontinentia urinae
Rheuma
Pyelonephritis
Rubeola
Scarlatina
Pertussis
Morbilli
Parotitis
Varicella
Scarlatina

Rheumatology

Rheumatoid arthritis
Arthrosis
Spondyloarthrosis
Spondyloarthritis
 ankylopoetica
Arthritis tuberculosa

MEDICAL APPLICATIONS cont.

Arthritis purulenta

Gynaecology and Obstetrics

Amenorrhoea
Hypomenorrhoea
Oligomenorrhoea
Hypermenorrhoea
Vulvitis
Herpes vulvae
Bartolinitis
Vaginitis
Endocervicitis
Ectopium cervici uteri
Endometritis
Adnexitis
Peritonitis tuberculosa
Toxicosis gravidarum
Pyelitis gravidarum
Pyelocistitis gravidarum
Ulcus puerperale
Hypogalactia
Rhagas mammae
Mammalgia
Neurology and Psychiatry

Neuritis facialis
Neuralgia nervi trigemini
Neuralgia nervi occipitalis
Neuralgia nervi
 glossopharingei
Neuralgia nervi oculomotori
Neuralgia nervi intercostalis
Neuralgia nervi cervicalis
Atrophy nervi optici
Polyneuritis
Causalgia
Phantom hallucination
Contractura
Sclerosis multiplex
 (disseminata)
Diencephalosis
Poliomyelitis
Tabes dorsalis
Encephalopatia traumatica
Atherosclerosis cerebri
Migraine
Myodistrophia progressiva
Myastenia gravis
 pseudoparalytica
Paralysis juvenilis

Neurosis
Hysteria
Psychosismaniaco-depressiva
Psychosis presenilica
Psychosis reactiva
Psychosis arteriosclerotica
Schizophrenia
Narcomania

Ophtalmology

Blepharitis
Dacryoadenitis
Dacryocystitis
Iritis
Iritis toxico-allergica
Iridocyclitis
Keratitis
Keratitis herpetica
Keratitis metaherpetica
Keratitis dendritica
Keratitis superficialis punctata
Keratitis sclerotica
Keratitis tuberculosa
Trachoma

MEDICAL APPLICATIONS cont.

Hordeolum

Otorhinolaryngology

Rhinitis acuta
Rhinitis chronica
Rhinitis atrophica
Rhinitis allergica
Sinusitis acuta
Sinusitis chronica
Pharyngitis chronica
Angina
Tonsillitis chronica
Combustio oesophagi
Laryngitis acuta
Laryngitis chronica
Otitis media acuta
Otitis media chronica
Catarrhus tubotympanalis
  acuta
Catarrhus tubotympanalis
  chronica
Eczema oticum

Surgical diseases

Endarteritis obliterica
Raynaud's disease
Thrombophlebitis
Abscessus
Gangrena
Erysipelas
Mastitis
Panaritium
Furunculosis
Carbunculus
Hydradenitis
Osteomyelitis
Periostitis
Epicondylitis
Myositis ossificans
  progressiva
Tendovaginitis
Tendovaginitis chronica
Tendoruptura
Bursitis acuta
Periarthrites
Lymphangitis acuta
Lymphangitis chronica

Lymphadenitis acuta
Lymphadenitis chronica
Lymphostasis
Contusio
Hematoma
Fractura
Infiltratum postoperative
Subluxatio
Prostatitis
Orchitis
Epididimitis
Hemorhoidae
Rhagas ani
Fistula ani

Occupational diseases

Silicosis
Occupational poisoning
  (inorganic or organic
  poisons)
Occupational disease
  produced by vibrations
Acne occupationalis

LASERS

The medical applications of lasers are well known. The International Non-Ionizing Radiation Workshop (Melbourne, 1988) devoted a few papers to this subject including some aspects of optical radiations in general, too (Jammet, 1988; Roy, 1988; Bosnjakovic 1988; McKinlay, 1988; Delpizzo and Cornelius, 1988; Villforth, 1988). Medical lasers and biological criteria and limits of their therapeutic effects were reported in 1989 in Erice by Court and Courant (Grandolfo et al. 1991). Many monographs and handbooks well described the medical applications of lasers (e.g. Goldmann et al. 1971). The low energy (soft) lasers in medicine for biostimulation of wound-healing; for treatment of ulcus cruris were introduced by Hungarian surgeon E. Mester. In years 1967-1991 several hundred papers and reviews were published (Miranda, 1981; Trelles, 1981; Mester et al. 1985). Most of these publications are describing the beneficial effects of soft lasers but a Swedish scientist considered that the clinical use of low energy lasers cannot be supported on scientific basis. This statement regards mainly the use of soft lasers in cosmetics (Harms-Ringdahl, 1988). There is no doubt about the beneficial effects of laser treatment (5 mW He-Ne) of ectopium cervici uteri (Kovacs and Tisza, 1984; Holland et al. 1983; Boloni et al. 1986). The treatment of rheumatoid arthritis by lasers was introduced in the early 1980's (Goldmann et al. 1982; Barabas er al. 1987; Bakos et al. 1987). New trials to develop laser applications in medicine are in progress and they are of great promise (e.g. photodynamic therapy of cancers).

PROTECTION AND STANDARDS OF OPTICAL RADIATIONS

The protection of medical applications of optical radiation (UV, visible, infrared and lasers) is based on the international statements published in several papers in the last 15 years (WHO EHC 14, 1979, WHO EHC 23, 1982, Duchene, 1988, WHO, 1989, IRPA Guidelines, Duchene at al. 1991, ILO 53, 1985, IRPA/INIRC Guidelines, 1989) Standards of optical radiations have been worked out mainly for lasers (IEC 825, 1984, 1987.). The above mentioned IEC standards are acceptable for national standards. Recently in Europe the IEC standards have become the source for the development of national standards (Szabo, 1990a,b, Driscoll, 1991, Standard of Hungary, 1991).

Protection against UV-light has been developed in the last decade (Sliney,

1980, Rosenthal, 1991, IRPA Guidelines, Duchene at al., 1991).

CONCLUSIONS

The up to date bibliography on medical applications of optical radiation is very important. Our knowledge must be increased in several ways:

1. Clearing up the valuable data published in earlier scientific literature.

2. Collection of experiences obtained in different countries and published in English, French, German, Russian and Spanish (in some cases we can find original data also in Italian, Danish, Swedish, Hungarian etc.).

3. Comparison of data of different continents (America, Australia, Europe, Asia and Africa, resp.).

4. Evaluation of results obtained in the countries of the world.

5. Estimation of future trends of development of medical applications of optical radiation.

REFERENCES

Bakos J., Szabo L.D., Barabas K., Sinay G., 1987. Biological effects of ND:Phosphate glass laser on human knee joint synovial membrane in vitro. Experimental Medicine. 39, 356-360. (in Hungarian)

Barabas K., Bakos J., Szabo L.D., Sinay G., 1987. "In vitro" effect of neodymium phosphate glass laser irradiation on the rheumatoid synovial membrane. Hungarian Rheumatology, Supplementum 58-62.

Boloni E., Batke I., Szabo L.D., Kovacs L., 1986. Effect of soft He-Ne laser on the metabolism of DNS and RNS in human portio uteri. Experimental Medicine, 38, 255-262. (in Hungarian)

Bosnjakovic B.F.M., 1988. Ultraviolet radiation. Risk limitation and protection of the public. Non-ionizing radiations, Editor: Repacholi M.H., IRPA/INIRC, Australia

Bozsoky S., Iranyi J., 1976. Physiotherapy Medicina, Budapest (in Hungarian)

Callies R., 1986. Rheumatologische Physiotherapie. VEB Fischer Verlag, Jena (in German)

Court L.A., Courant D., 1991. Medical Lasers and Biological Criteria and Limits of their Therapeutic Effects. In "Light, Lasers and Synchrotron Radiation" Grandolfo et al.

Delpizzo, V. and Cornelius, W.A., 1988. Lasers. Non-Ionizing Radiations, Editor: Repacholi, M.H., IRPA/INIRC, Australia

Driscoll C.M.H., 1991. Laser Standards. NRPB Radiological Protection Bulletin, No. 122, 10-13. Chilton, Didcot

Duchene A., 1988. Non-Ionizing Radiations, Editor: Repacholi M.H., IRPA/INIRC, Australia

Finsen R.T., 1899. Medelelser fra Finsens medicinska Lysinstitut

Goldmann J.A., Chapella I., Casex H.L., 1982. Lasers in Surgery and Medicine. 1, 93.

Goldman L., Rockwell R.J., 1971. Lasers in Medicine. Gordon and Breach, New York, London, Paris

Grandolfo M., Rindi A., Sliney D.H., 1991. Light, Lasers and Synchrotron Radiation. A Health Risk Assesment. Plenum Press, New York and London

Harms-Ringdahl M., 1988. SSI Project P-351.86, Stockholm (in Swedish)

Holland J., Csanyi P., Kokai A., Korosi L., Szabo L.D., Wagner K., Kovacs L., 1983. The effect of laser radiation on the prostaglandin receptor of erithrocytes and on protein synthesis of epithelium cervicis uteri. Proc. 7th ICRR sessions E: E8-04.

IEC-825, 1984. Radiation safety of laser products equipment classification requirement an user's guide. and Amendments to IEC-825, 1988. Geneva

ILO 53, 1985. Occupational Hazards from Non-ionizing Electromagnetic Radiation, Geneva

IRPA Guidelines on Protection against Non-ionizing Radiation, 1991. Ed.: Duchene A.S., Lakey J.R.A., Repacholi M.H., Pergamon Press, USA.

IRPA/INIRC Guidelines, 1989. Proposed Change to the IRPA 1985 Guidelines on Limits of Exposure to Ultraviolet Radiation. Health Physics 56, 971-972.

Jammet H.P., 1988. Comparison between radiological protection against ionizing radiation and non-ionizing radiation. Non-Ionizing Radiations, Editor: Repacholi M.H., IRPA/INIRC, Australia

Kovacs L., Tisza S., 1984. Neoplasma, 28, 351.

Lampert H., 1952. Physikalische Therapie. Verlag Steinkopff, Leipzig (in German)

McKinlay A.F., 1988. Visible (light) and infra-red radiation. Non-Ionizing Radiations, Editor: Repacholi M.H., IRPA/INIRC, Australia

Mester E., Mester A.F., Mester A., 1985. The biomedical effect of laser application. Lasers in Surgery and Medicine 5, 31-39.

Miranda R., 1981. La Biostimulatione Laser in Medicina. Athena, Modena (in Italian)

Moseley H., 1988. Non-ionizing Radiation. Medical Physics Handbooks 18. IOP Publ. Ltd. Bristol and Philadelphia

Nikolova L., 1983. Physiotherapy. Medicina, Sofia (in Russian)

Rosenthal F.S., West S.K., Munoz B., Emmett E.A., Strickland P.T., Taylor H.R., 1991. Ocular and Facial Skin Exposure to Ultraviolet Radiation in Sunlight: a Personal Exposure Model with Application to a Worker Population. Health Physics 61, 77-86.

Roy C.R., 1988. Ultraviolet Radiation: Sources, biological interaction and personal protection. Non-Ionizing Radiations, Editor: Repacholi M.H., IRPA/INIRC, Australia

Sliney D.H., Wolbarsht M., 1980. Safety with Lasers and Other Optical Sources. Plenum, New York

Sliney D.H., 1988. Measurement of optical radiations. Non-Ionizing Radiations, Editor: Repacholi M.H., IRPA/INIRC, Australia

Standard of Hungary, 1991. MSZ 16261. Radiation safety of laser products. Budapest (in Hungarian)

Szabo L.D., 1990.a. Diseases caused by non-ionizing radiations. In Occupational Diseases, Ed.: Timar M. OMIKK, Budapest (in Hungarian)

Szabo L.D., 1990.b. Safe use of medical lasers. In Clinical Application of Lasers, Ed.: Toth T., Medicina, Budapest, 162-168. (in Hungarian)

Trelles M., 1981. Soft Laser Therapy. Enar, Madrid (in Spanish)

Villforth J.C., 1988. Medical devices - Non-ionizing radiation standards. Non-Ionizing Radiations, Editor: Repacholi M.H., IRPA/INIRC, Australia

WHO EHC 14. 1979. Ultraviolet Radiation, Geneva

WHO EHC 23. 1982. Lasers and Optical Radiation, Geneva

WHO, 1989. Non-ionizing Radiation Protection, Ed.: Suess M.J., Copenhagen

PART V

ELF
and
STATIC FIELDS

BIOLOGICAL EFFECTS OF 50/60 HZ FIELDS

L. E. Anderson
Biology and Chemistry Department
Pacific Northwest Laboratory
Richland, Washington

INTRODUCTION

Until the last few decades, the natural background levels of atmospheric electric and magnetic fields were extremely low; however, in recent years they have increased dramatically. The industrialization and the electrification of society have resulted in the exposure of people, animals and plants to a complex milieu of elevated electromagnetic (EM) fields that span all frequency ranges. One of the most significant contributions to this changing electrical environment has been the technological advances associated with the growth of electrical power generation and transmission systems. In addition, EM field–generating devices have proliferated in industrial plants, office buildings, public transportation systems, homes and elsewhere.

EM fields, which may extend far beyond their sources, are mostly imperceptible to people. There has been considerable controversy as to whether 50/60 Hz fields could even cause significant biological effects, let alone pose a hazard to health. However, research and clinical experience have shown that biological effects from such fields are not precluded simply because they are not perceived. Recent data confirm some of the earlier reports that extremely low frequency (ELF) fields do cause changes in certain biological systems. Thus, it is both reasonable and timely to evaluate the interactions between the modern EM environment and living organisms and to investigate whether such interactions are beneficial or detrimental, transient or permanent. In the past two decades, research programs throughout the world have greatly expanded in scope and depth to address such issues. Significant progress has been achieved, both in defining the ways living organisms interact with ELF fields and in describing biological effects, both real and potential, from such fields. Much of this effort has been directed toward electric fields of power frequencies. However, frequencies other than 50 and 60 Hz have also been examined, and research has been expanded to include magnetic as well as electric fields. Although it is now clear that ELF EM fields do cause biological effects, the basis for those effects and the underlying mechanisms of interaction remain largely unknown, and the health implications for humans and animals have yet to be fully determined.

As in other areas of scientific investigation, the research being conducted on ELF bioeffects has been performed at several levels: human studies (primarily epidemiological), animal experiments, and cellular (mechanistic) studies.

Some of the earliest efforts to examine health–related issues of ELF fields were focused on the impacts of such fields in humans. Despite the obvious desirability of obtaining such data, they are the most complex and least complete. Additionally, experimental questions often cannot be investigated in humans. Therefore, many areas of biological investigation are more appropriately and efficiently conducted with animal models.

This paper specifically examines the biological effects of exposure to 50/60 Hz EM fields observed in in vivo (animal) studies. An attempt is made to evaluate experimental results and, insofar as possible, interpret them with respect to potential health implications. An overview of current concepts and possible mechanisms is given, and possible future directions of research are discussed.

REVIEW OF ANIMAL STUDIES

By far the largest body of information on biological effects of ELF fields has been obtained in experimental research on animals exposed to electric fields. Experiments have been performed primarily on rodents (rats and mice), but a wide variety of other subjects have also been used, including insects, birds, cats, dogs, swine, and nonhuman primates. A broad range of exposure levels has been employed and an equally large number of biological end points have been examined for evidence of possible electric and/or magnetic field effects. These multitudinous studies have been reviewed several times (e.g. Sheppard and Eisenbud, 1977; Anderson and Phillips, 1985; Anderson 1990). Summaries of important findings are presented here, arranged according to the biological systems that appear to be principally involved: neural and neuroendocrine systems (including behavior), reproductive systems (including fertility, growth and development), and other functions (including cardiovascular and blood chemistry, bone growth and repair, and immunology). A separate section addresses the issue of carconogenesis and mutagenesis.

Neural and Neuroendocrine Systems

Many of the biological effects observed in animals exposed to ELF fields appear to be directly or indirectly associated with the nervous system. This apparent relationship might be anticipated, since the nervous system is composed of tissues and processes that are unusually responsive to electrical signals. In addition, both the structure and function of this system are fundamentally involved in the interaction of an animal with its environment. The major features of this interaction; transmittal of sensory input from external stimuli, central processing of such information, and subsequent efferent innervation of tissues and organs, may provide the basis for explaining possible links between ELF exposure and observed biological consequences.

In early experimental studies, nervous system parameters were measured only occasionally, although many of the observed effects, primarily behavioral, were related to nervous system function. Before the late 1970s, studies on ELF exposure relating to nervous system function could generally be classified in three categories: assessments of activity or startle–response behavior, evaluations of stress–related hormones (such as corticosteroids), and general measurements of central nervous system responses (such as EEGs and interresponse times). Results were often contradictory, with claims of both effects and noneffects from ELF electric field exposure. However, because of the possible and suggested sensitivity of the nervous system to ELF fields, subsequent studies included a broader range of neurological assessments. Specific nervous system responses, in addition to behavior, began to be sought in experiments. This effort was mounted to determine the extent of ELF interaction with tissue and/or organ systems and also to investigate the mechanisms underlying the observed biological effects.

Behavior – Among the most sensitive measures of perturbations in a biological system are tests that determine modifications in the behavioral patterns of animals. This sensitivity is especially valuable in studying environmental agents of relatively low toxicity.

Behavioral studies in several species provide evidence of field perception and of the possibility that EM fields may directly alter behavior. The threshold of detection reported (Stern et al., 1983) in rats is between 4 and 10 kV/m. Human volunteers were able to detect a 9 kV/m 60 Hz field in certain postures (Graham et al., 1987). Thresholds for perception of the field in other animal species, reportedly in the 25 to 35 kV/m ranges, including mice (Rosenberg et al., 1983), pigs (Kaune et al., 1978), and pigeons and chickens (Graves et al., 1978). It appears that a change in other environmental factors, such as relative humidity has the potential to alter perception threshold values (Weigel et al., 1987b). Cutaneous sensory receptors that respond to a 60 Hz electric field have been identified in the cat paw (Weigel et al., 1987a). Whether such receptors exist in human skin is unknown.

An evaluation of the preference/avoidance behavior of animals for remaining in or out of the E–field has been conducted at several field strengths for 60 Hz electric fields. At 100 V/m, no effect of exposure, either in preference behavior or temporal discrimination, was evident in monkeys (deLorge, 1974). At 25 kV/m, rats preferred to spend their inactive period in the field, whereas at 75–100 kV/m they avoided exposure (Hjeresen et al., 1980). Swine (Hjeresen et al., 1982) remained out of the field (30 kV/m) at night but demonstrated few other observable behavioral changes. Alterations in activity have also been reported in animals exposed to ELF fields, including a transitory, increased activity response on initial exposure of rats or mice at 25 to 35 kV/m (Rosenberg et al., 1983).

Much of the behavioral work with nonhuman primates has been performed at very low field strengths (7 to 100 V/m), where essentially no effects of exposure were reported (NAS, 1977). At much higher field strengths (30 kV/m), Rogers et al. (1987) reported minor behavioral changes in exposed baboons that appear related to the animals' perception of the field. The observed effects do not seem to be permanent or deleterious. In studies examining the effect of ELF magnetic fields on behavior, many of the investigations carried out at low field intensities have shown behavioral alterations, primarily activity changes (Smith et al., 1977). In contrast, studies conducted at higher field intensities have shown no evidence of a field–associated effect on animal behavior (Davis et al., 1984).

In the experimental studies that have been conducted to determine whether ELF fields cause behavioral alterations, remarkably few robust effects have been demonstrated (Lovely, 1988). Effects that have been observed, usually arousal or activity responses, are probably due to the animal's detection and possible perception of the electric field.

<u>Biological rhythms</u> – Far from being static, living organisms exhibit marked dynamics in metabolism and function. Major elements of such dynamics are the endogenous rhythms of varying frequencies (such as ultradian, circadian, and infradian). These biological rhythms, which respond to exogenous

environmental cues, are normally a complex mix of phase–locked rhythms and have significant impacts on the physiological and psychological well–being of the organism. Biochemical processes, cellular communications, and functional systems are all intimately associated with endogenous rhythms, as is overall systemic response to the environment. Dysfunctions in these underlying rhythms can profoundly affect the organism and lead to a variety of biological effects.

A number of investigations have been conducted to examine the effects of ELF electromagnetic fields on natural biological rhythms. Following Wever's significant findings (Wever, 1971) on the influence of electromagnetic fields on humans, several studies have been performed. Researchers in Ehret's laboratory (Rosenberg, 1983; and Duffy and Ehret, 1982) used metabolic indicators to examine both circadian and ultradian rhythms in mice and rats exposed to 60 Hz electric fields. They observed no effects of exposure in rats, but they reported that the activity and rhythms of oxidative metabolism in male mice could be phase shifted by exposure.

Wilson et al. (1981) directly examined another aspect of circadian activity in rats by measuring the cyclical pineal production of indolamines and enzymes. A significant reduction in the normal nighttime rise of melatonin and biosynthetic enzymes in the pineal gland was observed in rats exposed to either 1.5 or 40 kV/m. Furthermore, the change in pineal indole response occurred only after at least 3 weeks of chronic exposure (Wilson et al., 1986). There was also a suggestion of phase–shifting in young rats exposed to 60 Hz fields (Reiter et al., 1988).

Sulzman and Murrish (1987) investigated the effects of ELF fields on circadian function in squirrel monkeys. In an examination of exposure to a range of electric field intensities (2.6, 26, and 39 kV/m), accompanied by a 100 μT magnetic field, they reported apparent intensity–related effects. None of the monkeys exposed to 2.6 kV/m showed any change in activity or feeding after 2 weeks of exposure. However, 33% of the monkeys exposed to 26 kV/m and 75% of those exposed to 39 kV/m had significant changes in their circadian cycles.

Although firm conclusions cannot yet be made regarding potential health impacts from ELF effects on circadian or biological rhythms, it is apparent that EM fields can alter the circadian timing mechanisms in mammals. Much work remains to be accomplished before the observed effects and their biological consequences are clearly understood. It seems probable that ELF effects on rhythms, particularly those mediated by neuroendocrine systems, could play an important role in other areas of observed bioeffects, such as behavior and development.

Neurochemistry/Neurophysiology

The relationship between the neurotransmitters norepinephrine and epinephrine and the physiological responses of stress and arousal is well established. As researchers began to look for potential biological effects of ELF electric fields, measuring these transmitters became one of the assessments used to examine the nervous system for evidence of a stress response in animals. This approach, which benefited from the ease of measuring these chemicals in serum, urine, or brain tissue, specifically addressed reports that ELF fields act as mild stressors (Marino and Becker, 1977). Unfortunately, potential methodological problems raised serious questions about the validity of the results from early studies (Michaelson, 1979). Experimental design and methodology problems have also contributed to contradictory results in some of the more recent studies.

In general, neurochemical data provide relatively weak evidence that exposure to electric fields in the power–frequency range may cause slight changes in nervous system function (see review by Anderson, 1990). The number of experiments is not large, and there are significant questions about the validity of several of the studies. Nevertheless, the findings support the hypothesis that ELF exposure alters internal rhythms, increases arousal in animals, and is transient in its effect.

Several laboratories have examined the morphology of brain tissue from animals exposed to ELF electric fields. Carter and Graves (1975) and Bankoske et al. (1976) exposed chicks to 40 kV/m E fields and saw no effects on central nervous system morphology. This finding was supported by those of Phillips et al. (1978), who examined rats exposed to 100 kV/m for 30 days. Again, no morphological evidence of an electric field effect was observed. In a study in Sweden (Hansson, 1981), dramatic changes in cell structure were reported in the cerebella of rabbits exposed to a 14 kV/m E field. However, these reported changes must be interpreted with caution. The animals were exposed outdoors and showed evidence of significant health deficits (whether resulting from the electric field, other environmental conditions, or some combinations of these factors is not clear). Furthermore, results from these studies are in conflict with experiments conducted by Portet and Cabanes (1988), in which no ultrastructural changes occurred in the cerebella of young rabbits exposed to 50 kV/m. These questions concerning neuroanatomical changes have yet to be resolved. However, the lack of obvious, significant functional deficits in the central nervous systems of thousands of animals exposed to date suggests that the dramatic morphological alterations in exposed rabbits may result from conditions unrelated to electric field exposure. The possibility of synergistic effects from the E field and a stressful environment cannot be ruled out.

Because the nervous system is by nature electrically sensitive, it has been assumed to be particularly sensitive to influence by external ELF fields. To some degree this assumption has been borne out by experimental results, although in the area of neurophysiology a confusing array of studies have claimed both effects and no effects of ELF field exposure. A case in point is the commonly used measure of general central nervous system activity, the EEG. Alterations in the EEG by electric and magnetic field exposures have been observed (Blanchi et al., 1973; Silney, 1979). Other studies have shown no effects (Graves et al., 1978; Silney, 1981).

In an assessment of a more specific electric "fingerprint" of the brain, the visual evoked response (VER), no effects of exposure were observed in adult or developing rats. Jaffe et al. (1983) assessed the VER in 114 rats exposed in utero through 20 days post partum. The dams, fetuses, and subsequent pups were exposed to a 65 kV/m, 60 Hz electric field. No consistent, statistically significant effects of exposure were observed. Wolpaw and associates (1987) examined evoked potentials in pig–tailed macaques exposed to combined electric and magnetic fields. As in Jaffe's studies, the VER and the auditory evoked potential showed no changes caused by exposure. However, an attenuation of the late components of the somatosensory evoked potentials was demonstrated in exposed animals. The authors suggest that these abnormalities may have been due to a particularly large numer of stimuli giving rise to a change in the mechanisms of attenuation.

Two other neurophysiological studies have had clear, replicable results. Jaffe et al. (1980) examined synaptic junctions from chronically exposed rats (60 Hz and 100 kV/m for 30 days). In these studies, presynaptic fibers were stimulated with a pair of above–threshold pulses. The height ratio of the resultant action potentials, observed as a function of the interspike interval, demonstrated an enhanced neuronal excitability in nerves from exposed animals. However, many other parameters tested in these nerves showed no changes in exposed animals. In a second experiment Jaffe et al. (1981) examined a wide range of physiological parameters of the peripheral nervous system and neuromuscular function. The only effect observed was slightly faster recovery from fatigue after chronic stimulation in one class of muscle; the soleus, slow–twitch muscle.

In summary, numerous studies have been initiated to determine how greatly an electrical environment containing electric or magnetic fields of ELF affects the nervous system. Many of the experiments have not confirmed any neuropathological effects, even after prolonged exposures to high–strength (100 kV/m) electric fields and high–intensity (5 mT) magnetic fields (Tenforde, 1985). As discussed previously, nervous system effects that have been observed include altered neuronal excitability, altered circadian levels of pineal hormones, and behavioral aversion to or preference for the field. In addition, in several instances where unconfirmed or controversial data exist, observed effects may or may not be real. Examples are changes in serum catecholamines or corticosteroids, morphology of brain tissue, and EEG waveforms. Possibly these and other putative effects are due to a direct interaction of the electric field with tissue or to an indirect interaction, such as a physiological response owing to detection or stimulation of sensory receptors by the field. The nature of the physical mechanisms involved in field–induced effects is obscure, and elucidating them is one of the urgent goals of current research.

The behavioral tests that most frequently showed an effect of exposure were those relating to detection of the field or to activity responses. Most other behaviors did not change with ELF field exposure. It should also be noted that influences of the nervous system on other biological systems are often mediated indirectly through neuroendocrine or endocrine responses.

Reproduction and Development

Developing organisms, including prenatal and postnatal mammals, are generally considered more sensitive to physical or chemical agents than are adult animals (Mahlum et al., 1978). This greater sensitivity, when it occurs, is thought to originate in subtle effects on the increased number and activity of processes and controls that guide the developing cellular interactions. A number of studies have examined the effects of ELF exposure on reproduction and development of both mammalian and nonmammalian species. These studies have been assessed in detail by other reviewers (e.g. Sikov, 1985) and are briefly highlighted here.

Most of the nonmammalian studies have been performed on chickens or pigeons. Many studies have indicated that electric field exposure of chicks at several field strengths, before and after hatching, did not significantly affect viability, morphology, behavior, or growth (e.g. Reed and Graves, 1984). However, in one series of experiments chicks exposed to 40 or 80 kV/m on days 1 to 22 after hatching showed significantly less motor activity during the week after removal from the field (Graves et al., 1978).

Unreplicated studies have given some indications that exposure of prenatal mammals to electric fields produces deleterious effects on postnatal growth and survival (e.g. Marino et al., 1980; Sikov et al., 1987). These results are countered by others in which rats, rabbits, or mice were exposed to 20, 50, 100, 200, or 240 kV/m and no effects on reproduction, survival, or growth and development were demonstrated (e.g. Portet and Cabanes, 1988; Sikov et al., 1984; Rommereim et al., 1987).

Conflict thus remains over results of studies investigating the potential for ELF electric or magnetic field exposure to affect reproduction and development. This confusion over results indicates the need for carefully designed, statistically sound experiments that will help clarify this important area of investigation.

Other Biological Functions

<u>Bone growth and repair</u> – One report of animals exposed to 60 Hz electric fields (McClanahan and Phillips, 1983) indicated that bone growth per se was not affected by exposure to 100 kV/m. However, this study, as well as an additional report (Marino et al., 1979), suggested that bone fracture repair was retarded in rats and mice exposed to 5 or 100 kV/m, 60 Hz fields but not to very low (1 kV/m) field strengths. In the report from McClanahan and Phillips, it is suggested that exposure affects the rate of healing but not the strength of the healed bone.

In contrast to the experiments with sinusoidal 60 Hz fields, research studies and clinical trials have been performed in which magnetic fields were used to treat bone fractures and arthroses in humans. The weak electrical currents induced in bone tissue by magnetic field pulses may enhance fracture repair by altering intracellular concentrations of calcium ions, thus modifying cellular metabolism and stimulating growth of the osteoblasts and chondrocytes (Luben et al., 1982). Why 60 Hz sinusoidal electric fields cause a retardation of fracture repair, whereas pulsed magnetic fields facilitate repair, is unclear.

<u>Cardiovascular system</u> – Cardiovascular function has been assessed by measuring blood pressure and heart rate and performing electrocardiographic

measurements. Early studies indicated as possible effects a decrease in heart rate and cardiac output in dogs exposed to 15 kV/m (Gann, 1972) and increased heart rates in chickens exposed to 80 kV/m (Carter and Graves, 1975). A more recent and comprehensive study in rats exposed to 100 kV/m showed no such effects of exposure, even when the animals were subjected to cold stress (Hilton and Phillips, 1980). Cerretelli and Malagute (1976) reported transient increases in blood pressure in dogs exposed to 50 Hz E fields greated than 10 kV/m. Hilton and Phillips (1980) were unable to confirm a report by Blanchi et al. (1973) of electrocardiographic changes in animals exposed to 100 kV/m.

Magnetic field exposure of dogs (50 Hz, 2 T) caused a stimulation of the heart in the diastolic phase, with salvos of ectopic beats appearing in the recordings (Silney, 1985). Humans exposed to combined electric and magnetic fields (9 kV/m, 20 µT) have demonstrated a longer cardiac interbeat interval than sham-exposed subjects (Graham et al., 1987).

Serum chemistry appears to be relatively unaffected by exposure to either ELF electric or magnetic fields (Ragan et al., 1983; Mathewson et al., 1977; Ragan et al., 1979). Hematological data, however, present a more confusing picture. With electric field exposure, white blood cell count was often elevated in populations of mice and rats (Ragan et al., 1983; Graves et al., 1979). With the exception of one report by Tarakhovsky et al. (1971), all the published studies on hematological effects of magnetic field exposure have shown no field-associated effects (e.g., deLorge, 1974; Fam, 1981; Beischer et al., 1973; Sander et al., 1982). The occasional positive or negative effects on the hematopoietic system must be carefully evaluated. Apparent sporadic effects may not be biologically or statistically significant; particularly when appropriate multivariate analyses are used to evaluate the wide range of hematological and serum chemistry parameters.

<u>Immunology</u> – There is some indication that exposure of animals to electric fields does not markedly affect the immune system. In a comprehensive investigation of the humoral and cellular aspects of the immune system, Morris et al. (1982) observed no effects of exposure at very low field strengths (150 to 250 V/m) in mice or rats. In subsequent experiments at higher field exposures (100 kV/m), no effects were seen in immune system response. However, Lyle et al. (1988) reported significant decrements in the cytolytic capacity of lymphocytes exposed to radiofrequency fields modulated at 60 Hz. Further work with 60 Hz electric fields alone also resulted in a suppression of T-lymphocyte cytotoxicity. A significant difference between the work reported from these two laboratories is that

Morris measured lymphocyte responses from exposed animals, whereas Lyle exposed lymphocytes in culture.

In contrast to the apparent lack of strong or consistent electric field influences on the immune system, immunoresponse to mitogens and antigens appears to be significantly susceptible to ELF magnetic fields (Conti et al., 1983).

Carcinogenesis and Mutagenesis

No effects suggesting a direct effect of electric field exposure on mutagenesis or carcinogenesis have been observed (Frazier et al., 1987). However, there is considerable research interest on this question due to an increasing number of epidemiological studies that suggest a possible association between ELF magnetic field exposure and cancer. As yet, only a few published laboratory studies, conducted in animals, bear directly on this question and there is an urgent need for such studies.

No studies to date have been reported in which spontaneous tumor development was followed in normal animals exposed to ELF fields. Another possible approach would be a cocarcinogenesis system in which EM exposure is used as a promoter following an initiating event (chemical or ionizing radiation).

MAJOR ISSUES AND RESEARCH GAPS

An overview of the literature suggests that exposure to ELF electric and magnetic fields definitely produces biological effects, although many indices of physiological status appear relatively unaffected by exposure. Most of the biological effects reported are quite subtle, and evidence suggests that, at least with short term exposure, these fields impose a relatively low potential hazard to biological systems. Still to be addressed are questions of longer term exposure and possible associated health risks.

Although the primary focus of health related issues currently appears to be carcinogenesis, care must be taken not to define the scope of a research agenda too narrowly. For example, areas associated with the nervous system have demonstrated the greatest number of effects, including altered neuronal excitability, neurochemical changes, altered hormone levels, and changes in behavioral responses. The fundamental science involved appears to be complex and may lead in a variety of directions. However, this area of research has the greatest potential for investigating interactions between EM fields and tissues in whole animal systems.

Knowledge about the mechanisms underlying observed biological effects is incomplete and remains a major challenge of this area of research. Further investigation is particularly needed to determine the mechanisms of effects that display nonlinear responses to field strength and frequency. A large number of models have been proposed to explain the electrochemical, cellular, and biochemical changes that occur as a result of ELF electric and magnetic field exposure. The need to define basic mechanisms of interaction has been enhanced because of emerging information on possible field associations with immunity, cancer, cell–to–cell communication, differentiation, and repair processes.

In vivo studies to date, have primarily concentrated on defining effects of exposure. In addition to such basic information, a much greater effort should be made to design and conduct studies that can evaluate the influence of electric and magnetic fields in cellular and animal systems, with emphasis on testing proposed mechanisms in systems sensitive to these fields.

No studies clearly demonstrate deleterious effects of ELF electric or magnetic exposure on mammalian reproduction and development, but several suggest such effects. Further efforts in this area of research should focus on the exposure of animals to the more complex fields of the work place.

Finally, in the area of carcinogenesis; clearly a large gap in our knowledge about possible EM involvement in this disease stems from the lack of any substantive animal research. Studies using accepted animal cancer models must be performed. These should include transplantation experiments wherein transformed cells are introduced into animals subsequently exposed to EM fields. Cocarcinogen type experiments would also be of value to examine the progression of cancer that had been initiated by known carcinogens (typically chemicals or ionizing radiation). A third type of study would determine   development of spontaneous tumors in normal animals or in animals predisposed to a specific type of cancer.

ACKNOWLEDGMENT

Portions of this chapter were previously published in "Biological Effects and Medical Applications of Electromagnetic Energy" Om Gandhi, Editor. Work supported in part by U.S. Department of Energy under Contract DE–AC06–76RLO–1830.

REFERENCES

Anderson, L.E. (1990). Biological effects of extremely low frequency and 60 Hz fields. In: Biological Effects and Medical Applications of Electromagnetic Energy, Gandhi, O., ed. Prentise Hall, Engelwood Cliffs, NJ., pp. 196–235.

Anderson, L.E., and Phillips, R.D. (1985). Biological effects of electric fields: an overview. In: Biological effects and dosimetry of static and ELF electromagnetic fields, Grandolfo, M., Michaelson, S.M., and Rindi, A. (eds.). Plenum Publishing Corp., New York, pp. 345–378.

Bankoske, J.W., McKee, G.W., and Graves, H.B. (1976). Ecological influence of electric fields. Interim Report 2, EPRI Research Project 129, EPRI Report No. EX–178. Palo Alto, CA, Electric Power Research Institute.

Beischer, D.E., Grissett, J.D., and Mitchell, R.R. (1973). Exposure of man to magnetic fields alternating at extremely low frequency, USN Report No. NAMRL–1180. Pensacola, FL, Naval Aerospace Medical Research Laboratory.

Blanchi, C., Cedrini, L., Ceria, F., Meda, E., and Re, G. (1973). Exposure of mammalians to strong 50–Hz electric fields: effect on heart's and brain's electrical activity. Arch. Fisiol. 70:33–34.

Carter, J.H., and Graves, H.B. (1975). Effects of high intensity AC electric fields on the electroencephalogram and electrocardiogram of domestic chicks: literature review and experimental results. University Park, Pennsylvania State University.

Cerretelli, P., and Malaguti, C., (1976). Research carried on in Italy by ENEL on the effects of high voltage electric fields. Rev. Gen. Elect. (spec. iss.):65–74.

Conti, P., Gigante, G.E., Cifone, M.G., et al. (1983). Reduced mitogenic stimulation of human lymphocytes by extremely low frequency electromagnetic fields. FEBS Lett. 162:156–160

Davis, H.P., Mizumori, S.J.Y., Allen, H., Rosenzweig, M.R., Bennett, E.L., and Tenforde, T.S. (1984). Behavioral studies with mice exposed to DC and 60–Hz magnetic fields. Bioelectromagnetics 5:147–164.

deLorge, J. (1974). A psychobiological study of rhesus monkeys exposed to extremely low–frequency low–intensity magnetic fields. USN Report NAMRL–1203. NTIS No. AD A000078, Springfield, VA., NTIS.

Duffy, P.M. and Ehret, C.F. (1982). Effects of intermittent 60–Hz electric field exposure: circadian phase shifts, splitting, torpor, and arousal responses in mice. In: Abstracts, 4th Annual Scientific Session, Biolectromagnetics Society, June 28–July 2, Los Angeles, Gaithersburg, MD., Bioelectromagnetics Society, p. 610.

Fam, W.Z. (1981). Biological effects of 60–Hz magnetic field on mice. IEEE Trans. Magnet. 17:1510–1513.

Frazier, M.E., Samuel, J.E., and Kaune, W.T. (1987). Viabilities and mutation frequencies of CHO–K1 cells following exposure to 60–Hz electric fields. In: Anderson, L.W., Weigel, R.J., and Kelman, B.J. (eds.). Interaction of biological systems with static and ELF electric and magnetic fields. Proceedings of the 23rd Annual Hanford Life Sciences Symposium. DOE Symposium Series CONF 841041, Springfield, VA., National Technical Information Service, pp. 255–268.

Gann, D.W. (1976). Biological effects of exposure to high voltage electric fields: final report. Electric Power Research Institute, RP98–02. Palo Alto, CA.

Graham, C., Cohen, H.D., Cook, M.R., Phelps, J., Gerkovich, M. and Fotopoulos, S.S. (1987). A double–blind evaluation of 60–Hz field effects on human performance, physiology, and subjective state. In: Anderson, L.E., Weigel, R.J., Kelman, B.J. (eds.). Interaction of biological systems with static and ELF electric and magnetic fields. Proceedings of the 23rd Annual Hanford Life Sciences Symposium. DOE Symposium Series CONF 841041, Springfield, VA., National Technical Information Service, pp. 471–486.

Graves, H.B., Carter, J.H., Kellmel, D., Cooper, L., Poznaniak, D.T., and Bankoske, J.W. (1978). Perceptibility and electrophysiological responses of small birds to intense 60–Hz electric fields. IEEE Trans. Power Appar. Syst. 97:1070–1073.

Graves, H.B., Long, P.D., and Poznaniak, D.T. (1979). Biological effects of 60–Hz alternating–current fields: a Cheshire cat phenomenon? In: Phillips, R.D. (ed.). Biological effects of extremely–low–frequency electromagnetic fields. Proceedings of the 18th Annual Hanford Life Sciences Symposium. CONF–781016, Springfield, VA., NTIS, pp. 184–197.

Hansson, H.–A. (1981). Lamellar bodies in Purkinje nerve cells experimentally induced by electric field. Exp. Brain Res. 216:187–191.

Hilton, D.I., and Phillips, R.D. (1980). Cardiovascular response of rats exposed to 60–Hz electric fields. Bioelectromagnetics 1:55–64.

Hjeresen, D.L., Kaune, W.T., Decker, J.R., and Phillips, R.D. (1980) Effects of 60–Hz electric fields on avoidance behavior and activity of rats. Bioelectromagnetics 1:299–312.

Hjeresen, D.L., Miller, M.C., Kaune, W.T., and Phillips, R.D. (1982). A behavioral response of swine to 60–Hz electric field. Bioelectromagnetics 2:443–451.

Jaffe, R.A., Laszewski, B.L., and Carr, D.B. (1981). Chronic exposure to a 60–Hz electric field: effects on neuromuscular function in the rat. Bioelectromagnetics 2:277–239.

Jaffe, R.A., Laszewski, B.L., Carr, D.B., and Phillips, R.D. (1980). Chronic exposure to 60–Hz electric field: effects on synaptic transmission and the peripheral nerve function in the rat. Bioelectromagnetics 1:131–147.

Jaffe, R.A., Lopresti, C.A., Carr, D.B., and Phillips, R.D. (1983). Perinatal exposure to 60–Hz electric fields: effects on the development of visual–evoked response in the rat. Bioelectromagnetics 4: 327–339.

Kaune, W.T., Phillips, R.D., Hjeresen, D.L., Richardson, R.L., and Beamer, J.L. (1978). A method for the exposure of minature swine to vertical 60–Hz electric fields. IEEE Trans. Biomed. Eng. 25:276–283.

Lovely, R.H. (1988). Recent studies in the behavioral toxicology of ELF electric and magnetic fields. In: O'Connor, M.E., and Lovely, R.H. (eds.). Electromagnetic fields and neurobehavioral function. New York, A.R. Liss, Inc., p. 2.

Luben, R.A., Cain, D.D., Chen, M.C.-y., Rosen, D.M., and Adey, W.R. (1982). Effects of electromagnetic stimuli on bone and bone cells in vitro: inhibition of responses to parathyroid hormone by low-energy low-frequency fields. Proc. Natl. Acad. Sci. USA 79:4180-4184.

Lyle, D.B., Ayotte, R.D., Sheppard, A.R., and Adey, W.R. (1988). Suppression of T-lymphocyte cytotoxicity following exposure to 60-Hz sinusiodal electric fields. Bioelectromagnetics 9:303-313.

Mahlum, D.D., Sikov, M.R., Hackett, P.L., and Andrew, F.D. (1978). Developmental toxicology of energy-related pollutants. CONF 771017, Springfield, VA., NTIS.

Marino, A.A., and Becker, R.O. (1977). Biological effects of extremely low-frequency electric and magnetic fields: a review. Physiol. Chem. Phys. 9:131-147.

Marino, A.A., Cullen, J.M., Reichmanis, M., and Becker, R.O. (1979). Fracture healing in rats exposed to extremely low-frequency electric fields. Clin. Orthop. 145:239-244.

Marino, A.A., Reichmanis, M., Becker, R.O., Ullrich, B., and Cullen, J.M. (1980). Power frequency electric field induces biological changes in successive generations of mice. Experientia 36:309-311.

Mathewson, N.S., Oosta, G.M., Levin, S.G., Diamond, S.S. , and Ekstrom, M.E. (1977). Extremely low frequency (ELF) vertical electric field exposure of rats: a search for growth, food consumption and blood metabolite alterations. Armed Forces Radiobiology Research Institute, Defense Nuclear Agency, Bethesda, MD., ADA 035954, Springfield, VA., NTIS.

McClanahan, B.J., and Phillips, R.D. (1983). The influence of electric field exposure on bone growth and fracture repair in rats. Bioelectromagnetics 4:11-20.

Michaelson, S.M. (1979). Analysis of studies related to biologic effects and health implications of exposure to power frequencies. Environ. Prof. 1:217–232.

Morris, J.E., and Phillips, R.D. (1982). Effects of 60–Hz electric fields on specific humoral and cellular components of the immune system. Bioelectromagnetics 3:341–348.

National Academy of Sciences. (1977). Biologic effects of electric and magnetic fields associated with proposed Project Seafarer, report of the Committee on Biosphere Effects of Extremely–Low–Frequency Radiation. Washington, DC., National Research Council.

Phillips, R.D., Chandon, J.J., Free, J.J., et al. (1978). Biological effects of 60–Hz electric fields on small laboratory animals. Annual Report, DOE Report No. HCP/T1830–3. Washington, DC., Department of Energy.

Portet, R.T., and Cabanes, J. (1988). Development of young rats and rabbits exposed to a strong electric field. Bioelectromagnetics 9:95–104.

Ragan, H.A., Buschbom, R.L., Pipes, M.J., Phillips, R.D., and Kaune, W.T. (1983). Hematologic and serum chemistry studies in rats and mice exposed to 60–Hz electric fields. Bioelectromagnetics 4:79–90.

Ragan, H.A. Pipes, M.J. Kaune, W.T., and Phillips, R.D. (1979). Clinical pathologic evaluations in rats and mice chronically exposed to 60–Hz electric fields. In: Phillips, R.D., et al. (eds.). Biological effects of extremely–low–frequency electromagnetic fields. Proceedings of the 18th Annual Hanford Life Sciences Symposium. CONF–781016, Springfield, VA., National Technical Information Service, pp. 297–325.

Reed, T.J., and Graves, H.B. (1984). Effects of 60–Hz electric fields on embryo and chick development, growth, and behavior, Vol. 1. Final Report, EPRI Project 1064–1. Palo Alto, CA., Electric Power Research Institute.

Reiter, R.J., Anderson, L.E., Buschbom, R.L., and Wilson, B.W. (1988). Reduction of the nocuturnal rise in pineal melatonin levels in rats exposed to 60–Hz electric fields in utero and for 23 days after birth. Life Sci. 42:2203–2206.

Rogers, W.R., Feldstone, C.S., Gibson, E.G., Polonis, J.J., Smith, H.D., and Cory, W.E. (1987). Effects of high–intensity, 60–Hz electric fields on operant and social behavior of nonhuman primates. In: Anderson, L.E., Weigel, R.J., and Kelman, B.J. (eds.). Interaction of biological systems with static and ELF electric and magnetic fields. Proceedings of the 23rd Annual Hanford Life Sciences Symposium. DOE Symposium Series, CONF 841041, Springfield, VA., National Technical Information Service, pp. 365–378.

Rommereim, D.N., Kaune, W.T., Buschbom, R.L., Phillips, R.D., and Sikov, M.R. (1987). Reproduction and development in rats chronologically exposed to 60–Hz electric fields. Bioelectromagnetics 8:243–258.

Rosenberg, R.S., Duffy, P.H., Sacher, G.A., and Ehret, C.F. (1983). Relationship between field strength and arousal response in mice exposed to 60–Hz electric fields. Bioelectromagnetics 4:181–191.

Sander, R., Brinkmann, J., and Kuhne, B. (1982). Laboratory studies on animals and human beings exposed to 50–Hz electric and magnetic fields. In: International Conference on Large High Voltage Electrical Systems (abstracts), No. 36–01. Paris, CIGRE.

Sheppard, A.R., and Eisenbud, M. (1977). Biologic effects of electric and magnetic fields of extremely low frequency. New York, New York Unviersity Press.

Sikov, M.R. (1985). Reproductive and developmental alterations associated with exposure of mammals to ELF (1–300 Hz) electromagnetic fields. In: Assessments and viewpoints on the biological and human health effects of extremely low frequency (ELF) electromagnetic fields. Washington, DC., American Institute of Biological Sciences, pp. 295–311.

Sikov, M.R., Montogomery, L.D., Smith, L.G, and Phillips, R.D. (1984). Studies on prenatal and postnatal development in rats exposed to 60–Hz electric fields. Bioelectromagnetics 5:101–112.

Sikov, M.R., Rommereim, D.N., Beamer, J.L., Buschbom, R.L., Kaune, W.T., and Phillips, R.D. (1987). Developmental studies of Hanford miniature swine exposed to 60–Hz electric fields. Bioelectromagnetics 8:229–242.

Silney, J. (1979). Effects of electric fields on the human organism. Institut zur Enforschung Electrischer Unfalle, Cologne, Medizinisch – Technischer Berichte, p. 39.

Silney, J. (1981).   Influence of low–frequency magnetic field on the organism. Proc. EMG, Zurich, pp. 175–180, Proceed 4th Symposium on Electromagnetic Compatibility, March 10–12, 1981.

Silney, J. (1985). The influence thresholds of the time–varying magnetic field in the human organism. In: Berhardt, J. (ed.). Proceedings of the Symposium on Biological Effects of Static and ELF Magnetic Fields. Neuherberg, May 13–15, 1985, BGA–Schriftenreibe, Munchen, MMv Medizin Verlag.

Smith, R.F., and Justesen, D.R., (1977). Effects of a 60–Hz magnetic field on activity levels of mice. Radio Sci. 12:279–285.

Stern, S., Laties, V.G., Stancampiano, D.V., Cox, C., and De Lorge, J.O. (1983). Behavioral detection of 60–Hz electric fields by rats. Bioelectromagnetics 4:215–247.

Sulzman, F.M., and Murrish, D.E. (1987). Effects of electromagnetic fields on primate circadian rhythms. Report to the New York State Power Lines Project, Albany, NY., Wadsworth Center for Laboratories and Research.

Tarakhovsky, M.L., Sambroskaya, Y.P., Medvedev, B.M., Zaderozhnaya, T.D., Okhronchuk, B.V., and Likhtenshtein, E.M. (1971).   Effects of constant and variable magnetic fields on some indices of physiological function and metabolic processes in albino rats. Fiziol. Zh. RSR–17:452– 459. (English translation JPRS 62865:37–46).

Tenforde, T.S. (1985). Biological effects of ELF magnetic fields. In: Biological and human health effects of extremely low frequency electromagnetic fields. Arlington, VA., American Institute of Biological Sciences, pp. 79–128.

Weigel, R.J., Jaffe, R.A., Lundstrom, D.L., Forsythe, W.C., and Anderson, L.E. (1987a). Stimulation of cutaneous mechanoreceptors by 60–Hz electric fields. Bioelectromagnetics 8:337–350.

Weigel, R.J., and Lundstrom, D.L. (1987b). Effect of relative humidity on the movement of rat vibrissae in a 60–Hz electric field. Bioelectromagnetics 8:107–110.

Wever, R., (1971). Influence of electric fields on some parameters of circadian rhythms in man. In: Menaber, M. (ed.). Biochronometry, Washington, DC., National Academy of Sciences, pp. 117–132.

Wilson, B.W., Anderson, L.E., Hilton, D.I., and Phillips, R.D. (1981). Chronic exposure to 60–Hz electric fields: effects on pineal function in the rat. Bioelectromagnetics 2:371–380.

Wilson, B.W., Chess, E.K., and Anderson, L.E. (1986). 60–Hz electric field effects on pineal melatonin rhythms: time course for onset and recovery. Bioelectromagnetics 7:239–242.

Wolpaw, J.R., Seegal, R.F., Lowman, R.I., and Satya–Murti, S., (1987). Chronic effects of 60 Hz electric and magnetic fields on primate central nervous system function. Final Report to the New York State Power Lines Project, Albany, NY., Wadsworth Center for Laboratories and Research.

VDT MEASUREMENTS AND ANIMAL STUDIES

Stuart M. Harvey

Ontario Hydro Research Division
Toronto, Ontario

Introduction

Most of the computer terminals and desktop computers presently in use display visual information on the screen of a device known as a video display terminal, or VDT. The VDT is similar to a small television set and employs the same type of cathode-ray-tube (CRT) technology. It is a common workstation feature in today's modern office.

The CRT, or picture tube, is essentially an evacuated glass bulb with an electron beam source at one end and a phosphor-coated surface at the other. The display is created by accelerating an intensity-modulated beam of electrons towards the phosphor-coated surface, and then deflecting the fast-moving beam back and forth, and up and down, across the surface. Light is produced by the phosphor in proportion to the intensity of the beam as it strikes each point on the surface. The CRT is an electromagnetically operated device: electric fields are used to accelerate the electron beam, and magnetic fields are used to deflect it.

During the 1970's, a period that initiated a revolution in office technology based on the computer, concerns were generated about the possibility of harmful radiation from the video terminals. Since that time, all types of electromagnetic emissions from gamma rays, through x-rays and microwave radiation, to static fields, have been looked for, analysed, and assessed (Bergqvist, 1984; Walsh et al., 1985; Marriott and Stuchly, 1986; Elliott et al., 1986). Today, the electromagnetic environment of the VDT is arguably better known than that of any other office machine. It is considered by most scientists to be completely benign, with the possible exception of the residual electric and magnetic fields, spanning a frequency range from 0 to a few hundred kHz, that are produced outside of the VDT housing by the beam acceleration and deflection systems of the CRT.

In this paper, I will look at the physical and engineering aspects of VDT field characterization and of testing for adverse health effects with small animals. Owing to the complex nature of these subjects, a brief review cannot do justice to the technical information contained in original reports and papers. The reader is encouraged to consult the references for more detailed descriptions.

Field Measurement and Characterization

The electric and magnetic fields produced by the VDT may be detected within one or two meters of the surface with simple sensors and conventional display instruments. A quick survey will reveal that the fields resemble in their spatial distribution the fields of classical electric and magnetic dipoles: the field lines are not, in general, radially or azimuthally directed but return to the unit, producing an inverse third power dependence on distance from the source conductors for distances that are larger than the dimensions of the conductors. The electric and magnetic fields are generated by different sources within the unit, and must be measured as independent quantities. There is no detectable radiated power.

The fields have both static and time varying components. The time varying components have a complex spectrum containing the power frequency (60 Hz in Canada and the U.S.), the vertical refresh frequency (typically between 60 and 70 Hz), the horizontal raster or flyback frequency (typically between 16 and 22 kHz), some electron beam modulation frequencies, and their harmonics. Fields having frequencies between 30 and 300 Hz are usually referred to as extremely low frequency (ELF) fields, and those having frequencies between 3 and 30 kHz are usually referred to as very low frequency (VLF) fields. Figure 1 shows representative waveforms for VLF electric and magnetic fields.

Owing to the complex spatial and time variations, and the possibility of direct interaction between the sensor and the source circuits, quantitative field measurements within a few tens of cm from the unit present some technical difficulties. Some of these difficulties are reviewed in the following sections.

<u>Measurement of magnetic fields</u>

It is customary to measure the time-varying magnetic fields with multi-turn pickup coils (Figure 2a). The frequency response of the coil or signal conditioning equipment must be tailored to yield true waveforms or true rms (root-mean-square) field values and it is usually necessary to use two coils, one for the ELF frequency band and one for the VLF band. Static fields,

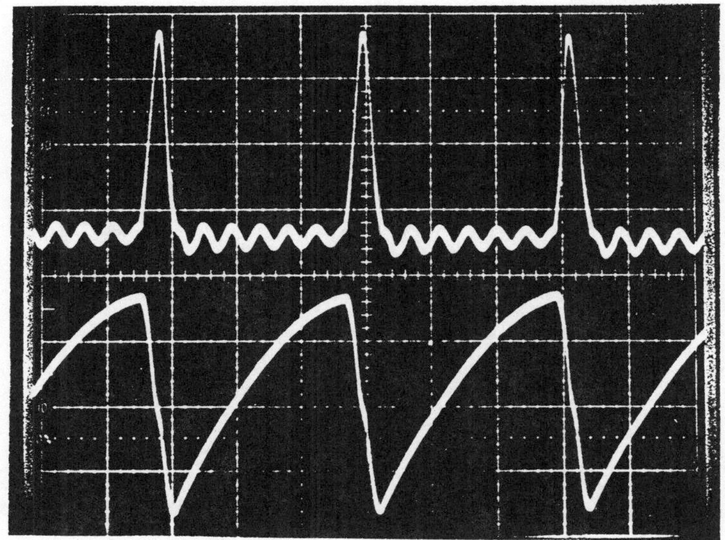

Figure 1: VLF electric (top) and magnetic (bottom)
field waveforms.

if present, are unlikely to be detectable in the presence of the static field
from the earth, and have not been reported.

The size and the impedance of the coil are important factors when
measurements are to be made close to the VDT case. A coil with a 5 cm
radius located at a distance of 10 cm from a dipole source will "read" the
field at an actual distance of about 8 cm, a field value difference of about
50%. The source and the pickup coil form the primary and secondary
windings of an air-cored transformer and, if appreciable current is drawn
from the pickup coil, the field may be affected through transformer action.

A single coil needs to be rotated to find the major field axis, and the
harmonic content of the field will be found to be dependent on the coil
orientation. The need for rotation can be removed by combining the outputs
of three concentric and mutually orthogonal coils, as provided by some
commercially available instrumentation. However, the resultant field value
may be different from the value recorded by a properly oriented single coil.

Notwithstanding the above, true magnetic fields can be measured with
adequate precision if properly calibrated instrumentation is used at a distance
of 30 cm or more from the cover of the VDT.

Measurement of Electric Fields

It is customary to measure the electric fields with capacitive sensors (Figure 2b) having either a dipole or a parallel plate configuration. As a broad frequency response can be obtained without great difficulty, a single sensor may be used for both the ELF and the VLF bands and even, if necessary, for the static fields as well. The electric fields have a complex waveform and true waveform display or true rms field amplitude readings are essential. Sensors that employ simple half-wave diode detection of the field-generated voltage will read incorrectly.

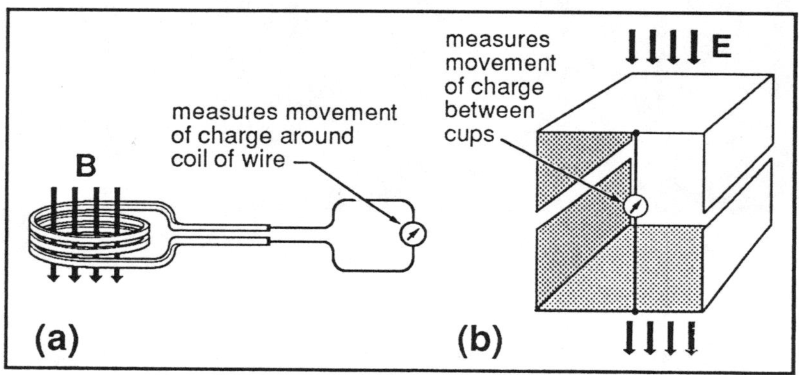

Figure 2: (a) magnetic field sensor, (b) electric field sensor.

All capacitive sensors modify, or perturb, the electric field being measured. This is not a problem for spatially uniform fields or for radiative far fields, where the perturbation can be taken into account either theoretically or by appropriate calibration procedures. In the complex fringing field environment close to the VDT, however, no single calibration factor will yield the true unperturbed fields. Two alternative approaches have been adopted to circumvent this difficulty.

In the first approach, the perturbing effect of the sensor is made as small as possible in order to approximate free-space field conditions, and the measurements are then reported as the field at the sensor position without correction. As with pickup coils, the size of the sensor, the orientation, and the coupling to the source of the field can influence results close to the VDT case.

In the second approach, the sensor is configured to produce specific perturbations similar to those produced by the body of the VDT operator, and the measurements are then reported as representing the fields that actually occur at the operators body. While properly proportioned conductive mannequins would be ideal for this purpose, parallel plate sensors with large (typical dimension about 30 cm) grounded reference planes are preferentially used. Recorded fields will be higher than true free-space field values. However it is possible, at least in principle, to calculate correction factors for measured sensor voltages and currents that would yield an equivalent uniform unperturbed field suitable for comparison with radiation exposure guidelines (Harvey, 1983a; Guy, 1987).

The path between measured fields, actual fields, and equivalent operator exposure is a complex one for very localized electric fields such as those produced by display terminals. Because it is difficult to navigate, I refer to this problem area as the "Bermuda Triangle" of VDTs (Figure 3).

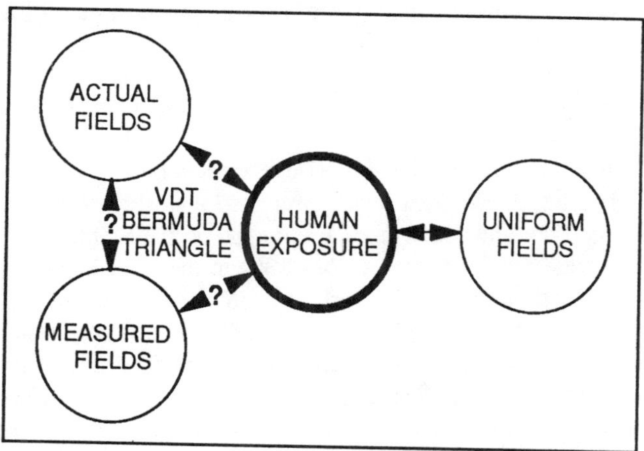

Figure 3: VDT electric field "Bermuda Triangle".

Results

Many studies of the electric and magnetic fields produced by VDTs have been reported. Over the past ten years, a convention has been established of recording field levels at a distance of either 30 or 50 cm from the case of the unit. The 30 cm distance is the closest practical point of approach, given existing sensors, and may be considered to represent a "worst-case" situation under normal operating conditions. A distance of 50 cm from the display screen is more representative of an operator's normal position, but low field strengths may present a measurement problem in some locations.

Application of the inverse third-power law noted above, combined with estimated distances of the sources from the surface of the case, indicates that the field strengths at 50 cm should be approximately 25 to 35% of those at 30 cm. This is confirmed by experiment in laboratories with low background fields.

The individual measurements of static and time-varying fields, as might be expected, show substantial variations among units and between investigators. They are too numerous to itemize here. Instead, the maximum ELF and VLF field values recorded at a distance of 30 cm from the VDT display screen in 13 reputable studies carried out over the past ten years are presented in Table I. It is gratifying to note that, allowing for product variability and measurement uncertainty, the reported range of electric and magnetic field values has remained quite consistent over this period.

<u>Proposed IEEE Standard</u>

Standardized measurement procedures for VDT electric and magnetic fields are proposed in an unapproved draft report circulated under the auspices of the Institute of Electrical and Electronics Engineers, or IEEE (Moongilan, 1991). A parallel-plate ground plane sensor of 30 cm diameter is specified for electric field measurements and an assembly of three mutually perpendicular coils, each having a nominal radius of 5.6 cm, is specified for magnetic field measurements. Measurement locations are specified to lie at defined positions on circles of radius (L/2+50) cm, where L is the distance from front to back of the unit. Front-of-screen measurements are therefore located at a distance of 50 cm from the screen, and a measurement location with low background field levels is required. If these proposals are adopted, then reported VDT emissions may be compared with greater confidence.

Electric and magnetic field strengths considered to be achievable at the measurement locations using current technology are shown in Table II.

Animal Studies

The field exposures that are produced by the VDT at the operator's normal position are well below those associated with any known adverse health effect, and are comparable to common electric and magnetic field exposures

Table I

Maximum Electric and Magnetic Fields Measured at 30 cm from the Screen.

Reference	ELF Fields		VLF Fields	
	E (V/m)	B (μT)	E (V/m)	B (μT)
BRH, 1981	-	-	2.4	0.05
Harvey, 1982, 1984	65*	-	35*	-
Stuchly, 1983a,b	-	0.27	5	-
Harvey, 1983b	-	0.7	-	0.20
Roy, 1984	-	-	2.7	0.10
Paulsson, 1984	-	-	-	0.22
Guy, 1984	-	-	40*	0.18
Juutilainen, 1986	-	0.44	-	-
Guy, 1987	-	-	10.2	0.10
Jokela, 1989	165	0.72	4.8	0.14
Tofani, 1991	-	0.6	-	0.06

Notes: *Electric Fields marked with a * are measured with reference to a grounded plane and are uncorrected for field perturbations.*

Field strengths at a distance of 50 cm from the unit may be estimated by applying a factor of 0.25 to 0.35 to the tabulated values.

due to other environmental influences (Walsh et al., 1991; Kavet and Tell, 1991). Nevertheless, adverse pregnancy outcomes have become popularly associated with VDT operation, while a few laboratory and epidemiological studies have raised new but as yet unverified concerns about the safety of even very-low-level ELF and VLF magnetic field exposures.

Birth defects and ELF/VLF magnetic fields were linked by Delgado and co-workers (Delgado et al., 1982; Ubeda et al., 1983) in a study of the effect of pulsed magnetic fields on developing chick embryos (the pulse refers to the actual field, not a modulation envelope). Teratogenic effects were

TABLE II

Achievable Field Strengths on circles of radius (L/2+50) cm (IEEE).

Equipment Class	ELF Fields		VLF Fields	
	E (V/m)	B (μT)	E (V/m)	B (μT)
IEC Class I	50	0.25	10	0.025
IEC Class II	250	0.25	10	0.025

observed for particular combinations of pulse shapes, with rise times in the μs range, and amplitudes of 1 μT and above. Subsequent efforts to replicate Delgado's findings have had mixed results. In the most ambitious international attempt at replication to date, six laboratories combined in what has become known as the "Henhouse" project (Berman et al., 1990). Although only two of the six laboratories were able to find a significant teratogenic effect, the overall results of the experiments are considered to be statistically significant.

Cancer and ELF magnetic fields have been linked in some epidemiological studies of childhood illness in homes with background 60 Hz field levels of a fraction of 1 μT (Wertheimer and Leeper, 1979; Savitz et al., 1988). The incidence of certain types of cancer was significantly elevated in homes considered to have higher magnetic field exposures as a result of the electrical wiring configurations near the homes. Although actual measurements have failed to substantiate the purported link with ELF magnetic fields, alternative explanations for the association between the illness and the wiring configurations have not at this time displaced the magnetic field hypothesis.

Controlled laboratory experiments with mamallian species could help to resolve questions relating to the effect of low-level ELF or VLF magnetic field exposures on human health. Because of the widespread use of 50 Hz or 60 Hz electrical power throughout the world, ELF studies are proceeding independently of any concerns about VDTs. Our interest in this paper will therefore be on experiments that employ VLF magnetic fields of the type produced by the display terminals, and whose endpoint is the influence on fetal development in laboratory animals.

Animal Experiments with VLF Fields

In 1986, Sweden's Karolinska Institute reported the preliminary results of an animal experiment in which pregnant mice were exposed to pulsed magnetic fields designed to replicate those used by Delgado and co-workers, and to VLF fields designed to replicate those produced by VDTs (Tribukait et al., 1987). The VLF fields were described as 20-kHz sawtooth waves, with rise and fall times of 45 μs and 5 μs respectively. Field amplitudes of 1 μT and 15 μT peak-to-peak were used (corresponding to rms amplitudes of about 0.3 and 4.5 μT respectively for the VLF fields) and the exposure extended from day 0 to day 14 of the pregnancy. The investigators reported that no teratogenic effects were observed due to exposure to the pulsed fields, but that fetuses among animals exposed to the 15 μT VLF fields showed a significant increase in external malformations.

This report re-ignited concerns that VDT emissions could adversely influence human fetal development, and initiated at least three attempts at replication; one in Sweden, and two in Canada.

The second Swedish study was carried out at the Swedish Agricultural University in Uppsala (Frölén et al., 1987, 1989). Pregnant mice (a different strain) were again exposed to 15 μT VLF magnetic fields with rise and fall times of 45 and 5 μs, but the exposure duration was extended to 19 days. The investigators reported no effect on fetal malformations, but a significant increase in placental resorptions among the exposed animals, with the latter effect persisting even if the onset of exposure was delayed for up to five days.

The first Canadian study was performed by Health and Welfare Canada (Stuchly et al., 1988). In this experiment, pregnant Sprague-Dawley rats were exposed to sawtooth VLF magnetic fields having rise and fall times of 44 μs and 12 μs, respectively, and field amplitudes of 5.7, 23, and 66 μT peak-to-peak (corresponding to rms amplitudes of about 1.7, 6.9, and 20 μT). The animals were exposed for seven hours per day starting two weeks prior to pregnancy and extending to the 22nd day of pregnancy. The investigators reported no significant teratogenic effects from field exposure except for a increase in the number of minor skeletal anomalies in the highest exposure group, attributed to common biological "noise".

The second Canadian study was performed at the University of Toronto, with technical assistance and sponsorship from Ontario Hydro and IBM (Wiley et al, 1990). The design of this experiment specifically addressed experimental uncertainties noted in earlier studies. It was the first experiment to be conducted in a completely blind manner and the first to rigorously control or randomize experimental variables among the exposed and control animals. Pregnant mice (a third strain) were exposed to sawtooth VLF magnetic fields with rise and fall times of 45µs and 5µs respectively and with amplitudes of 3.6, 17, and 200 µT peak-to-peak (1.1, 5.1, and 60 µT rms). The animals were exposed for 20 to 21 hours per day for the first 18 days of pregnancy. The investigators report no teratogenic effects due to exposure to the fields.

A more detailed comparative review of the four experiments may be found in the literature (Wiley, 1990; Kavet and Tell, 1991). In terms of field strengths, the laboratory animals in these experiments were exposed to whole-body VLF magnetic fields ranging from 1.5 to 300 times the maximum VLF field measured at 30 cm from the VDT screen (Table I). In terms of currents induced in body tissue, calculated scaling factors derived on the basis of relative body sizes (Stuchly et al., 1988) yield experimental exposures ranging from 0.1 to 20 times the "worst-case" VDT operator exposure. Over these ranges, the experiments have shown no detectable influence of VLF magnetic fields on reproductive outcomes in laboratory rodents that is reproducible across species, strains, or laboratories.

Suggestions for Future Experiments

Do the experiments described in this paper suggest weak biological effects of VLF magnetic field exposure that are only manifested under specific but undefined experimental conditions, or are the reported associations simply the result of biological "noise" or experimental artifacts?

Many experimenters are now searching for possible biological effects of exposure to low-level magnetic fields. Such effects have proven to be quite elusive; indicating either that they are very weak, or that the experimental conditions responsible for them are, at best, poorly understood. Frequently, observations or working hypotheses having important implications for human health have to be based on results that fail to achieve conventional levels of statistical significance.

Under such conditions, there is a need to minimize biological variability, to thoroughly test experimental protocols, and to verify to the greatest possible extent the absence of any experimental artifacts or confounders that may affect the outcome of an experiment at, or even below, the statistical noise level. As a fact of life, verification requires that a potential confounder, if it cannot be eliminated, be tested with the same or higher level of scientific rigor as the agent which the experiment is designed to test.

The coils and the power equipment used to generate the magnetic fields to which animals are exposed are potential sources of experimental artifacts and confounders. Careful design and testing are essential pre-requisites for any animal study involving magnetic fields, particularly in the frequency range above about 16 kHz, where human sensory inputs lack sensitivity, instrument readings may be unreliable, and spurious system responses are possible. The following factors are among those unavoidably associated with magnetic field generators.

- Heat: heat is produced by current flowing in coils, transformers, and other conductors. The amount of differential heat production between the exposed and control (or sham) animals that will produce a biological effect at the statistical noise limit of the experiment will determine experimental tolerances.

- Vibration and sound: equipment designed to produce ac magnetic fields will vibrate because of the forces between wires. Mandatory physical isolation between animal cages and all field generating equipment will alleviate the problem. Rats and mice are very sensitive to sounds beyond human audibility such as those produced by switching power supplies, oscilloscopes, and desktop computers, that may not be detected with conventional microphones.

- Chemicals: Solvents, glues, and other materials can evolve chemicals over time, at a rate that depends on the temperature and the work history of the apparatus. Careful choice of materials, identical environments for exposed and control animals, and good ventilation, are important.

- Stray fields and spurious signals: A rigorous examination of the electric and magnetic fields and other influences experienced by the exposed and control animals will help to reveal environmental or equipment-generated artifacts and will clarify the meaning of "sham" exposure. Indirect or inferential methods of measurement are always suspect.

In addition, field generating equipment may introduce indirect biases due to an influence on room illumination, cage occupancy or ventilation, behaviour of laboratory personnel, induced voltages and currents on lab equipment, or unspecified sensory inputs to the animals. It is the task of the experimental scientist to take these factors into account when designing the exposure facility and when interpreting the experimental data.

Conclusions

Because of concerns for the health of video display terminal operators, electromagnetic radiation and other forms of emissions from VDTs have been subjected to close scrutiny. Electromagnetic emissions, both ionizing and non-ionizing, have been measured in considerable detail over the past ten years, and can now be considered to be well characterized. The spatial and temporal variation of the electric and magnetic fields produced by VDTs in the VLF and ELF ranges are well understood, and the range of field strengths has been established. Standardized measurement techniques and guidelines for achievable field strengths at various locations around the unit are features of a proposed IEEE standard for VDT emissions.

The field strengths experienced by VDT operators are comparable to the electric and magnetic field strengths produced by other common sources, and have no known adverse health effects. However, concerns that exposure to the VLF magnetic fields could influence pregnancy outcomes have continued, largely as a result of two experiments performed with living animals. The first experiment, performed with chick embryos, suggested an association between pulse-type magnetic fields and embryonic development in avian species. The second experiment, performed with laboratory rodents, suggested that VLF magnetic fields similar to those produced by display terminals could influence fetal development in mammals.

Three attempts to replicate the rodent experiment have been made, with whole body field exposures up to 300 times the field strength that might be experienced by a VDT operator, but no influence on fetal development in rodents that is reproducible between laboratories has been found at the level of statistical significance. If VLF magnetic fields of the type produced by VDTs do have an adverse influence on pregnancy outcomes, the effect is not sufficiently robust in laboratory rodents to survive confounding by biological variability or experimental artifacts. In other words, it is below the experimental "noise level".

Magnetic field exposure systems are potential sources of experimental artifact. Careful design and testing are pre-requisites for lowering the inter-laboratory noise level for biological experimentation.

References:

Bergqvist, U.O. (1984) Video Display Terminals and Health: A technical and Medical Appraisal of the State of the Art. Scandinavian Journal of Work, Environment, and Health, 10:Supplement 2.

Berman, E. et al. (1990) Development of Chicken Embryos in a Pulsed Magnetic Field, Bioelectromagnetics, 11:169.

BRH (1981) An Evaluation of Radiation Emission from Video Display Terminals, U.S. Department of Health and Human Services publication FDA 81-8153, Rockville, Maryland.

Delgado, J.M. et al. (1982) Embryological Changes Induced by Weak, Extremely Low Frequency Electromagnetic Fields, J. Anat. 134:533.

Elliott, G., Roy, C.R. and Gies, H.P. (1986) Video Display Terminals and Radiation Emissions- Current Status, Radiation Protection in Australia, 4:123.

Frölén, H. et al. (1987) Repetition of a Study of the Effect of Pulsed Magnetic Fields on the Development of Fetuses in Mice, SSI Report 346.86 (English translation), Swedish National Institute of Radiation Protection, Stockholm, Sweden.

Frölén, H. and Svedenstål, B.M. (1989) The Effect of a Pulsed Magnetic Field on Fetal Development in the Mouse: Complementing and Expanding on a Previously Reported Investigation, SSI Report 493.88 (English Translation), Swedish National Institute of Radiation Protection, Stockholm, Sweden.

Guy, A.W. (1984) Health Hazards Assessment of Radio Frequency Electromagnetic Fields Emitted by Video Display Terminals, Report prepared for IBM Office of the Director of Health and Safety, University of Washington, Seattle, WA.

Guy, A.W. (1987) Measurement and Analysis of Electromagnetic Field Emissions from 24 Video Display Terminals in American Telephone and Telegraph Office Washington, D.C., Report prepared for the National Institute for Occupational Safety and Health, University of Washington, Seattle, WA.

Harvey, S.M. (1982) Characteristics of Low Frequency Electrostatic and Electromagnetic Fields Produced by Video Display Terminals, Ontario Hydro Research Report 82-528-K.

Harvey, S.M. (1983a) Analysis of Operator Exposure to Electric Fields from Video Display Units, Ontario Hydro Research Report 83-503-K.

Harvey, S.M. (1983b) Characterization of Low Frequency Magnetic Fields Produced by Video Display Units, Ontario Hydro Research Report 83-504-K.

Harvey, S.M. (1984) Electric Field Exposure of Persons Using Video Display Units, Bioelectromagnetics, 5:1.

Jokela, K., Aaltonen, J. and Lukkarinen, A. (1989) Measurement of Electromagnetic Emissions from Video Display Terminals at the Frequency Range from 30 Hz to 1 MHz, Health Physics, 57:79.

Juutilainen, J. and Saali, K. (1986) Measurements of Extremely Low Frequency Magnetic Fields around Video Display Terminals, Scand. J. Work Environ. Health, 12:609.

Kavet, R. and Tell, R.A. (1991) VDTs: Field Levels, Epidemiology, and Laboratory Studies, Health Physics, 61:47.

Marriott, I.A. and Stuchly, M.A. (1986) Health Aspects of Work With Visual Display Terminals, Journal of Occupational Medicine, 28:833.

Paulsson, L.E., Kristiansson, I. and Malmström, I. (1984) Stralning fran Dataskarmar (Radiation from Data Screens), Statens Stralskyddsinstitut Report a84-08 (in Swedish), ISSN 0281-1359, Stockholm.

Moongilan, D. et al. (1991) Standard Procedures for the Measurement of Electric and Magnetic Fields from Video Display Terminals (VDTs) from 5 Hz to 400 kHz, P1140 Draft 7, Unapproved, Institute of Electrical and Electronics Engineers, New York.

Roy, C.R. et al. (1984) Measurement of Electromagnetic Radiation Emitted from Visual Display Terminals (VDTs), Radiation Protection in Australia, 2:26.

Savitz, D.A. et al. (1988) Case-Control Study of Childhood Cancer and Exposure to 60-Hz Magnetic Fields, Am. J. Epidemiol., 128:21.

Stuchly, M.A., Lecuyer, D.W. and Mann, R.D. (1983a) Extremely Low Frequency Electromagnetic Emissions from Video Display Terminals and Other Devices, Health Physics, 45:713.

Stuchly, M.A. et al. (1983b) Radiofrequency Emissions from Video Display Terminals, Health Physics, 45:772.

Stuchly, M.A. et al. (1988) Teratological Assessment of Exposure to Time-Varying Magnetic Field, Teratology 38:461.

Tofani, S. and D'Amore, G. (1991) Extremely-Low-Frequency and Very-Low-Frequency Magnetic Fields Emitted by Video Display Units, Bioelectromagnetics 12:35.

Tribukait, B., Cekan, E and Paulsson, L.E. (1987) Effects of Pulsed Magnetic Fields on Embryonic Development in Mice, In: Work with Display Units 86, Knave, B. and Wideback, P.G. eds, Elsevier Science Publishers, North-Holland.

Ubeda, A. et al. (1983) Pulse Shape of Magnetic Fields Influences Chick Embryogenesis, J. Anat. 137:513.

Walsh, M.L. et al. (1985) Hazard Assessment of Video Display Units: Final Report (2 Volumes), Safety Services Department, Ontario Hydro, Toronto.

Walsh, M.L. et al. (1991) Hazard Assessment of Video Display Units, AM. Ind. Hyg. Assoc. J., 52:324.

Wertheimer, N and Leeper, E. (1979) Electrical Wiring Configurations and Childhood Cancer, Am. J. Epidemiol., 109:273.

Wiley, M. et al. (1990) Magnetic Field Rodent Reproductive Study (MFRRS), Ontario Hydro Report No. HSD-91-3, Ontario Hydro, Toronto.

MEASUREMENTS OF ELF FIELDS

D.A. Agnew

Ontario Hydro
Health and Safety Division
Pickering, Ontario

1.0   INTRODUCTION

The increasing concern regarding the possible adverse health effects associated with the exposure to extremely low frequency (ELF) electric and magnetic fields has stimulated activities in instrumentation development and electric and magnetic field measurement. The measurements include those performed to address the concerns of workers and members of the public, and those that are carried out in support of scientific studies designed to answer questions about the nature and sources of the fields and their effect on human health.

In 1987 Ontario Hydro decided to carry out measurements of electric and magnetic fields in response to requests from its workers and members of the public (Mu89). The majority of the requests (500) have been for measurements in residences, often to address concerns raised when citizens see media reports on the EMF issues and/or when becoming aware of the close proximity of their home to a visible part of the transmission or distribution system. The remaining requests have come from industrial (50) and commercial (25) firms, schools and other institutions (30) and farms (6). The procedure for residences normally involves taking point-in-time measurements of the electric and magnetic field strength both inside and outside the house using survey-type meters according to a prescribed protocol (Mu90). For residential owners, as well as other customers, measurements are made at requested locations to address specific concerns.

All early human health studies and some recent ones have used surrogates of ELF exposure rather than field measurements. For example, occupational studies frequently use job title (Co83, Mi85, Ro91), while studies of adult or childhood cancer and residential exposures have employed features of the distribution system, such as distance to transformer stations and high voltage power lines (Mc86, Co89) and the wiring configuration of the distribution system (We79). The lack of exposure measurement information has proved to be limiting in the evaluation of study results (De91).   More recent studies, (Sa88, Lo91) including a number now in progress (Wa92), use a variety of measurement techniques in order to assign electric and magnetic field exposure to study subjects. Techniques include point-in-time measurements with survey meters in areas occupied by subjects, data recorders which collect and store electric and magnetic field strength information for a few locations over a period of hours or days, and personal monitors which are worn by the subjects and record the electric and/or magnetic field strength as the subject goes about his/her daily routine.

Studies that identify the electric currents responsible for the magnetic fields in residences, schools and other environments, and the computation of those fields (EP89, Ma90), have provided us with a clearer insight into the types of measurements which are appropriate for assessing human exposure and provide some basis for mitigation should it be necessary.   These studies generally employ point-in-time measurements of magnetic fields, and the logging of electrical currents and magnetic field strengths at specified locations for periods of hours or days.

Until now the main interest has been in measuring power frequency (50/60 Hz) electric and magnetic fields.   However, epidemiology studies to date have not revealed a statistically significant correlation between elevated levels of exposure to power frequency fields and an increased risk of cancer.   There is now a growing interest in examining other metrics, such as ELF harmonics, transients, and intermittent exposures.   However, at least in terms of the total exposure assessment of subjects in scientific studies, it is premature to emphasize field characteristics other than the time weighted average 50/60 Hz fields and those that can be monitored with existing exposure meters.   This paper will deal only with the measurement of power frequency electric and magnetic fields.

2.0   MEASUREMENT OF ELECTRIC FIELD STRENGTH

2.1   Electric Field Meters

The electric field meter consists of two parts, the probe or field sensor, and the detector which consists of a signal processing circuitry with an analog or digital display.   The probe can take different forms depending on the type of measurement required and the specific manufacturer.   For survey-type measurements one requires a meter that is portable, allows measurements above the ground plane and does not require a known ground reference. A commonly used survey-type meter is the free-body (or dipole) meter. Electro-optical meters (IE87b) also have the required characteristics to perform field surveys. Ground reference meters are not useful for survey measurements but can be used to measure the field strength at ground level and to provide a calibrated field under overhead lines.

2.1.1   Free-Body Meter

The probe of the free-body meter is made of an electrical conductor and can be of any shape, although in North America it is generally in the shape of two rectangular boxes with side dimensions ranging from 7-20 cm. When the probe is aligned in a uniform sinusoidal electric field, for example E_0 $\sin\omega t$, an electric charge oscillates between the two halves of the probe, and the current is given by (IE87)

$$I = \frac{dQ}{dt} = k\omega\epsilon_0 E_0 \cos\omega t \qquad (1)$$

where ω is the angular frequency and k can be thought of as a field strength meter constant, and is determined by calibration.

The detector measures the steady state induced current or charge oscillating between the conducting halves of the probe. It can be located inside or construction as an integral part of the probe. There are free-body meters designed for remote display of the electric field strength. In this case, part of the detector circuitry is contained in the probe and the rest, including the display, is housed separately. The two components are connected by a fibre-optic link. The detector is normally calibrated to read the root mean value of the power frequency field. At power frequency the accuracy of the detector depends on the stability of the components at a given temperature and humidity, and is generally high ($<0.5\%$ uncertainty) (IE87). The response of the detector to harmonic components in the electric field depends on the electronic design of the detector. The frequency response of the meter should be provided by the manufacturer, but can be determined by injecting a known current at various frequencies onto the two halves of the probe (see 2.2.3).

Because the background electric field is perturbed in the vicinity of the operator, the free-body meter is attached to a long, electrically insulated handle so that it can be positioned away from the body. If the meter is held in a vertical electric field at a height of 1 meter above the ground and at a distance of 1.5 meters from a grounded operator, the meter will read as much as 9% below the unperturbed or background field strength (Di78). At a distance of 2 meters the shielding by the operator reduces the reading to less than 5% below the background field strength. Free-body meters are insensitive to height above ground down to a distance of 30 cm. At clearances of less than 30 cm, the meter can be expected to give erroneous results due to the redistribution of ground charge (Di78).

2.1.2 Electro-Optical Meters

Electro-optical meters, like free-body meters can be used for field survey measurements. The probe is a dielectric crystal a few centimetres in size, and so can be used to measure the electric field strength much closer to sources of electric fields, without disturbing their charge distributions, than can most free-body meters. The electric field induces an optical birefringence in the dielectric crystal, the magnitude of which is proportional to the electric field strength. The result is a reduction in the intensity of the polarized light that is emitted after passing through the crystal (IE87b). The light source is in the detector. It is transmitted to the probe and back to the detector by optical fibres.

2.1.3 Ground-Reference Meters

This type of meter is used to measure the electric field at ground level. The probe can consist of a single conducting plate and a flat guard ring, all mounted on a thin insulating layer. Alternately, two parallel plates separated by an insulating layer can be used. In the latter case the bottom layer is grounded. In both cases the sensing layer is connected to the detector with a shielded cable (IE87b).

For a sensing surface of area A, the induced root mean square (rms) current is given by:

$$I = A\omega\epsilon_0 E \qquad\qquad (2)$$

where ϵ_0 is the permittivity of vacuum and E is the uniform rms electric field strength (IE87b). This apparatus can be placed under overhead lines to produce uniform fields of known strength for the purpose of calibrating survey-type meters (IE87).

2.2 Calibration of Electric Field Meters

Electric field meters can be calibrated using a parallel plate structure, a ground plate with a flat guard ring that is placed under overhead high voltage lines, and a current injection circuit. Calibration checks should be made prior to and after any extended period of use.

2.2.1 Calibration of Survey Meters Between Parallel Plates

A uniform electric field can be produced between two parallel plates provided that the spacing between the plates is sufficiently small relative to the plate dimensions (Sh77). A parallel plate apparatus which is suitable for calibrating a free-body type meter is shown in Figure 1 (IE87b). The spacing of the plates is 0.75 meters and the plates are 1.5 meters square. The plates are energized with a centre tap transformer to minimize distortions in the electric field from nearby objects, and current limiting resistors are placed in the transformer output leads as a safety measure (10 $M\Omega$ resistors of adequate voltage rating are satisfactory for voltages up to 10 kV). The edges of the plates should be at least 0.5 meters from nearby objects (wall, floors, etc.), otherwise graded guard rings can be used to isolate the space between the plates from surrounding perturbations (IE87). With this arrangements the calibration field is within 1% of the uniform field value V/d (V=voltage between plates, d-plate separation).

The survey-type meter is calibrated by placing it in the centre of the spacing between the plates, supported by the insulating handle normally used during measurements. The maximum value is recorded after aligning the meter to within 10^0 of the vertical electric field. At least three calibration points should be obtained for each range of the meter. The meter readings should be within 5% of the calculated electric field strength between the plates. Additional details can be found in reference IE87. For measuring devices of smaller dimensions, plates with a smaller dimension and spacing can be used (Sh7).

2.2.2 Ground Plate Under a High Voltage Line

Another method to create a uniform electric field of known magnitude employs a conducting plate, as described in 2.1.3 (IE87). If the grounded plate is placed under a high voltage overhead line, the electric field strength at the surface of the plate can be calculated from equation 2. The electric field strength meter to be calibrated is attached to the insulating handle and placed 1 meter above the plate where the field is assumed to be approximately uniform.

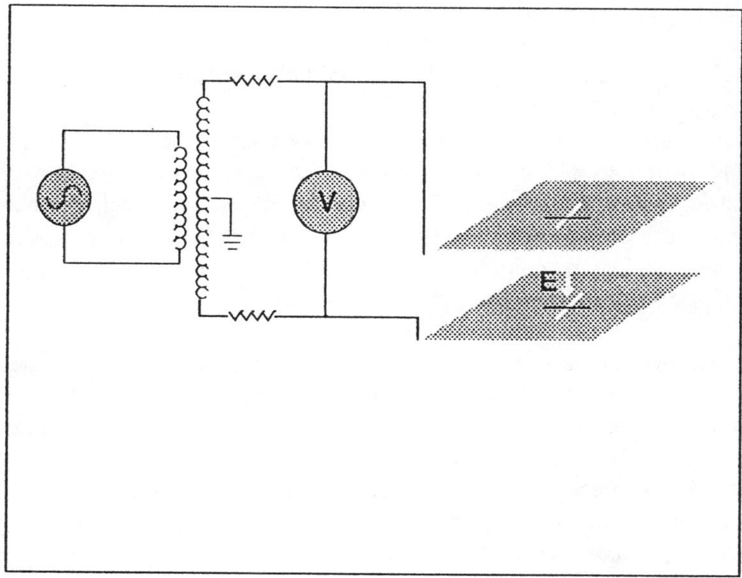

Figure 1 Parallel Plates Used for Electric Field Meters

2.2.3   Current Injection

If a free-body meter has been previously calibrated so that the field strength
meter constant, k, is know, then injecting a known current onto the probe
sensing plates can be used to check the calibration of the meter (see
equation 1). Current injection can also be used to measure the frequency
response of the meter by injecting known currents at various frequencies.
If current injection is used, care must be taken to ensure adequate shielding
from ambient sources such as interior lighting, power supplies, etc. For
further information on current injection see IE87 and IE87b.

2.3   Electric Field Strength Measurement Procedures

At present, published Standard Procedures for measuring electric fields
address primarily the overhead power line environment (IE87, IE87b). This
likely reflects the fact that the first reports of adverse health effects from
electric and magnetic fields were from electrical workers employed near

very high voltage electrical transmission equipment (Ko72). However, now a large fraction of the measurements are made in the public sector, primarily in residential areas.

2.3.1   Measurements Near High Voltage Power Lines

If possible, measurements should be taken on relatively level ground, avoiding tall grass, shrubs and trees. The effects of vegetation on the electric field strength can be significant. In general, field enhancement occurs near the top of isolated vegetation and field attenuation occurs near the sides. The magnitude of the perturbation can depend on the water content of the vegetation.

To measure the background (unperturbed) electric field strength, the free-body meter should be held 1 meter above ground level (use of other heights should be reported), and at a distance from the body of the operator to achieve the desired accuracy (see 2.1.1). If the measurements are taken near an overhead line that is expected to dominate the electric field environment, the handle of the meter should be held parallel to the line, and the meter rotated to obtain the maximum electric field strength value. Otherwise the operator must determine the orientation of the maximum value of the electric field, ensuring that he/she is positioned in the lowest field possible in order to minimize perturbation of the electric field (Di78).

When measuring the background electric field near a nonpermanent object, such as a vehicle, the meter to object distance should be at least 3 times the height of the object (IE87). Background measurements near a permanent object should be taken at least 1 meter from the object.

All relevant details of the electric field survey should be documented (IE87), such as environmental conditions (temperature, humidity, ground cover), transmission line parameters (such as line voltage, current, conductor geometry), measurement locations, and instrument used.

2.3.2   Measurement at Residential and Non-Residential Sites

The Ontario Hydro measurement program (Mu90) requires that a sketch is made of the layout of the lot and the location of the house or other relevant buildings. Indicated on the sketch are features such as the driveway, doors, deck, electrical service connection and its point of entry, transformers and distance to rights-of-way or conductors of nearby transmission or distribution lines. Indicated on the sketch are the locations of field measurements.

Locations of electric field measurements taken outdoors vary according to the nature of the site (Mu89). For residential sites, readings are taken on patios or decks, near front, rear or side entrances, in the middle of front, rear or side yards, near lot boundaries (particularly if bordering a right-of-way or other utility property), at the end of the driveway or front walk, at service entry points, and near ground mounted, or under pole mounted transformers. For non-residential sites, readings are taken to give information on electric and magnetic field profiles under transmission or distribution lines at locations occupied by workers, near equipment of concern to the individual(s) who requested measurements, etc.

Inside residences the electric field is usually vertical, or nearly so. Thus, although rotating the meter will lead to some changes in the reading obtained, such changes usually are not large. It should be noted, however, that there will be occasions when the maximum value will be observed at an orientation of the meter far off the vertical.

Inside readings are taken typically in the living room, dining room, kitchen, family room, one or more bedrooms, in one or two locations in the basement, and at requested locations that address specific concerns. Measurements are usually taken in the middle of the rooms with the meter held about 1 meter above the floor and at least 1 meter away from all appliances. Measurements may be made at other locations within the room to address various specific concerns, however, it is usually assumed that the field measured in the middle of the room is a reasonable value to attribute to the room as a whole.

2.3.3   Parameters Affecting Accuracy of Measurements

The measurement accuracy during practical field measurements of electric field strength using commercially available free-body meters is typically within 10%, although this can be reduced under more controlled conditions (IE87b). A number of factors can contribute to the measurement error and they include:

- difficulty in positioning the meter
- reading errors
- temperature and humidity effects
- observer proximity effects
- leakage current along the support handle

- the presence of significant harmonic content in the field
- sensitivity of meter to ambient magnetic fields
- mechanical imbalance of movement of analog meter
- measurement with probe in non-uniform field

Further information on these factors can be found in IE87b.

3.0 MEASUREMENT OF MAGNETIC FIELD STRENGTH

3.1 Magnetic Field Meters

Magnetic field meters consist of a probe, or field sensor, and the detector which consists of a signal processing circuitry and an analog or digital display. The probe may take the form of an electrically shielded coil of wire or a Hall Effect semiconductor device.

3.1.1 Magnetic Field Meters with Shield Coil Probe

When the plane of a loop of wire with a cross sectional area, A, is placed perpendicular to an alternative magnetic field $B_0 \sin\omega t$, an electromotive force (EMF) is developed in the loop, and a current will flow in response to the time-rate-of-change in magnetic flux density. It can be shown using basic EM theory that the electromotive force is given by:

$$EMF = -\omega B_0 A \cos\omega t \qquad (3)$$

For a loop of many turns the EMF given above will be generated in each loop, and the voltage output of the probe increased accordingly.

Magnetic field meters with a single coil probe are often used for point-in-time measurements. The sensing coils are usually air core and appear in a variety of sizes. In order to accurately measure the magnetic field, the operator must position the plane of the coil to be perpendicular to the direction of the magnetic field, or the semi-major axis of the ellipse generated by the rotating magnetic field vector in the case of multiple phase magnetic field sources.

Three orthogonal coils are often used in data recorders where further characterization of the magnetic field is required. Data recorders are used to collect information on electric and/or magnetic fields for extended periods of time in specific work locations such as rooms of houses, different areas in a workplace, or along a lateral profile under a transmission line. The instruments can be operated unattended during the period of data collection. To characterize the field the coils measure the three orthogonal components of the magnetic field. It should be noted that, in general, vectorially summing the rms values of the three components will not give the same answer as measuring the rms magnetic field strength along the semi-major axis of the field ellipse. For linearly polarized fields the error is 0%. The worst case is for circularly polarized fields where the maximum error is approximately 40% (IE90).

3.1.2   Magnetic Field Meters With Hall Effect Probe

A Hall Effect sensor is a thin strip of semiconducting material through which a constant control current is passed (Figure 2). When a magnetic field is introduced at right angles to the face of the semiconducting strip, the action of the Lorentz force on the charge carriers produces a small voltage at the contacts that are placed on the sides of the strip. The voltage produced is equal to:

$$V_H = \frac{R_H}{d} I_c B \tag{4}$$

where V_H is the Hall voltage
R_H is the Hall coefficient
d   is the thickness of the semiconductor
I_c is the control current
B   is the magnetic flux density

Unlike the shielded coil probe, the response of the Hall Effect probe is proportional to the magnetic flux density, B, and not the time-rate-of-change, dB/dt. Hall Effect gausemeters that can measure magnetic flux densities from DC to 50 kHz are available. As with the shielded coil probe the Hall Effect probe must be properly positioned with respect to the magnetic field to obtain a maximum reading.

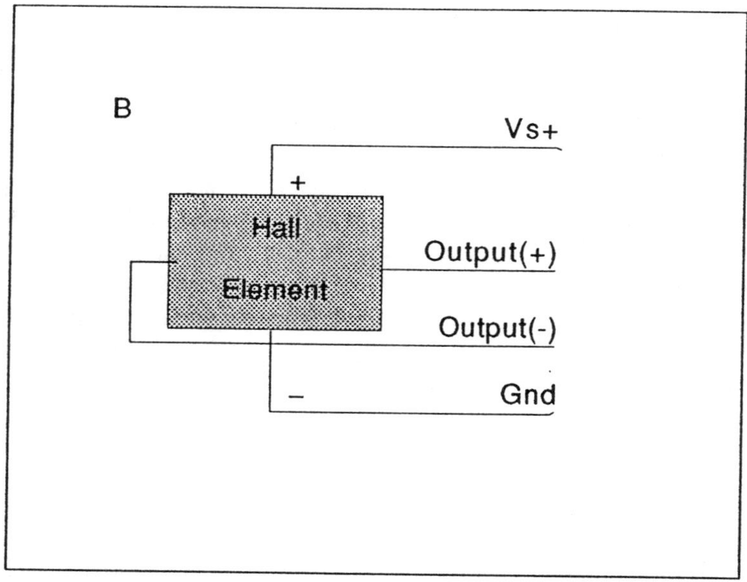

Figure 2 Hall Effect Sensor

3.1.3   Frequency Response of Magnetic Field Meters

Most instruments have a band pass frequency response characteristic (IE90). There is a low frequency below which the response of the meter drops off monotonically. The purpose is to prevent anomalous readings as a result of physical movement of the instrument in the earth's geomagnetic field. There is also a high frequency cutoff above which the meter response drops off monotonically. This occurs because either an imperfect sensor core material is used or because a high frequency filter is built into the instrument. Between the two cutoffs there are different types of frequency responses possible.

Instruments where the response is proportional to the frequency of the magnetic field are said to have a **linear** response. An example is a meter with a shielded coil probe and no narrow band filter of integrator. If the response of the meter is independent of field frequency the response is **flat (integrated)**. A meter with a shielded coil probe where the signal is passed through an integrator circuit would be an example. Instruments with a filter with a narrow band centred on the frequency of interest are said to have a narrow band response. In most cases the frequency is centred on 50/60 Hz but in some cases it is switchable to allow measurement of harmonic content.

3.2 Calibration of Magnetic Field Meters

Calibration of magnetic field meters is performed by placing the probe in a uniform field of known magnitude and direction. A simple method that produces acceptable results is to place the probe at the centre of a square coil of wire (IE87, IE90). The dimensions of the sides of the coil should be at least 1 meter and the energized power supply should be nearly free ($<1\%$) of harmonic distortion.

To calibrate a meter, place the probe in the centre of the square coil. Align the probe so as to get a maximum reading. The response of the meter should be within 5% of the calculated magnetic field strength to be considered accurate (IE87).

3.3 Magnetic Field Meter Measurement Procedures

ELF magnetic fields are not significantly perturbed by the body of the operator, and so the meter can be hand held. Nearby vegetation will not significantly affect the readings.

Measurements are normally made by holding the probe about 1 meter above the ground or floor surface. When using a single shielded coil or hall Effect probe, vary the orientation of the probe to obtain the maximum response. To provide a more complete description of the field at a point of interest, one can measure the maximum and minimum fields with their orientations in the plane of the field ellipse. Non-permanent objects containing magnetic material or nonmagnetic conductors should be at least three times the largest dimension of the object away from the point of measurement in order to measure the unperturbed field value. The distance between the probe and permanent magnetic objects should not be less than 1 meter in order to accurately measure the ambient perturbed field (IE87).

For additional information on measurement procedures near high voltage lines, and at residential and other sites, see the procedures for measuring electric field strength in sections 2.3.1 and 2.3.2.

3.3 Parameters Affecting Accuracy of Measurements

Many of the difficulties described in section 2.3 that apply to electric field measurements are not serious considerations when making magnetic field measurements. Positioning the meter, reading errors, proximity effects of the observer or nearby conducting objects, leakage of electrical current along support handles and the non-uniformity of fields are of little or no importance. The harmonic content of the magnetic field, the temperature and humidity sensitivity of the meter, and the mechanical balance of the meter movement are possible sources of error. It is essential that the probe is properly shielded against ambient electric fields.

4.0 PERSONAL EXPOSURE MONITORS

Personal exposure monitors are self contained monitoring devices, sufficiently small and lightweight that they can be worn on the body without significantly interfering the normal daily activities. They have made possible the measurement of cumulative electric and magnetic field exposure for individuals in occupational or residential environments. With most personal monitors, sample measurements can be taken at selected rates (second to minutes) for periods that, depending on the sampling rate, range from 24 hours to 7 days. Data stored in the monitor is normally downloaded to a PC for analysis and display. The options available to the user during data collection and data analysis vary considerably with the manufacturer. Several monitors come with a display that shows the operational status and most recent field measurement values.

5.0 REFERENCES

Co83   Coleman M, Bell J and Skeet R. Leukemia incidence in electrical workers. Lancet i: 982-983 (1983).

Co89   Coleman MP, Bell CMJ, Taylor HL and Primic-Zakelj M. Leukemia and residence near electric transmission equipment: a case-control study. Br. J. Cancer 60:793-798 (1989).

De91   Dennis JA, Muirhead CR and Ennis JR. Epidemiological studies of exposure to electromagnetic fields: II. Cancer. J. Radiol. Prot. 11:13-25 (1991).

Di78   DiPlacido J, Shih CH and Ware BJ. Analysis of the proximity effects in electric field measurements. IEEE Transactions on Power Apparatus and Systems. Vol. PAS-97, Nov/Dec (1978).

IE87b   IEC International Standard. Measurement of power-frequency electric fields. International Electrotechnical Commission. Publication 833 (1987).

EP89   EPRI, Technical Report EL-6509, Pilot Study of Residential Power Frequency Magnetic Fields, September, 1989.

IE87   IEEE Standard procedures for measurement of power frequency electric and magnetic fields from AC power lines. ANSI/IEEE Std 644-1987

IE90   IEEE Magnetic Fields Task Force of the AC Fields Working Group of the Corona and Field Effects Subcommittee of the Transmission and Distribution Committee. An evaluation of instrumentation used to measure AC power system magnetic fields. Paper presented at the IEEE/PES 1990 Summer Meeting, Minneapolis, Minnesota, July 15-19, 1990.

Ko72   Korobkova VP, Morozov YA, Stolarov MS and Yakub YA. Influence of the electric field in 500 and 750 kV switchyards on maintenance staff and means for its protection. Conference Internationale des Grands Reseaux Electrique a Haute Tension. p 23-26 (1972).

Lo90   London SJ, thomas DC, Bowman JD, Sobel E, Cheng TC and Peters JM. Exposue to residential electric and magnectic fields and risk of childhood leukemia. Am. J. epidemio. $\underline{134}$:923-37 (1991).

Ma90   Mader DL, Barrow DA, Donnelly Ke, Scheer RR and Sherar MD. A simple model for calculating residential 60 Hz magnetic fields. Bioelectromagnetics $\underline{11}$:283-296 (1990).

Mc86   McDowall ME. Mortality of persons resident in the vicinity of electricity transmission facilities. Br. J. Cancer $\underline{53}$:271-279 (1986).

Mi85   Milham S. Mortality in workers exposed to electromagnetic fields. Environ. Health Persp. <u>62</u>:297-300 (1985).

Mu89   Muc A.M. Summary Of Electric And Magnetic Field Measurements To 1989 06 16. Ontario Hydro Information Report HSD-IR-89-2. August 4, 1989.

Mu90   Muc A.M. Procedures For Conducting Power Frequency Electric And Magnetic Field Measurements In Residences. Ontario Hydro Health and Safety Division Report HSD-TS-90-27, September, 24, 1990.

Ro91   Robinson CF, Lalich NR, Bkurnett CA, Sestito JP and Fine LJ. Electromagnetic field exposure and leukemia mortality in the United States. J. Occup. Med. <u>33</u>:160-161 (1991).

Sa88   Savitz DA, Wachtel H, Barnes FA, John EM and Tvrdik JG. Case-control study of childhood cancer and expsoure to 60 Hz magnetic fields. Am. J. Epidemiol. <u>128</u>:21-38 (1988).

Sh77   Shih CH, DiPlacido J and Ware BJ. Analysis of parallel plate simulation of the transmission line electric field as related to biological effects laboratory studies. IEEE Transactions on Power Apparatus. Vol PAS-96, No3, May/Jun (1977).

Wa92   Walsh ML. Summary of Canadian Research in NIR. Second International Workshop: Non-Ionizing Radiation. Vancouver, BC. May 10-14 (1992).

We79   Wertheimer N and Leeper E. Electrical wiring configurations and childhood cancer. Am. J. Epidemiol. <u>109</u>:273-284 (1979).

RESEARCH IN CANADA INTO THE HEALTH
EFFECTS OF POWER FREQUENCY
ELECTRIC AND MAGNETIC FIELDS

M.L. Walsh and K.E. Donnelly

Ontario Hydro
Health and Safety Division
757 McKay Road
Pickering, Ontario
L1W 3C8

1.0   INTRODUCTION

The objective of this paper is to provide the reader with an appreciation of
the scope and nature of the research being conducted into the human health
effects of exposures to power frequency (60 Hz plus harmonics) electric and
magnetic fields (EMF). This review is limited to the Extremely Low
Frequency (ELF) area, excluding dc, and to major research initiatives that
are known to the authors. We also do not discuss the engineering or source
characterization studies. Overviews are provided on selected studies.

Canadian research can be divided into epidemiological human studies,
laboratory animal studies, *in vitro* investigations of mechanisms, and
exposure assessment. Because of the findings of some epidemiological
studies, most research has focused on the possible relationship between
cancer development and exposure to the fields. The estimated total cost of
presently committed EMF programs for the next three to five years is
approximately $10.8M (U.S.).

The strength of the Canadian programs lies primarily in the emphasis on
Quality Assurance in the studies which are funded by major sponsors. For
example, Ontario Hydro's Risk Assessment program is periodically reviewed
by a panel of the Royal Society of Canada; The Royal Society of Canada is
similar to the National Academy of Sciences in the United States and the
Royal Society in the United Kingdom. This review process was favourably
discussed at the recent hearings on the U.S. Environmental Protection

Agency's report on possible EMF carcinogenesis. Favourable outcomes have been both scientific improvements to the Ontario Hydro sponsored studies and significantly enhanced credibility of the program with the public and government officials.

2.0 EPIDEMIOLOGY STUDIES

There are four epidemiology studies being conducted: two are occupationally focused while, two are investigating the potential link between childhood leukaemia and EMF exposures.

2.1 Occupational Epidemiology Studies

2.1.1 Hydro–Québec, Ontario Hydro and Electricité de France Occupational Epidemiological Study on the Long Term Effects of Exposure to 50 Hz and 60 Hz Electric and Magnetic Fields.

The principal investigators are:

- Gilles Thériault, – School of Occupational Health, McGill University, Montreal, Québec;

- Anthony B. Miller, – Faculty of Medicine, University of Toronto, Toronto, Ontario; and

- M. Goldberg – Université de Paris

The sponsors are Hydro Québec, Ontario Hydro and Electricité de France.

Target Completion Date: December 1992

This study was initiated in September 1988. The primary objective of the study is to test the hypothesis of an association between cancer and exposure to EMF. Study design is case–control, within a cohort consisting of employees from the three electrical utilities, and totalling 173,292 men. All cancers are being considered with specific emphasis on haematopoietic cancers, brain and skin cancers. The total number of cancers expected is 4,373.

Lifetime exposure to EMF, accounting for each individual's job history, will be assessed through job exposure matrices at each utility. Exposure of current employees is being measured using Positron and IREQ dosimeters. Over 40 job categories are being assessed. Exposure to potential occupational confounders will be included, and historical changes in EMF exposure are being addressed with the help of former employees and companies' experts.

2.1.2 Brain Cancer and Occupation: A Case/Control Study with Reference to Electrical Occupations.

The principal investigators are:

- Richard Gallagher, Pierre R. Band, Mary McBride – British Columbia Cancer Agency, Vancouver, British Columbia

The study is sponsored by British Columbia Hydro.

Results from several recent epidemiologic studies have suggested that workers in "electrically related" occupations have an elevated risk of brain cancer. A case–control study was recently completed comparing the occupational history of 301 brain cancer cases and 9,972 "other cancer" controls diagnosed in British Columbia from 1983 to 1989.

No statistically significant elevated risks for brain cancer were detected among workers employed as electricians, power station operators, linemen, electric or electronic assemblers and repairmen or welders, although odds ratios for welders exceeded unity.

2.2 Paediatric Leukaemia Epidemiology Studies

2.2.1 Case/Control Study of Paediatric Leukaemia in Southern Ontario

The principal investigators are:

- Anthony B. Miller, Lois M. Green, University of Toronto; Marc L. Greenberg, Hospital for Sick Children, Toronto; and David A. Agnew, Ontario Hydro, Toronto, Ontario.

The study is sponsored by the Canadian Electrical Association, the Electric Power Research Institute (EPRI), and Ontario Hydro.

This population–based case–control investigation, with geographic focus confined to Southern Ontario, is designed to assess the relationship between paediatric leukaemia and exposure to electric and magnetic fields.

Cases will be ascertained from the records of the Hospital for Sick Children for a seven–year period, January 1, 1984 to December 31, 1990. Cases may be of either sex, may be either alive or dead, and will be of the ages 1 to 14 years at time of diagnosis. Controls will be frequency matched to the cases on sex and age.

In 1990 a pilot study was undertaken to compare a number of methods by which exposure to electric and magnetic fields might be assessed, and determine their suitability for use with children. This study provided estimates of the uncertainty inherent in different methods in a residential context. This information has provided an objective basis for the development of a measurement protocol, including measurements at previous residences, which will improve upon those which have been used in previous epidemiological investigations.

Exposure to electric and magnetic fields will be measured according to several different methods, as dictated by cooperation and availability of the child and his/her family. These include personal dosimetry of the child or a sibling using the <u>Positron</u> instruments and measurements of magnetic fields in the internal and external residential environment. The Wertheimer and Leeper wiring code will be used to provide comparability with other epidemiologic research in this area.

Other factors which will be evaluated in the case control investigation are ionizing radiation, parental occupation, medications, infectious agents, immunologic status, physical and chemical agents.

2.2.2   A Case/Control Study of Childhood Leukaemia

The principal investigators are:

- Mary McBride and Richard Gallagher, British Columbia Cancer Agency, Vancouver, British Columbia

The study is sponsored by the Canadian Electrical Association, Electric Power Research Institute and Health and Welfare, Canada.

A prospective, population–based case–control study, investigating the possible association between exposure to low frequency EMF and childhood leukaemia risk, is currently underway in 5 cities across Canada. EMF exposure will be measured by 48–hour personal dosimetry, correlated with personal activity/location diaries. Indirect assessment, using spot measurements and wiring codes, will also take place. Information on other exposures will be collected using a standardized interview with subject parents. This protocol called for the selection or development of a number of specialized data collection instruments.

A series of field tests and pilot studies have been performed in consultation with experts from epidemiology and other disciplines, to assess the feasibility of the protocol. Issues of validity and reliability of the instruments, and of subject compliance, were resolved and ongoing quality control procedures were developed. It is intended that the results of the Ontario study (section 2.2.1 above) and this study will form the basis of a Canadian national study.

3.0 *IN VIVO* STUDIES

3.1 Evaluation of the Potential Carcinogenicity of 60 Hz Magnetic Fields in F334 Rats

The principal investigators are:

- Rosemonde Mandeville, Daniel Oth, Gilles Lussier, Jean–Paul Descôteaux, Eduardo Franco, Martin Lis. Institut Armand–Frappier, University of Québec, Laval des Rapides, Québec. S. Harvey, D. Agnew, Ontario Hydro, Toronto, Ontario

This study is funded by Health and Welfare Canada, Hydro Québec, Ontario Hydro and Institut Armand–Frappier.

A number of *in vivo* animal studies have produced evidence on the effects of EMF on neuronal and neuroendocrine systems. Moreover, epidemiologic studies seem to point to a putative increased risk of brain cancers in workers exposed to magnetic fields and their children. These data have clearly established the nervous system as a salient mediator of EMF effects in vertebrates. This apparent relationship might be anticipated since the nervous system is composed of tissues and cells that are unusually responsive to electrical and magnetic signals. In addition, both the structure and function of this system are fundamentally involved in the interaction of an animal with its environment. To address the potential carcinogenic effect of EMF *in vivo*, a study containing two distinct elements is being conducted.

Total Carcinogenesis

The effect of magnetic fields per se on tumour development will be observed in female rats exposed to 60 Hz magnetic fields from birth until 2 years (104 weeks) of age. Tumour incidence in 45 organs as well as haematopoietic cancer will be determined.

Brain Tumour Promotion Study

The effect of 60 Hz magnetic fields on the promotion of brain tumours induced in female Fischer rats by the carcinogen N–Ethyl–N–nitroso–urea (ENU).

Animals will be exposed to 60 Hz magnetic fields of varying intensities ranging from 0 (Sham exposure) to 2, 20, 200 and 2000 μT. Exposure will be performed for 20 consecutive hours (10 hours in the dark and 10 hours in the light). In these studies, all measurements of exposures will be carefully controlled and all confounders closely monitored. At sacrifice, blood will be drawn for immunological competence testing and for the evaluation of the presence of leukaemia. The statistical design of these experiments should allow for the detection of a doubling in the incidence of tumours in rats exposed to 60 Hz magnetic fields of different intensities, with a power of 95% in both elements.

3.2   Tumour Copromotion in the Mouse Skin by 60 Hz Magnetic Fields

The principal investigators are:

- Jack R.N. McLean, Maria A. Stuchly, Michael Goddard and David W. Lecuyer, Health and Welfare Canada, Ottawa; Ronald E.J. Mitchel, Atomic Energy of Canada Limited, Chalk River, Ontario.

This study is funded by Health and Welfare Canada.

This study is an extension of one completed in 1990 which suggested that exposure of DMBA–initiated TPA–promoted Sencar mice to a 60 Hz (2 mT) magnetic field, could affect the immune system and the development of skin papillomas. In that study, the exposure of mice to a magnetic field appeared qualitatively to increase the rates of growth and progression of papillomas. A dose–response curve for TPA in DMBA–initiated Sencar mice was initially established. At 0.3 µg of TPA/wk, it was found that 23% of mice developed papillomas (mean yield, 1.0 pap/mouse, SEM ± 0.4), with minor skin irritation observed on some mice.

In this study, the dorsal skin of DMBA–initiated Sencar mice will be treated weekly for 21 weeks with 0.3 µg of TPA. Groups will be sham or field exposed at 60 Hz (2 mT) for 6 hrs/day 5 days per week. To establish statistical significance at p=0.05 and a power of 0.8, 54% of field exposed mice would have to develop papillomas (59% for a power of 0.90). TPA will not be applied from week 21 through week 55 but mice will continue to be sham or field exposed. At termination, mice will undergo complete histological assessment with the mean yield of carcinomas and proportion of mice with carcinomas being important endpoints.

3.3 University of Toronto Evaluation of the Potential Carcinogenicity of 60 Hz 20 mT Fields in Mice.

The principal investigator is:

- Michel Wiley, University of Toronto, Department of Anatomy, Toronto, Ontario

This study is funded by Ontario Hydro.

The Institute Armand–Frappier study, the Health and Welfare Study and other planned carcinogen screenings will be carried out at exposures not exceeding 2 mT. To ensure that the screening studies are comparable to more traditional chemical carcinogen screening studies a level mimicking the sublethal dose needs to be tested. Twenty mT was selected as an exposure level that may be considered an analogue for sublethal exposures and could be cost–effectively achieved.

Exposure devices to house experimental and control animals under identical environmental conditions have been designed and are now under development. Mice were selected as the test animal in preference to rats because of the logistic difficulties that would be met in constructing apparatus capable of housing the required number of larger animals. Either species is typically used in such screenings.

The strain of choice is the mouse B6C3F1; this strain is to be used in the NIH/NTP screening study, and a good database on natural occurrence of tumours already exists. Forty (40) mice will be exposed to the field for 20 hours per day over a one–year period. Exposure will be initiated at conception.

4.0   *IN VITRO* MECHANISTIC STUDIES

4.1   Effect of 60 Hz Magnetic Fields on the Steady State mRNA Levels of C–FOS and C–MYC Oncogenes in NIH–3T3 Cells

The principal investigators are:

- Claire Guilbault, Michel Bourdage and Yves Langelier, Institut de cancer de Montréal, Hôpital Notre–Dame, Montréal, Québec and Institut de Recherche de l'Hydro Québec, Varennes, Québec

This study is funded by Hydro Québec.

Recent reports by Goodman and Czerska have suggested that the mRNA steady state levels of certain oncogenes were altered in cells exposed to a 60 Hz magnetic field. A constant parameter in these studies was the use of exponentially growing cells. As the expression of a subset of oncogenes such as c–fos, c–jun and c–myc is restricted to G1–cells stimulated to enter the S phase of the cell cycle, the effect of a 60 Hz field on NIH–3T3 cells arrested in G1 by serum starvation was examined. After 36 hours in medium containing 0.5% fetal calf serum, cells were exposed for different periods of time to magnetic fields of intensity from 25 to 450 µT. Sham exposed cells were used as controls. Northern blots of total cellular RNA were hybridized with c–fos and c–myc probes labelled by nick–translation. The autoradiographic signals were quantified by laser densitometry and corrected to take into account differences occurring during the RNA transfer. Preliminary results indicate that a 60 Hz magnetic field can produce a very weak increase in the c–fos mRNA level.

4.2   Effect of ELF Magnetic Fields on Cell Motility

The principal investigators are:

- Ingrid Spadinger, Branko Palcic, Cancer Imaging, British Columbia Cancer Research Centre, Vancouver, and David Agnew, Ontario Hydro, Toronto, Ontario

This study is funded by Ontario Hydro and BC Hydro.

The effect of exposure to various frequencies and intensities of ELF magnetic fields on mammalian cell motility have been studied. Swiss 3T3 fibroblasts, which exhibit motile behaviour *in vitro*, were exposed for 2–hour intervals to sinusoidally varying magnetic fields ranging in frequency from 10 to 72 Hz and in amplitude from 10 to 200 µT. A large number of frequency–amplitude combinations were tested, with emphasis on frequency effects. Frequencies were tested at 2 Hz intervals throughout most of the stated range. Additional frequencies were also tested, if they corresponded to the estimated geomagnetic cyclotron resonance frequencies for ions such as K^+, Mg^{2+} or Ca^{2+}. Field strengths of 20 and 80 µT were applied at each of the frequencies tested. Additional field strengths were applied for selected frequencies such as 60 Hz (power distribution frequency) and 22 Hz (the estimated cyclotron resonance frequency for Ca^{2+}).

Cell motility was measured using an automated image cytometry device to locate and record the positions of selected cells at regular time intervals (usually every 10 minutes). Up to 100 cells were tracked per experiment, with a typical experiment lasting for approximately 10 hours. Two–hour exposure periods were bracketed by 2–hour "quiet" intervals, which acted as controls. Two exposure conditions were tested per experiment. Various motility parameters (e.g., speed, acceleration, persistence and direction of movement) could be calculated from the recorded data.

No reproducible effects of ELF magnetic fields on cell speed have been detected for any of the frequency amplitude combinations tested. Other motility parameters are currently being analyzed, but the data so far indicate that ELF fields do not produce measurable effects of 3T3 cell motility *in vitro*.

4.3   Cancer Implications of the Effects of EMF on (1) Blood–Brain Barrier Permeability in the Rat and (2) OPOID Function in Mouse and Snail

The principal investigators are:

- Frank S. Prato, K. Kavaliers, K–P Ossenkopp, R.R. Shivers, J.R.F. Frappier, D.J. Drost, St. Joseph's Health Centre, London, Ontario

This study is funded by MRC, Upjohn Pharmaceutical Company, National Sciences and Engineering Research Council, Canada.

An increase in blood brain barrier (BBB) permeability after exposure of anaesthetized rats to a Magnetic Resonance Imaging procedure has been reported previously by the authors. Experiments to date indicate that the extremely low frequency magnetic field exposure may cause this increase. This observation may be important since increases in brain tumours (astrocytomas) have been reported in human populations chronically exposed to EMFs, whereas brain tumours, such as meningiomas, which are not protected by the BBB, are not reported to increase. This suggests the hypothesis that incidence of cancer is linked to an increased exposure to blood–borne chemical carcinogens.

Over the last decade, the authors have reported that increased analgesia in mice induced by exogenous opiates, such as morphine, can be attenuated by exposure to EMF. Experiments have indirectly implicated intracellular free calcium regulatory mechanisms as the site of an EMF interaction which attenuates morphine induced analgesia. Specifically, we have evidence to suggest that EMF may activate protein kinase C[3]. These *in vivo* experiments provide a rationale for the investigation of EMF effects on intracellular calcium *in vitro*. This suggests the hypothesis that EMF effects on such cellular mechanisms may impact on carcinogenesis by altering tumour promotion and/or immunity.

4.4   *In Vitro* Cytosolic Free Ca^{2+} Measurement During Magnetic Resonance Imaging Gradient Field Exposure

The principal investigators are:

- Jeffrey J.L. Carson, University of Western Ontario, Frank S. Prato, D.J. Drost and S.J. Dixon, St Joseph's Health Centre, London, Ontario

This study is funded by The Medical Research Council.

It has been hypothesized that many reported bioelectromagnetic phenomena are mediated by a primary effect on the concentration of cytosolic free calcium ($[Ca^{2+}]_i$). Using a fluorescence assay on HL–60 cells, the authors have shown that $[Ca^{2+}]_i$ increases by approximately 30 nM above resting levels (100 to 200 nM) after Magnetic Resonance Imaging (MRI) gradient magnetic field exposure. To investigate the kinetics of this gradient magnetic field–mediated response, a system which will measure $[Ca^{2+}]_i$ during exposure is being constructed. Cytosolic free calcium concentration will be measured by loading the cytoplasm of HL–60 cells with the fluorescent probe, indo–1. Indo–1 fluorescence will be detected with a spectrofluorimeter and calibrated to the concentration of free calcium in the cytoplasm. With this apparatus we will measure the kinetics of the $[Ca^{2+}]_i$ changes in HL–60 cells during exposure to gradient magnetic fields typical of magnetic resonance imaging.

4.5 Magnetic Field Effects on Cell Proliferation in Spheroid Cell Cultures

The principal investigators are:

- G.P. Arron and L.A. Bremner, Ontario Hydro, Toronto, Ontario; M.D. Sherar, Ontario Cancer Institute, Princess Margaret Hospital, Toronto, Ontario

This study is sponsored by Ontario Hydro.

It has been reported that combined dc and ELF ac magnetic fields consistent with calcium ion cyclotron resonance affects the growth rate of cells. Similar experiments were conducted using multicell spheroids of V79 (Chinese hamster lung fibroblasts) cells to investigate whether intercellular communication has a role in the biological effects of ELF magnetic fields. The growth rate of the spheroids growing in media at ph 7.4 (normal growing conditions) or 6.8 (about 65% of normal growth rate) was monitored for!5 days when exposed to 16 Hz ac magnetic field and a controlled dc field of 21 μT.

Previous work with other systems (e.g., Ca^{2+} metabolism in rat thymocytes and Na, K–ATPase activity) has shown that sensitization or activation of the systems (by treatment with the mitogen Concanavalin A or reducing the K^+ concentration, respectively) makes them more responsive to field exposure. It was anticipated that growing the spheroids at pH 6.8 would sensitize them, in that the activity of the Na^+/H^+ antiporter would be affected, and they would not be growing at their maximal rate. At both pH 7.4 and 6.8 exposure had no significant effect on the growth of the spheroids.

Further experiments at ph 7.4 and 6.8 using magnetic fields consistent with hydrogen ion cyclotron resonance conditions are being conducted.

5.0 RISK ASSESSMENT STUDIES

5.1 Methodology Development for the Risk Assessment of Potential Carcinogenic Effects Resulting from Exposure to Magnetic Fields

The principal investigator is:

- R. Stephen McColl, Institute for Risk Research and Department of Health Studies, University of Waterloo, Waterloo, Ontario.

The study is sponsored by Ontario Hydro.

The objective of the *EMF Risk Assessment Methodology* project is to develop specific analytical procedures for assessing the nature and extent of possible cancer risk to humans arising from exposure to extremely low frequency electromagnetic fields (EMF), and to characterize these risks in a societal context.

At present, there is no satisfactory theoretical foundation for understanding possible cause and effect relationships for EMF as a carcinogenic agent. Therefore, to assess the level of risk related to EMF exposure, it will be necessary to develop a novel methodological framework for the risk assessment of EMF. The methodological framework will be used to compile, evaluate, and integrate many diverse sources of risk assessment data pertaining to EMF carcinogenicity.

The utility of the proposed risk assessment methodology will be evaluated by a case study approach, prior to its application to the EMF studies. The carcinogenic chemical formaldehyde will serve as the prototype agent exemplifying an incompletely resolved risk assessment problem.

The methodological framework will subsequently be employed to analyze carcinogenicity data obtained from the EMF studies undertaken by Ontario Hydro, as well as data from other ongoing EMF studies, to produce initial estimates of possible cancer risks from EMF exposure. To minimize methodological bias, the risk assessment framework will be presented for external review before the individual EMFRAP study results are analyzed for risk estimation purposes. The methodological framework will include detailed scientific protocols for hazard identification, risk estimation, exposure assessment, and risk characterization of EMF.

In order to assure that the conceptual framework and risk assessment methodology developed by the project meet the criteria of scientific validity and practical utility for the risk assessment of EMF exposure, a scientific advisory panel will be assembled which will include representatives from the National Radiological Protection Board, the Electric Power Branch Institute, Health and Welfare Canada, and Ontario Hydro.

STATIC MAGNETIC FIELDS: A SUMMARY OF BIOLOGICAL INTERACTIONS, POTENTIAL HEALTH EFFECTS, AND EXPOSURE GUIDELINES

T. S. Tenforde

Life Sciences Center
Battelle, Pacific Northwest Laboratories
Richland, Washington 99352 (U.S.A.)

INTRODUCTION

Interest in the mechanisms of interaction and the biological effects of static magnetic fields has increased significantly during the past two decades as a result of the growing number of applications of these fields in research, industry and medicine (Stuchly, 1986; Tenforde, 1986). A major stimulus for research on the bioeffects of static magnetic fields has been the effort to develop new technologies for energy production and storage that utilize intense magnetic fields (e.g., thermonuclear fusion reactors and superconducting magnetic energy storage devices).

Interest in the possible biological interactions and health effects of static magnetic fields has also been increased as a result of recent developments in magnetic levitation as a mode of public transportation. In addition, the rapid emergence of magnetic resonance imaging as a new clinical diagnostic procedure has, in recent years, provided a strong rationale for defining the possible biological effects of magnetic fields with high flux densities (Budinger and Lauterbur, 1986). In this review, the principal interaction mechanisms of static magnetic fields will be described, and a summary will be given of the present state of knowledge of the biological, environmental, and human health effects of these fields. Extensive reviews of these subjects have been published in the last several years (Frankel, 1986; Tenforde, 1985a, 1985b, 1988, 1989a, 1990, 1991; Tenforde and Budinger, 1986; World Health Organization, 1987).

1 INTERACTION MECHANISMS

Three classes of physical interactions of static magnetic fields with bio-logical systems are well established on the basis of experimental data: (1) electrodynamic interactions with ionic conduction currents; (2) magnetomechanical effects, including the orientation of magnetically anisotropic structures in uniform fields and the translation of paramagnetic and ferromagnetic materials in magnetic field gradients; and (3) effects on electronic spin states of the reaction intermediates in certain types of charge transfer processes. Each of these physical interaction mechanisms, along with relevant experimental data, will be described in the following paragraphs.

1.1 Electrodynamic Interactions

Ionic currents interact with static magnetic fields as a result of the Lorentz forces exerted on moving charge carriers. This electrodynamic interaction gives rise to an induced electric field, $E_i = -v \times B$, where v is the velocity of current flow and B is the magnetic flux density. For ion flows through channels in cell membranes, the interaction of a static magnetic field is extremely weak. It has been estimated, for example, that a static magnetic field in excess of 24 Tesla (T) would be required to produce a Lorentz force on nerve ionic currents equal to one-tenth the force they experience from the local electric field of the nerve membrane (Wikswo and Barach, 1978). The absence of effects of static fields up to 2 T on nerve bioelectric properties has been demonstrated experimentally (Gaffey and Tenforde, 1983).

In the case of bulk flow of an electrically conductive fluid such as blood, significant electrical potentials are induced magnetically at field levels well below 1 T. It is a direct consequence of the Lorentz force exerted on moving ionic currents that blood flowing through a cylindrical vessel of diameter, d, will develop an electrical potential, ψ, equal to $| E_i | d = | v | | B | d \sin \phi$, where ϕ is the angle between B and the velocity vector v. It has been demonstrated that an induced electrical potential associated with pulsatile aortic blood flow in the presence of a static magnetic field can be detected in the electrocardiogram (ECG) at the locus of the T-wave signal. As reviewed by Tenforde (1989a, 1991), this feature of aortic blood flow potentials has been demonstrated for rats, rabbits, dogs, baboons and monkeys. Figure 1 shows the three-lead ECG of a 5-kg *Papio* baboon prior to and during exposure to a 1.5-T static magnetic field. The largest superimposed electrical potential occurs at the T-wave locus in the ECG, which corresponds temporally to the opening of the aortic valve during pulsatile ejection of blood from the left ventricular chamber of the heart. The augmentation of the T-wave signal that is observed during magnetic field exposure is completely reversed upon removal of the field.

In large animal species such as baboons, monkeys and dogs, the aortic blood flow potential can be detected in the ECG at field levels above approximately 0.1 T, and is a linear function of field strength up to 1.0 T (Tenforde et al., 1983, 1985). At higher field levels, the total electrical potential at the T–wave locus in the ECG increases more rapidly as a function of magnetic field strength, possibly as a result of the superposition of additional, weaker flow potentials which cannot be detected at field strengths below 1.0 T. Based on the timing of valve displacements during the cardiac cycle (see Figure 1), the magnetically induced flow potential associated with pulsatile ejection of blood into the pulmonary artery may also contribute to the total ECG signal at the locus of the T–wave during exposure to very large magnetic fields. It is also evident from the ECG recordings shown in Figure 1 that other magnetically induced flow potentials can be detected during exposure to strong magnetic fields. The temporal sequence of these flow potentials relative to cardiac valve displacements indicates that they may be associated with rapid movements of blood during the filling and emptying phases of the heart cycle. A similar conclusion has been drawn from studies with *Macaca* monkeys and beagle dogs, in which a combination of phonocardiography and echocardiography were used to study the temporal sequence of cardiac valve displacements in relation to the timing of signals recorded in the ECG during magnetic field exposure (Tenforde, 1989a).

The electrodynamic interaction between an applied magnetic field and a flowing electrolyte solution such as blood also creates a net volume force within the fluid. This force is equal to $\mathbf{J} \times \mathbf{B}$, where $\mathbf{J} = -\,\sigma\,\mathbf{v} \times \mathbf{B}$ is the ionic conduction current density resulting from the induced electric field within the flowing solution, and σ is the electrical conductivity. For a moving fluid within a static magnetic field, the magnetohydrodynamic consequence of this electrical force is a reduction in the flow velocity of the fluid. As an experimental test of the strength of magnetohydrodynamic interactions within the circulatory system, a combination of arterial blood flow velocity measurements and intra–arterial blood pressure measurements were carried out in beagle dogs and *Macaca* monkeys exposed to static magnetic fields with flux densities up to 1.5 T (Tenforde, 1989a). In accord with theoretical predictions (Tenforde, 1985a), these experimental results demonstrated that magnetohydrodynamic interactions in a 1.5–T field do not produce a measurable alteration in blood flow dynamics.

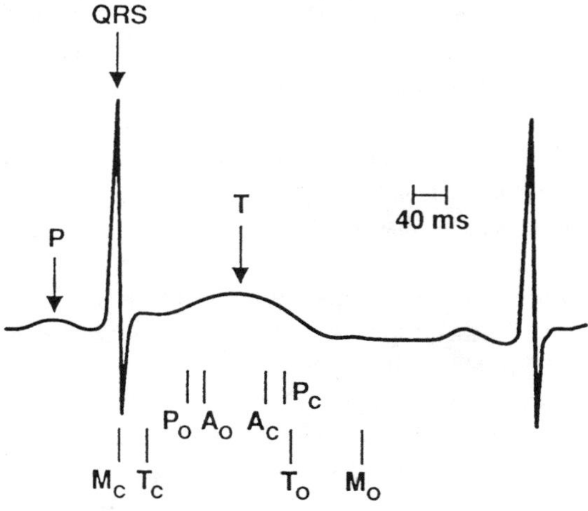

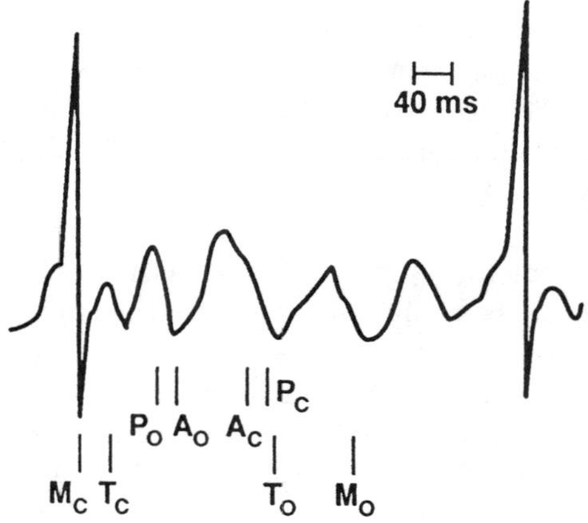

Figure 1. Electrocardiograms are shown for a female *Papio* baboon immediately prior to and during exposure to a 1.5–T static magnetic field. The estimated times of opening (subscript "o") and closing (subscript "c") of the mitral (M), tricuspid (T), pulmonary (P) and aortic (A) valves are denoted by vertical bars. (From Tenforde et al., 1985, Figure 1)

1.2 Magnetomechanical Interactions

There are two basic mechanisms through which static magnetic fields exert mechanical forces and torques on objects. In the first type of magnetomechanical interaction, rotational motion of a substance occurs in a uniform field until it achieves a minimum energy state. The second mechanism involves the translational force exerted on a paramagnetic or ferromagnetic substance placed in a magnetic field gradient. These two types of interaction will be discussed separately.

Macromolecules and structurally ordered molecular assemblies with a high degree of magnetic anisotropy will experience a torque in a uniform magnetic field and rotate until they reach an equilibrium orientation that represents a minimum energy state. Macromolecules such as DNA that exhibit this property generally have a cylindrical symmetry, and magneto-orientation occurs as a result of the anisotropy of the diamagnetic susceptibility tensor along the axial and radial coordinates. The extent to which these molecules orient is a function of their magnetic interaction energy, U, relative to the Boltzmann thermal energy, kT, where k is the Boltzmann constant (1.382×10^{-23} J/K) and T is the Kelvin temperature. The interaction energy depends upon the product of B^2, the molecular volume, and the difference in magnetic susceptibility along directions parallel and perpendicular to the applied field. The ratio U/kT is referred to as the "order parameter," and reflects the extent of molecular orientation in an applied magnetic field.

For individual macromolecules, the order parameter is much less than unity at field levels that can be easily achieved in the laboratory, and the extent of orientation of individual molecules in strong magnetic fields is very small. For example, optical birefringence measurements on calf thymus DNA in solution have demonstrated that a field of 13 T is required to produce 1% orientation of the molecules (Maret et al., 1975). In contrast, there are several examples of molecular assemblies that can be completely oriented by fields on the order of 1 T (Tenforde, 1985b). These assemblies behave as structurally coupled units in which the summed magnetic anisotropy is large, thus giving rise to a large magnetic interaction energy. Examples of molecular aggregates that exhibit magneto–orientation include retinal rod outer segments, muscle fibers, photosynthetic systems (chloroplast grana and *Chlorella* cells), purple membranes of *Halobacteria*, and filamentous virus particles. An example of an intact cell that can be oriented magnetically is the deoxygenated sickled erythrocyte. It has been shown that these cells, in which the deoxygenated hemoglobin is paramagnetic, will align in a 0.35–T static field with the long axis of the sickled cell oriented perpendicular to the magnetic flux lines (Murayama, 1965). This equilibrium orientation results from the stacking of the planar haem moieties parallel to the long axis of the sickled erythrocyte, with the net magnetic moment oriented perpendicular to the long axis.

Despite the fact that magneto–orientation of biologically important structures such as retinal rod outer segments can be demonstrated by optical techniques when these units are suspended in an aqueous medium (Becker et al., 1978), extensive electrophysicological studies have failed to reveal any influence of this effect on visual functions *in vivo*. A series of electroretinogram (ERG) recordings were made from a *Macaca* monkey before, during and after exposure to static magnetic fields with flux densities up to 1.5 T. Neither the A–wave (receptor field potential) or the B–wave (postsynaptic potential) elicited by flashes of white light were altered during magnetic field exposure. Similar results were obtained in ERG studies on three monkeys and six cats (Tenforde, 1989a; Tenforde et al., 1985). These data indicate that magneto–orientational forces exerted by static magnetic fields with flux densities up to 1.5 T have no significant influence on visual phototransduction processes *in vivo*. The most likely reason for this lack of effect is the motional restriction imposed on retinal photoreceptors by virtue of being embedded in a rigid structural matrix within the intact retina.

A second major type of magnetomechanical interaction is the translation of paramagnetic and ferromagnetic substances in static magnetic field spatial gradients. Denoting the magnetic susceptibility of an object as χ and the volume as V, the force, F(z), experienced in a linear magnetic field gradient, dB/dz, is equal to the product of the net magnetic moment and the field gradient: $F(z) = (\chi VB/\mu_o)(dB/dz)$, where μ_o is the magnetic permeability of free space $(4\pi \times 10^{-7} \text{ H/m})$. The forces exerted on paramagnetic and ferro-magnetic substances by strong static magnetic field gradients provide the physical basis for a number of useful biological and biochemical processes (Frankel, 1986; Tenforde, 1991). Examples of the application of magnetic forces include the targeting of drugs encapsulated in magnetic microcarriers, the separation of deoxygenated erythrocytes from whole blood, and the separation of antibody–secreting cells from a suspension of bone marrow cells.

In contrast to their useful applications in biology, the forces exerted by strong magnetic field gradients can pose a significant physical hazard in the workplace and in magnetic resonance imaging facilities. There is a risk of large tools and other metallic objects becoming projectiles in the proximity of a high–field magnet. In addition, significant magnetic forces are exerted on many types of implanted medical devices, including aneurysm clips, dental amalgam, prostheses, and pacemaker cases (Tenforde and Budinger, 1986).

1.3 Magnetic Field Effects on Electronic Spin States

Several classes of organic chemical reactions can be influenced by static magnetic fields greater than approximately 1 mT as a result of effects on the electronic spin states of the reaction intermediates (Schulten, 1982). One example of such reactions that has been studied extensively is the photo–induced charge transfer reaction in bacterial photosynthesis (Hoff, 1981). This reaction involves a radical pair intermediate state through which electron transfer occurs to the ultimate acceptor molecule, a ubiquinone–iron complex. Under natural conditions the electron transfer occurs within 200 ps following flash excitation of the bacteriochlorophyll. However, chemical reduction of the acceptor molecules extends the lifetime of the intermediate state to about 10 ns. With an extended lifetime, the singlet state of the radical pair intermediate evolves into a triplet state via the hyperfine interaction mechanism. In the presence of an external magnetic field greater than approximately 10 mT, however, the triplet channels are completely blocked and the resulting yield of triplet product is expected to decrease by two thirds. This predicted blocking of triplet channels by a weak magnetic field has been confirmed experimentally using laser pulse excitation and optical absorption measurements (Michel–Beyerle et al., 1979).

It should be emphasized that the magnetic field effect on the photo–induced electron transfer in photosynthesis occurs only when the photosynthetic system is placed in an abnormal state by chemical reduction of the electron acceptor molecules. The possibility cannot be excluded, however, that similar magnetic field effects may occur in other radical–mediated biological processes under naturally occurring conditions. For example, it has been proposed that an anisotropic Zeeman interaction with a radical–mediated reaction system could provide a basis for geomagnetic direction finding (Schulten et al., 1978). Several types of enzymatic reactions also involve radical intermediate states that may exhibit sensitivity to static magnetic fields (Tenforde, 1985b).

2 ORGANISMS WITH UNIQUE SENSITIVITY TO MAGNETIC FIELDS

Various types of organisms have been demonstrated to possess sensitivity to extremely weak magnetic fields, comparable in strength to the geomagnetic field. In several instances, there is direct experimental evidence indicating that this magnetic sensitivity is linked to direction–finding ability. The two basic mechanisms of magnetoreception are (1) magnetic induction of weak electrical signals in specialized sensory receptors, and (2) magnetomechanical interactions with localized deposits of single–domain magnetite crystals (Tenforde, 1989b). These two mechanisms of geomagnetic field detection are described in the following paragraphs.

2.1 Elasmobranch Fish

A well–known example of electrodynamic interactions involving weak magnetic fields is the electromagnetic guidance system of elasmobranch fish, a class of marine animals that includes sharks, skates and rays. The heads of these fish contain long jelly–filled canals with a high electrical conductivity, known as the ampullae of Lorenzini. As an elasmobranch swims through the lines of flux of the geomagnetic field, small voltage gradients are induced in its ampullary canals. These induced electric fields can be detected at levels as low as 0.5 μV/m by the sensory epithelia that line the terminal ampullary region (Kalmijn, 1982). The polarity of the induced field in an ampullary canal depends upon the relative orientation of the geomagnetic field and the compass direction along which the fish is swimming. As a consequence, the weak electric fields induced in the ampullae of Lorenzini provide a sensitive directional cue for the elasmobranch fish.

2.2 Magnetotactic Bacteria

An example of a cellular structure in which significant magnetic orientational effects occur in response to the geomagnetic field is the magnetotactic bacterium (Blakemore, 1975). Approximately 2% of the dry mass of these aquatic organisms is iron, which has been shown by Mössbauer spectroscopy to be predominantly in the form of magnetite, Fe_3O_4 (Frankel et al., 1979). Magnetite crystals are synthesized within the magnetotactic bacterium ("biogenic magnetite"), and they are arranged as a chain of approximately 20 to 30 single domain crystals encapsulated in a membrane structure (Gorby et al., 1988). The orientation of the net magnetic moment is such that magnetotactic bacteria in the northern hemisphere migrate towards the north pole of the geomagnetic field, whereas strains of these bacteria that grow in the southern hemisphere move towards the south magnetic pole (Blakemore et al., 1980; Kirschvink, 1980). Magnetotactic bacteria that have been found at the geomagnetic equator are nearly equal mixtures of south–seeking and north–seeking organisms (Frankel et al., 1981). Because of the polarities of their magnetic moments, the magnetotactic bacteria in both the northern and southern hemispheres migrate downwards in response to the vertical component of the geomagnetic field. It has been proposed that this downward–directed motion, which carries the bacteria into the bottom sediments of their aquatic environment, may be essential for the survival of these microaerophilic organisms (Blakemore et al., 1980). In support of this hypothesis it has been shown that the population density of magnetotactic bacteria decreases abruptly as the vertical component of the magnetic field at the mud/water interface approaches zero (Chang and Kirschvink, 1989).

2.3 Avian Navigation

The effects of the static geomagnetic field on the navigation of avians have been studied extensively (Tenforde, 1991). In early experiments by Keeton (1971), measurements were made of orientation and homing ability in groups of pigeons to which small bar magnets were attached to the back at the base of the neck, producing a static field of about 45 μT at the head. Nonmagnetic brass bars of comparable weight were attached to the control group of birds. The results of these experiments illustrated a significant disorientation of birds wearing the bar magnets compared to controls when they were released under overcast skies.

A remarkable recent finding by Moore (1988) has challenged the widely accepted view that the geomagnetic field influences avian navigation. In an evaluation of unpublished data collected by the late W. T. Keeton during the period 1971–1979, he found no evidence for statistically significant effects of bar magnets attached to the backs of pigeons on either the consistency or the accuracy of their initial orientation during flight under overcast skies. These findings are in direct contrast to the results of Keeton's earlier studies conducted in 1969 and 1970, in which statistically significant decreases in both the accuracy and consistency of pigeon orientation were observed in response to an altered magnetic field environment produced by an attached bar magnet. Moore (1988) concluded that it is conceivable that pigeons can detect magnetic fields, but that some unknown factor masked or blocked the effect in Keeton's later studies. An alternative explanation proposed by Moore is that the difference in results between the two sets of experiments may indicate that pigeons do not detect magnetic fields and that the positive outcome of the earlier studies by Keeton resulted from some unknown source of bias or as a result of random chance alone. Regardless of the explanation, the remarkable divergence between the results of Keeton's 1971–1979 experiments and his earlier studies raises a severe challenge to the concept that the geomagnetic field provides a back–up compass under overcast skies.

2.4 Magnetite and Magnetoreception in Animals

It was first observed by Lowenstam (1962) that the teeth of mollusks contain a high concentration of iron in the form of magnetite. Evidence has subsequently been obtained that the interaction of the geomagnetic field with these iron–containing crystals may influence the kinetic movements of mollusks (Ratner, 1976; Kirschvink and Lowenstam, 1979). The important role of magnetite deposits in geomagnetic field detection has now been shown for a large variety of biological organisms, including the magnetotactic bacteria discussed above (Kirschvink et al., 1985). Following the demonstration of biogenic magnetite in bacteria, sensitive magnetometer measurements have detected the presence of localized magnetite domains in a variety of animal species, including pigeons, bees, dolphins, tuna, salmon, turtles, and humans (Tenforde, 1991). In several of these species, there is an apparent sensitivity to the geomagnetic field, which confers direction–finding ability. Baker (1980) has claimed that humans can also use the geomagnetic field for orientation and direction–finding. However, tests of this hypothesis have let to negative results (Gould and Able, 1981; Fildes et al., 1984).

The possible influence of ambient magnetic fields on the direction–finding ability of bees is also an unresolved issue. Experimental evidence obtained by Lindauer and Martin (1968) indicated that the directional information conveyed in the bee's waggle dance is influenced by the geomagnetic field. This finding, and other evidence (Kirschvink, 1981), have raised the possibility that bees use the earth's magnetic field as a secondary directional cue when the sun is not visible. However, experiments by Dyer and Gould (1981) on bee navigational patterns indicated that these animals possess a "memory" of the sun's position at different times of the day, which serves as a back–up orientational cue when the sky is overcast. The results of these experiments suggest that magnetic field sensitivity may play only a minor role in direction–finding by bees. In contrast, recent behavioral experiments by Walker and Bitterman (1985, 1989) have demonstrated that the foraging behavior of honeybees can be influenced by ambient magnetic fields. This finding has been confirmed and initial evidence for a ferromagnetic transduction mechanism was recently reported (Kirschvink and Kobayashi–Kirschvink, 1991). Currently available evidence suggest that bees can sense weak magnetic fields, but do not necessarily use this information as a directional cue.

3 LABORATORY STUDIES ON STATIC MAGNETIC FIELD EFFECTS

As discussed in the preceding section, several species of marine animals and various other species possess an inherent sensitivity to static magnetic fields with flux densities as low as that of the geomagnetic field. In higher organisms, however, the only effect that is well established at the present time is the magnetic induction of electrical potentials in the central circulatory system. There are also numerous instances in which contradictory results have been reported in the literature.

During the past decade, a large number of studies have been conducted in which the biological effects of static magnetic fields were examined under well–controlled laboratory conditions, including the use of precise dosimetry, large numbers of experimental subjects, quantitative biochemical and physiological end points, and careful control of ambient environmental conditions that could influence the experimental results. Based on laboratory studies involving field levels of 1 T or higher, the following important biological processes appear not to be altered by exposure to static magnetic fields at high flux densities (Tenforde, 1985b, 1988): (1) cell growth and morphology, (2) DNA structure and gene expression, (3) reproduction and development (pre– and postnatal), (4) visual functions, (5) nerve bioelectric activity, (6) cardiovascular dynamics, (7) hematological indices, (8) immune responsiveness, (9) physiological regulation and circadian rhythms, and (10) animal behavior.

Although the majority of published studies have not shown significant behavioral or physiological effects of strong magnetic fields, occasional reports have appeared in the literature on the response of various organ and tissue systems to relatively weak magnetic fields. For example, it has been reported that the electrical activity of rodent and avian pineal cells can be altered by artificial changes in the strength and direction of the local geomagnetic field (Semm, 1983). In addition, evidence has been obtained that pineal melatonin content and serotonin–N–acetyltransferase activity in rodents can be modified by changes in the ambient magnetic field strength (Welker et al., 1983; Lerchl et al., 1990). Weak magnetic fields have also been reported to abolish the circadian rhythmicity of Purkinje cell responses to pineal melatonin in pigeons (Demaine and Semm, 1986). Other studies have indicated that circadian variations may exist in the sensitivity of turtle retinal cells to static magnetic fields with flux densities as low as 2 to 3 mT (Raybourn, 1983). In these and other observations of apparent biological effects of weak magnetic fields, the types of interaction mechanisms involved have not as yet been elucidated. Further research will be required to verify the existence and the nature of these interactions, and to determine their potential implications for human health.

4 HUMAN HEALTH EFFECTS

Several epidemiological studies in the United States and Europe have shown no adverse health effects associated with occupational exposure to static magnetic fields at National Laboratories (Budinger et al., 1984), and in chemical separation plants that use electrolytic cells operated with large DC currents (Marsh et al., 1982; Barregård, 1985). Two studies on workers in aluminum plants, where large static magnetic fields are present near prebake anode cells (Tenforde, 1986), have demonstrated an increased mortality from leukemia and various other types of cancer in comparison with the general population (Milham, 1979; Rockette and Arena, 1983). However, the possible influence of potentially carcinogenic factors other than magnetic fields was not adequately addressed in these studies. In addition, a large study on French aluminum workers showed their cancer mortality and mortality from all causes not to differ significantly from that observed for the general male population of France (Mur et al., 1987).

Another important aspect of potential human health effects is the influence of static magnetic fields on the operation of medical electronic devices (Tenforde and Budinger, 1986). Of particular concern is the fact that static magnetic fields exceeding 1.7 mT can produce closure of a reed relay switch used in modern cardiac pacemakers, thereby causing the pacemaker to revert to an asynchronous mode of operation (Pavlicek et al., 1983). As a consequence, persons wearing pacemakers should not enter areas where the static magnetic field levels exceed 1 mT.

5 MAGNETIC FIELD EXPOSURE GUIDELINES

Several sets of guidelines limiting human exposure to static magnetic fields in the workplace have been proposed during the past two decades. Until recently, the most widely used guidelines have been those established at the Stanford Linear Accelerator in California in 1970. These guidelines limit whole–body or head exposure to 20 mT during the entire workday, and to 0.2 T for short intervals of several minutes duration. The limits for exposure of the arms and hands are 10 times greater than those for the whole body or head. An occupational limit of 20 mT for whole–body exposure to static magnetic fields was also adopted in the United Kingdom (National Radiological Protection Board, 1986).

A less conservative set of exposure guidelines for static magnetic fields was implemented in 1985 at the Lawrence Livermore National Laboratory in California. These guidelines limit whole–body exposure to a time–weighted–average field strength of 60 mT measured at the torso or 0.6 measured at the extremities. The rationale for the whole–body limit of 60 mT was based on a calculation of the field level that would induce a maximum electrical potential in the aortic vessel of 1 mV (Miller, 1987). From earlier research with experimental animals described in this paper, it was concluded that magnetically induced potentials of this magnitude should not produce adverse effects on cardiac performance or hemodynamic parameters. The Lawrence Livermore National Laboratory guidelines prohibit individuals with cardiac pacemakers from entering areas where the magnetic field level exceeds 1 mT. This field strength was also set as a cautionary warning level for individuals with aneurysm clips or other implanted prosthetic devices. The maximum field level to which any worker can be exposed was set at 2 T. The American Conference of Governmental Industrial Hygienists (ACGIH) recently adopted a set of occupational exposure guidelines for static magnetic fields that are identical to those used at the Lawrence Livermore National Laboratory (ACGIH, 1991). The International Radiation Protection Association is in an advanced stage of development of new guidelines for exposure of workers and the general public to static magnetic fields (Repacholi, 1991).

6 ACKNOWLEDGMENTS

The skillful assistance of Claudine A. Baldwin in the preparation of this paper is gratefully acknowledged. Research support was received from the U.S. Department of Energy under Contract DE–AC06–76RLO 1830 with the Pacific Northwest Laboratory. The Pacific Northwest Laboratory is operated for the U.S. Department of Energy by the Battelle Memorial Institute.

7 REFERENCES

American Conference of Governmental Industrial Hygienists (1991) Threshold Limit Values and Biological Exposure Indices for 1991–1992. American Conference of Governmental Industrial Hygienists, Cincinnati, Ohio.

Baker, R.R. (1980) Goal orientation by blindfolded humans after long distance displacement: Possible involvement of a magnetic sense. Science, 270:555–557.

Barregård, L., Jarvholm, B. and Ungethum, E. (1985) Cancer among workers exposed to strong static magnetic fields. Lancet, 2(8460):892.

Becker, J.F., Trentacosti, F. and Geacintov, N.E. (1978) A linear dichroism study of the orientation of aromatic protein residues in magnetically oriented bovine rod outer segments. Photochem. Photobiol., 27:51–54.

Blakemore, R. (1975) Magnetotactic bacteria. Science, 190:377–379.

Blakemore, R.P., Frankel, R.B. and Kalmijn, A.J. (1980) South–seeking magnetotactic bacteria in the southern hemisphere. Nature, 286:384–385.

Budinger, T.F. and Lauterbur, P.C. (1984) Nuclear magnetic resonance technology for medical studies. Science, 226:288–298.

Budinger, T.F., Bristol, K.S., Yen, C.K. and Wong, P. (1984) Biological effects of static magnetic fields. In: Proc. Society of Magnetic Resonance in Medicine, Third Annual Meeting, New York, 4–6 August 1984, pp. 113–114.

Chang, S.R. and Kirschvink, J.L. (1989) Magnetofossils, the magnetization of sediments, and the evolution of magnetite biomineralization. Ann. Rev. Earth Planet. Sci., 17:169–195.

Demaine, C. and Semm, P. (1986) Magnetic fields abolish nycthemeral rhymicity of responses of purkinje cells to the pineal hormone melatonin in the pigeon's cerebellum. Neurosci. Lett., 72:158–162.

Dyer, F.C. and Gould, J.L. (1981) Honey bee orientation: A backup system for cloudy days. Science, 214:1041–1042.

Fildes, B.N., O'Loughlin, B.J. and Bradshaw, J.L. (1984) Human orientation with restricted sensory information: No evidence for magnetic sensitivity. Perception, 13:229–236.

Frankel, R.B. (1986) Biological effects of static magnetic fields. In: Handbook of Biological Effects of Electromagnetic Fields (Eds.: C. Polk and E. Postow). CRC Press, Boca Raton, Florida, pp. 169–196.

Frankel, R.B., Blakemore, R.P. and Wolfe, R.S. (1979) Magnetite in freshwater magnetotactic bacteria. Science, 203:1355–1356.

Frankel, R.B., Blakemore, R.P. Torres de Araujo, F.F. and Esquival, D.M.S.

(1981) Magnetotactic bacteria at the geomagnetic equator. Science, 212:1269–1270.

Gaffey, C.T. and Tenforde, T.S. (1983) Bioelectric properties of frog sciatic nerves during exposure to stationary magnetic fields. Radiat. Environ. Biophys., 22:61–73.

Gorby, Y.A., Beveridge, T.J. and Blakemore, R.P. (1988) Characterization of the bacterial magnetosome membrane. J. Bacteriol., 170:834–841.

Gould, J.L. and Able, K.P. (1981) Human homing: An elusive phenomenon. Science, 214:1061–1063.

Hoff, A.F. (1981) Magnetic field effects on photosynthetic reactions. Quart. Rev. Biophys., 74:599–665.

Kalmijn, A.J. (1982) Electric and magnetic field detection in elasmobranch fish. Science, 218:916–918.

Keeton, W.T. (1971) Magnets interfere with pigeon homing. Proc. Natl. Acad. Sci. (USA), 68:102–106.

Kirschvink, J.L. (1980) South–seeking magnetic bacteria. J. Exp. Biol., 86:345–347.

Kirschvink, J.L. (1981) The horizontal magnetic dance of the honeybee is compatible with a single–domain ferromagnetic magnetoreceptor. BioSystems, 14:193–203.

Kirschvink, J.L. and Lowenstam, H.A. (1979) Mineralization and magnetization of chiton teeth: Paleomagnetic, sedimentologic and biologic implications of organic magnetite. Earth Planet. Sci. Lett., 44:193–204.

Kirschvink, J.L. and Kobayashi–Kirschvink, A. (1991) Is geomagnetic sensitivity real? Replication of the Walker–Bitterman magnetic conditioning experiment in honey bees. Am. Zool., 31:169–185.

Kirschvink, J.L., Jones, D.S. and MacFadden, B.J. (Eds.) (1985) Magnetite Biomineralization and Magnetoreception in Animals: A New Bio-magnetism, Plenum Press, New York.

Lerchl, A., Noraka, K.O., Stokkam, K.–A. and Reiter, R.J. (1990) Marked rapid alterations in nocturnal pineal serotonin metabolism in mice and rats exposed to weak intermittent magnetic fields. Biochem. Biophys. Res. Comm., 169:102–108.

Lindauer, M. and Martin, H. (1968) Die Schwereorientierung der Bienen unter dem Einfluss des Erdmagnetfelds. Vergl. Physiol., 60:219–243.

Lowenstam, H.A. (1962) Magnetite in denticle capping in recent chitons (*Polyplacophora*). Geol. Soc. Am. Bull., 73:435–438.

Maret, G., VonSchickfus, M., Mayer, A. and Dransfeld, K. (1975) Orientation of nucleic acids in high magnetic fields. Phys. Rev. Lett., 35:397–400.

Marsh, J.L., Armstrong, T.J., Jacobson, A.P. and Smith, R.G. (1982) Health effect of occupational exposure to steady magnetic fields. Am. Ind. Hyg. Assoc. J., 43:387–394.

Michel–Beyerle, M.E., Scheer, H., Seidlitz H., Tempus, D. and Haberkorn, R. (1979) Time–resolved magnetic field effect on triplet formation in photosynthetic reaction center of *Rhodopseudomonas sphaeroides* R–26. FEBS Lett., 100:9–12.

Milham, S., Jr. (1979) Mortality in aluminum reduction plant workers. J. Occup. Med., 21:475–480.

Miller, G. (1987) Exposure guidelines for magnetic fields. Am. Ind. Hyg. Assoc. J., 48:957–968.

Moore, B.R. (1988) Magnetic fields and orientation in homing pigeons: Experiments of the late W.T. Keeton. Proc. Natl. Acad. Sci. (USA), 85:4907–4909.

Mur, J.M., Moulin, J.J., Meyer–Bisch, C., Massin, N., Coulon, J.P. and Loulergue, J. (1987) Mortality of aluminum reduction plant workers in France. Int. J. Epidemiol., 16:257–264.

Murayama, M. (1965) Orientation of sickled erythrocytes in a magnetic field. Nature, 206:420–422.

National Radiological Protection Board (1986) Advice on the Protection of Workers and Members of the Public From the Possible Hazards of Electric and Magnetic Fields With Frequencies Below 300 GHz: A Consultative Document. Oxon, England.

Pavlicek, W., Geisinger, M., Castle, L., Borkowski, G.P., Meaney, T.F., Bream, B.L. and Gallagher, J.H. (1983) The effects of nuclear magnetic resonance on patients with cardiac pacemakers. Radiology, 147:149–153.

Ratner, S.C. (1976) Kinetic movements in magnetic fields of chitons with ferromagnetic structures. Behav. Biol., 17:573–578.

Raybourn, M.S. (1983) The effects of direct–current magnetic fields on turtle retinas *in vitro*. Science, 220:715–717.

Repacholi, M.H. (1991) Static magnetic fields and health. In: Abstracts of the Sixteenth Annual Conference of the Australian Radiation Protection Society, Melbourne, Australia, 30 September – 2 October 1991, p. 65.

Rockette, H.E. and Arena, V.C. (1983) Mortality studies of aluminum reduction plant workers: Potroom and carbon department. J. Occup. Med., 25:549–557.

Schulten, K. (1982) Magnetic field effects in chemistry and biology. In: Festkorperprobleme XXII: Advances in Solid State Physics (Ed. P.G. Aachen), Vieweg, Braunschweig, West Germany, pp. 61–63.

Schulten, K., Swenberg, C.E. and Weller, A. (1978) A biomagnetic sensory mechanism based on magnetic field modulated coherent electron spin motion. Zeit. Phys. Chem., 111:1–5.

Semm, P. (1983) Neurobiological investigations on the magnetic sensitivity of the pineal gland in rodents and pigeons. Comp. Biochem. Physiol., 76A:683–689.

Stuchly, M.A. (1986) Human exposure to static and time–varying magnetic fields. Health Phys., 51:215–225.

Tenforde, T.S. (1985a) Mechanisms for biological effects of magnetic fields. In: Biological Effects and Dosimetry of Static and ELF Electromagnetic Fields (Eds. M. Grandolfo, S.M. Michaelson and A. Rindi) Plenum Press, New York, pp. 71–92.

Tenforde, T.S. (1985b) Biological effects of stationary magnetic fields. In: Biological Effects and Dosimetry of Static and ELF Electromagnetic Fields (Eds. M. Grandolfo, S.M. Michaelson and A. Rindi) Plenum Press, New York, pp. 93–127.

Tenforde, T.S. (1986) Magnetic field applications in modern technology and medicine. In: Biological Effects of Static and Extremely Low Frequency Magnetic Fields (Ed. J.H. Bernhardt) MMV Medizin Verlag, Munich, West Germany, pp. 23–35.

Tenforde, T.S. (1988) Interaction mechanisms, biological effects and biomedical applications of static and extremely–low–frequency magnetic fields. In: Proc. Twenty–Second Annual Meeting of the National Council on Radiation Protection and Measurements: Nonionizing Radiations and Ultrasound, 2–3 April 1986, NCRP Press, Bethesda, Maryland, pp. 181–217.

Tenforde, T.S. (1989a) Biological responses to static and time–varying magnetic fields. In: Electromagnetic Interaction With Biological Systems (Ed. J.C. Lin) Plenum Press, New York, pp. 83–107.

Tenforde, T.S. (1989b) Electroreception and magnetoreception in simple and complex organisms. Bioelectromagnetics, 10:215–221.

Tenforde, T.S. (1990) Biological effects of static magnetic fields. Int. J. Appl. Electromag. Mat., 1:157–165.

Tenforde, T.S. (1991) Interaction mechanisms and biological effects of static magnetic fields. Automedica, in press.

Tenforde, T.S. and Budinger, T.F. (1986) Biological effects and physical safety aspects of NMR imaging and *in vivo* spectroscopy. In: NMR in Medicine: Instrumentation and Clinical Applications (Eds. S.R. Thomas and R.L. Dixon) American Association of Physicists in Medicine, New York, pp. 493–548.

Tenforde, T.S., Gaffey, C.T., Moyer, B.R. and Budinger, T.F. (1983)

Cardiovascular alterations in *Macaca* monkeys exposed to stationary magnetic fields: Experimental observations and theoretical analysis. Bioelectromagnetics, 4:1–9.

Tenforde, T.S., Gaffey, C.T. and Raybourn, M.S. (1985) Influence of stationary magnetic fields on ionic conduction processes in biological systems. In: Proc. Sixth Symposium and Technical Exhibition on Electromagnetic Compatibility (Ed. T. Dvorák) Zurich, Switzerland, 5–7 March 1985, pp. 205–210.

Walker, M.M. and Bitterman, M.E. (1985) Conditioned responding to magnetic fields by honeybees. J Comp. Physiol., A157:67–71.

Walker, M.M. and Bitterman, M.E. (1989) Conditioning analysis of magnetoreception in honeybees. Bioelectromagnetics, 10:261–275.

Welker, H.A., Semm, R.P. Willig, R.P., Commentz, J.C. Wiltschko, W. and Vollrath, L. (1983) Effects of an artificial magnetic field on serotonin N–acetyltransferase activity and melatonin content of the rat pineal gland. Exp. Brain Res., 50:426–432.

Wikswo, J.P., Jr. and Barach, J.P. (1978) An estimate of the steady magnetic field strength required to influence nerve conduction. IEEE Trans. Biomed. Eng., BME–27:722–723.

World Health Organization (1987) Environmental Health Criteria 69: Magnetic Fields. World Health Organization, Geneva, Switzerland.

PART VI

POPULATION STUDIES
and
STANDARDS

EPIDEMIOLOGICAL STUDIES: WHAT DO THEY TELL US?

Harvey Checkoway, Ph.D.

University of Washington
Department of Environmental Health
Seattle, Washington 98195

INTRODUCTION

During the past several decades, epidemiology has assumed an increasingly prominent role in the identification and quantification of health risks from environmental hazards. In fact, the extent of epidemiology's gain in recognition can even be seen from a casual perusal of lay literature; epidemiologic findings are referenced more often than animal studies, and nearly as often reports of therapeutic advances. This trend is due in part to a growing need for risk information derived from human experience, rather than merely from animal experimentation. To the epidemiologist, the objectives, merits, and shortcomings of the discipline are usually apparent, yet these may be unclear, if not fully mysterious, to the reviewer unversed in epidemiologic terminology and methods. Accordingly, the purpose of this chapter is to illustrate some of the principal uses and interpretations of epidemiologic research. The central focus of this review is to examine what can be learned from epidemiologic studies, and how one might best synthesize the findings with information derived from other disciplines, namely toxicology and clinical medicine.

The points to be made in this review will be illustrated with examples taken from the published literature on environmental risk factors for disease. I have chosen examples other than studies pertaining to electromagnetic fields because this issue is examined in depth in other papers at this symposium. It should also be pointed out that no attempt will be made to review comprehensively epidemiologic study designs, data sources, or methods of data analysis. Readers interested in pursuing these topics are advised to consult standard epidemiology reference texts (MacMahon and Pugh, 1970; Rothman, 1986; Kelsey et al., 1986; Monson, 1990).

HISTORICAL DEVELOPMENT OF EPIDEMIOLOGY

In order to address the question at hand: "What do epidemiolgical studies tell us?," I think it useful to review selectively the historical background of the field by mentioning some highlights that have established epidemiology as a significant public health science. This brief review should offer a backdrop for the scope of epidemiologic applications.

Until the the early to middle twentieth century, epidemiolgists devoted most of their efforts to characterizing the risks of communicable diseases in populations. As such, epidemiolgists often recorded the rates of disease according to time, geography, and population groups within geographic regions. This is known as <u>descriptive epidemiology</u>. Elucidation of the route of transmission of infectious agents, e.g., cholera, was a principal benefit of the research. Descriptive epidemiology has other valuable applications, including the generation of causal hypotheses, based on the population distributions of disease incidence, and uses in the planning of health care facilities. Also, it should be mentioned that hypotheses that emerge from in-depth studies addressing etiologic questions may be examined on a macroscopic level to determine whether the population patterns of disease risk are consistent with presumed causative associations. For example, it is of interest to examine international trends of coronary heart disease rates to determine, albeit crudely, whether the countries with relatively high fat dietary profiles have the greatest risks, as would be suggested from a large body of clinical and epidemiologic research.

Epidemiologists devote perhaps the greatest fraction of their time and effort to the study of disease etiology. The origins of causal hypotheses often are clinical observations of apparent excesses or "clusters" of disease in small patient groups who share a distinctive exposure. An often cited example is the recognition of liver angiosarcoma in 3 workers at a vinyl chloride plant (Creech and Johnson, 1974); the extreme rarity of the disease and the characteristic exposure, vinyl choride, was sufficient evidence to trigger formal epidemiologic studies which confirmed the association (Waxweiler et al., 1976). Case cluster identification and follow-up is usually a less productive source of etiologic insight, particularly when the health outcome of interest is not rare and the suspected causative exposure(s) is poorly defined. Thus, for example, an explanation for the occurrence of 10 cases of hypertention in residents of a community thought to have contamined drinking water sources may never be forthcoming using epidemiologic techniques.

Epidemiology and toxicology are to a great extent companion disciplines. Their goals are identical: the recognition of hazards and elucidation of disease induction mechanisms, which can ultimately lead to disease prevention. Of course, epidemiology and toxicology differ in their approaches. Nelson (1988) has summarized some of the these differences in approach and interpretability of findings (see Table 1). Certainly the most profound difference (other than the obvious choice of species for study) is the toxicologist's freedom to control and quantify the administration of exposure. Ethical considerations preclude controlled experimentation in epidemiology, except in the case of randomized therapeutic trials. For the most part, epidemiology is an observational science in which the investigator neither controls exposures nor eliminates other sources of disease. Instead, the epidemiologist measures the subsequent effects of toxicants in relation to the extent of exposure.

Table 1
Contrasts of Toxicology and Epidemiology*

Issue	Toxicology	Epidemiology
• Control of variables —exposure —environment —confounders	Excellent	Poor
• Identifying causal factors	Excellent	Poor
• Size of population	Limited	May be large
• Sensitivity	Poor	Poor
• Genetic diversity	Intentionally narrow	Broad
• Intercurrent disease	Controllable	Not controllable
• Elucidation of disease mechanisms	Accessible but of uncertain relevance	Not accessible usually but directly relevant

* Source: Nelson N (1988)

Perhaps an ideal sequence of hazard recognition would be the accumulation of experimental data supporting an effect, followed by confirmation of similar effects in humans and further experimental research to elucidate the mechanisms of toxicity. Of course, there needs to be an initial motivation to examine effects of specific agents. This often comes from case reports of illness, as mentioned earlier. Alternatively, experimental or epidemiologic evidence from studies of similar substances may trigger investigations. A good example of the latter is the interest in examining respiratory disease and cancer risks related to fiberglass and other man-made mineral fibers which have replaced asbestos as insulating materials (Kilburn, 1982).

EPIDEMIOLOGIC APPROACHES FOR DISEASE INVESTIGATION

In this section, I will attempt to illustrate the approaches commonly used in epidemiology to characterize population disease risks and to identify hazardous exposures and their mechanisms of effect. The features of study design and potential sources of bias will be highlighted with some illustrative the examples.

Descriptive and Correlational Studies

Descriptive studies of disease occurrence may reveal etiologic clues, especially when time trends of change in incidence can be associated with changes in environmental conditions. The data in Table 2 are incidence rates of two neurological diseases, Parkinson's dementia and amyotrophic lateral sclerosis, in the populations of two of the Mariana Islands, Guam and Rota Reed et al., 1987. The causes of these conditions have remained unclear, although there have been strong suspicions that a toxic component in the cycad seed, which formed an important part of the traditional native Chamorro diet, was a causative factor. These data indicate that both conditions have declined in incidence on Guam, but not so on Rota. Interpreting these incidence patterns is complicated by the absence of more specific dietary and environmental exposure information, yet the trends suggest that there has been the elimination of some etiologic factor(s) on Guam. In contrast, a constant incidence rate over time would suggest that the environment, if associated with disease, has not changed, or alternatively that other factors, such as genetics, may be at play. This example demonstrates that the value of descriptive epidemiologic analyses, as well as the limitations for identifying specific etiologic factors.

Table 2

Time Trends of Incidence of Amyotrophic Lateral Sclerosis (ALS) and Parkinsonism-Dementia (P-D) on Guam and Rota*

	Average annual incidence rates per 100,000		
Time period	ALS on Guam	P-D on Guam	P-D on Rota
1968-73	32	43	38
1974-78	8	77	41
1979-83	0	25	39

*Source: Reed D, et al. (1987)

An extension of the purely descriptive approach is the attempt to correlate disease rates with environmental characteristics. Typically, such studies examine geographic patterns of disease occurrence in relation to environmental exposures. An example of this approach is demonstrated with data from a study in Quebec in which the prevalence rates of Parkinson's disease in rural regions of the province were correlated with the use of pesticides (Barbeau et al., 1986). Visual inspection of these data reveals a positive correlation between Parkinson's disease and pesticide exposure. The motivation for this study was the prior recognition of Parkinsonism among intravenous drug users exposed to a contaminant in "synthetic heroin," that is structurally similar to some pesticides, especially paraquat (Langston et al., 1983).

Table 3
Prevalence of Parkinson's Disease and Pesticde Use in Rural Quebec*

Region	1983 Prevalence of Parkinson's disease per 1,000	1982 Pesticide Use (Kg/yr)
1	0.62	66,511
2	0.80	170,268
3	0.89	1,561,189
4	0.73	221,867
5	0.83	266,180
6	0.33	21,398
7	0.13	25
8	0.32	932
9	0.20	0

*Source: Barbeau A, et al. (1986)

There are some significant limitations of correlational studies, not the least of which is the great potential for improper exposure attribution. In particular, such studies attribute to individuals exposures based on the aggregate profile of the geographical unit. Thus, it is possible to observe a spurious association between disease risk and exposure if the persons truly at greatest risk within the high rate areas were not, in fact, exposed to the agent in question. If in the preceding example, the highest Parkinson's disease rates occurred among persons not exposed to pesticides, yet residing in regions where the average pesticide use was high, then the observed association would not be meaningful. Moreover, the agent chosen for study may be truly associated with disease risk, but not in a causative manner. This might occur if the putative environmental risk factor were highly correlated with the true etiologic agent. However, it should be appreciated that distinguishing between etiologic and merely statistical associations is a problem that is by no means unique to descriptive correlational studies.

Analytic Studies

Identifying specific risk factors requires obaining exposure data for individuals and relating this information to disease risk. This strategy forms the basis of <u>analytic epidemiology</u>. It should be recognized that, although it is necessary to obtain individuals' exposure information, conclusions reached pertain to risks in populations. For example, merely knowing that a particular person has elevated blood lipid levels is not necessarily predictive of his or her risk for coronary heart disease. However, the vast body of literature demonstrating higher coronary heart disease risks in populations with elevated lipid concentrations, relative to populations with normal lipid values, offers support for an etiologic link.

In the example of Parkinson's disease and pesticides (Barbeau et al., 1986), an apparent association was indicated by the data. A more detailed examination of this issue was made in a recent case-control study of Parkinson's disease risk factors (Stern et al., 1991). The data are displayed in Table 4. In this study, the investigators asked Parkinson's disease cases and controls (persons free of Parkinson's disease) about past exposures to a variety of environmental and lifestyle factors. There is no evidence in this study that either insecticides or herbicides were related to risk, although the elevated risks associated with rural residence, especially among older patients, indicate that there may indeed be other as yet unidentified risk factors in rural environments. In comparing the evidence from this study with the previous correlational analysis, care should be taken to avoid over-generalizing the discrepancies because of the differences in study design, specificity of exposure characteriza-tion (more complete in the Stern study), and differences in location and hence populations; the Barbeau study was conducted in Quebec and the Stern study was conducted in Pennsylvania and New Jersey.

Table 4
Environmental and Lifestyle Exposures in Parkinson's Disease*

| Exposure | Age at onset of PD | | | |
| | < 40 yr | | > 60 yr | |
variable	RR	(95% CI)	RR	(95% CI)
Well under ≥ 1 yr	0.9	(0.4-2.1)	1.2	(0.6-2.7)
Rural residence ≥ 1 yr	1.2	(0.6-2.5)	2.4	(1.0-5.4)
Insecticides	0.6	(0.2-1.7)	0.8	(0.3-2.1)
Herbicides	0.9	(0.5-1.7)	1.3	(0.7-2.4)
Head injury	3.0	(1.2-7.6)	2.3	(0.9-6.1)
Smoking (ever)	0.8	(0.4-1.6)	0.5	(0.2-1.0)
Education ≤ 12 yr	1.0	(0.5-2.1)	2.7	(1.2-6.2)

*Source: Stern M, et al. (1991)

ESTABLISHING CAUSATION

A prominent feature of epidemiologic research is the collection of large quantities of data. In conducting a study, an epidemiologist ordinarily has a limited number of prior hypotheses or suspicions about which exposures are risk factors. There are numerous ways in which one can formulate an hypothesis. Prior experimental or epidemiologic evidence is clearly the most attractive basis for hypothesis formulation, but other avenues may be equally productive in some instances. Some hypotheses may emerge from a <u>post hoc</u> inspection of data. In fact, the timing of hypothesis formation in relation to the actual conduct of the study may have no bearing on the validity of the hypothesis; in some cases serendipitous findings may be more meaningful than results based on prior expectation. Given that all researchers are obligated to interpret their data, even if only to say that a particular finding has no clear explanation, the challenge becomes one of distinguishing causal from artifactual associations.

There has been substantial attention devoted to the issue of determining causation in epidemiologic research. Some of the discussion has become philisophical and esoteric. Fortunately, there are practical ways to address this question. In 1965, Hill published a useful set of guidelines for assessing causation from epidemiologic research findings. Included among these are: biological plausibility, strength of associations, freedom from bias, consistency within the study, consistency with data from other studies of the same issue, demonstration of a dose-response relationship, and the beneficial effect of removing or reducing exposure. These guidelines will be discussed briefly.

<u>Biological Plausibility</u>

Whether an observed association is biologically plausible depends largely on the state of knowledge at the time of the study. Certainly, 50 years ago few would have imagined that electromagnetic fields, for example, could be carcinogenic or teratogenic. However, during the past 10 years there has been some research that has pointed toward a health hazard from electromagnetic fields, and this development alone causes some to consider the plausibility of effects. A more vivid example is the case of benzene, which before the early 1960s was known to cause central nervous system and bone marrow toxicity, but has since been shown to be a potent leukemogen at high exposure levels (Austin et al., 1988). The leukemogenic potential of benzene is not only biologically plausible, but is a widely accepted phenomenon.

Strength of Association and Consistency

It seems logical that the largest observed effects are most likley to be real. If for example, lung cancer occurs 10 times more often among cigarette smokers than non-smokers, whereas the risk ratio for some other agent is, say, 2, then there is reason to assume that the effect of smoking is more important, even if not necessarily more valid (real), than the effect of the other agent.

On the other hand, some associations that appear very strong initially do not stand the test of time. An observed relative risk of 5 or 6, which is generally considered to indicate a strong effect, may be based on small numbers in a given study, and may not be seen subsequently, whereas a seemingly weaker effect (e.g. relative risk of 1.5) may persist in subsequent studies. The latter relationship should thus receive more credibility.

Bias in Epidemiologic Research

It is customary in epidemiology to classify study biases into three main categories: selection bias, information bias, and confounding. An example of selection bias would be a comparison between chronic bronchitis cases and normal controls of their past histories of exposure to community air pollution, where the cases were identified from urban communities and the controls from rural areas. The source of the cases, urban areas, increases their likelihood of having experienced greater exposures than controls. The avoidance of selection bias can be achieved by selecting subjects on the basis exposure status, but irrespective of health status in a cohort or cross-sectional study, or on the basis of disease status, but irrespective of exposure status in a case-control study.

Information bias can take several forms. The most important type in environmental epidemiology is misclassification of exposure. If one cannot correctly classify subjects as exposed or non-exposed in the simplest case, or if exposure level assignment is erroneous in a dose-response study, then the net effect will be an inability to observe true associations. This problem is akin to distinguishing the "signal from the noise"; as the noise level (misclassificaton) increases the signal (exposure-disease association) becomes less discernible. Reducing exposure assignment error can be a difficult task for very practical reasons. It is impossible to monitor humans for toxic exposures throughout their lifetimes, and even if lifetime monitoring were achievable, new information invariably will introduce added uncertainty about which exposures warrant monitoring. A case in point is the carcinogenicity of vinyl chloride which was only recognized years after its introduction into the workplace.

Confounding is the mixing of the effects of the main study exposure variables with the effects of extraneous factors that are also capable of causing disease. For example, attributing chronic neurological impariment to an industrial solvent exposure may be erroneous if workers with the highest solvent exposures were also the heaviest consumers of alcohol. In order for some extraneous exposure to be a confounder, it must be an independent risk factor for the health outcome of interest (i.e., it can cause disease even in the absence of the study exposure), and it must occur more commonly among exposed than non-exposed persons.

Eliminating or minimizing confounding is an important aspect of all research. Toxicologists can usually eliminate confounding by selecting in-bred strains of animals and controlling the environment. These luxuries are seldom if ever available to epidemiologists. The options for controlling confounding in an epidemiologic study include restricting the study to persons free of the presumed confounder (e.g., non-smokers), matching the comparison groups with respect to the confounder and making comparisons separately among persons with and without the confounder, or using statistical correction techniques to the data. Each approach has merits for certain situations, yet one can never be fully confident that all confounding has been eliminated from an epidemiologic study. There are simply too many unknown causes of disease that may be related to the exposures of concern.

Dose-Response

It is intuitive that causal evidence is strengthened if it can be shown that disease risk increases in proportion to the amount of exposure. This phenomenon has been seen in numerous instances of generally acknowledged relationships. A linear, or at least monotonically increasing, gradient of risk according to exposure level is a standard dose-response paradigm. However, not all dose-response relationships follow predictable patterns. Dose-response relationships that are non-linear or that have opposing effects at different ends of the dose spectrum may reflect true underlying biological processes but are difficult to interpret from epidemiologic data. Part of the difficulty in detecting and interpreting epidemiologically-derived dose-response trends is directly attrributable to flaws in measuring exposures. Here again, the signal-to-noise ratio can pose a serious problem.

Intervention and Disease Reduction

Although the demonstration of benefit from exposure removal or reduction is the most convincing argument for causation, this is generally only seen in instances of acute, reversible effects. Demonstrating reductions of long-term sequelae of chronic exposures requires prolonged periods of observation. Moreover, concurrent changes in other environmental conditions may confound the apparent trends of risk abatement.

INTERPRETING EPIDEMIOLOGIC DATA

Several examples taken from the occupational and environmental epidemiology literature are presented to demonstrate the strengths and limitations of epidemiologic evidence. In considering these, it is helpful to keep in mind the Hill's guidelines for causal inference.

Example 1: Radon and Lung Cancer Among Chinese Tin Miners

The data from a case-control study of approximately 100 lung cacer cases and 100 controls are shown in Table 5 (Qiao et al., 1989). The association between radon in underground mines and lung cancer had been demonstrated convincingly in numerous prior studies, thus establishing a prior biologically plausible hypothesis. The investigators report a strong dose-reponse trend with increasing radon levels (working level months, WLM), and the risk estimates have been adjusted statistically for confounding from age, smoking and arsenic exposure. On balance, it is reasonably clear that this study offers convincing evidence in support of a strong effect of cumulative radon exposure on lung cancer risk.

Table 5
Radon and Lung Cancer Among Chinese Tin Miners*

WLM[†]	Cases	Controls	RR[‡]	(95% CI)
0	7	45	1.0	(−)
1–240	24	31	4.8	(1.6–14.8)
241–541	39	14	15.0	(4.4–50.8)
541–1,762	37	17	9.5	(2.7–33.1)

[†]Working level months
[‡]Relative risk adjusted for age, smoking, arsenic exposure
*Source: Qiao Y-L, et al. (1989)

In addressing possible mechanisms of effect, the investigators further examined the influences of the components of dose, duration and dose rate (Table 6). It appears that duration of radon exposure has a stronger influence on risk than dose rate, although the effect of dose rate is also prominent. Distinguishing unique contributions to disease risk from duration and dose rate (or intensity of exposure) is difficult to achieve in most epidemiologic studies because exposure information is often not sufficiently detailed to permit separate assessments.

Table 6
Lung Cancer Among Chinese Tin Miners by Duration
and Rate of Radon Exposure*

Duration of exposure (yr)	RR[†]	(95% CI)	Rate of radon exposure (WLM/yr)	RR[†]	(95% CI)
0	1.0	(–)	0	1.0	(–)
1–19	5.8	(1.9–17.6)	1–22.8	7.5	(2.5–23.0)
20–55	20.0	(4.9–82.7)	22.9–57.4	4.2	(1.2–14.3)

* Source: Qiao Y-L, et al. (1989)
† Relative risk adjusted for age, smoking, and arsenic exposure

Example 2: Childhood Leukemia and Paternal Occupational Radiation Exposure

There has been growing concern about the potential for increased cancer risks among residents living near nuclear installations. This concern heighted following a recent publication by Gardner et al. (1990) demonstrating an association between fathers' occupational radiation exposures at the Sellafield nuclear plant in West Cumbria, U.K. and leukemia in their children. The only exposure data available pertained to external penetrating ionizing radiation; internal radionuclide contamination could not be taken into account. The results regarding pre-conception exposure are summarized in Table 7.

Table 7
Leukemia and Paternal Radiation Dose Before Conception Among Residents
Near the Sellafield Nuclear Plant*

Dose (mSv)	Cases	Controls	RR	(95% CI)
0	38	262	1.00	–
1–4	3	18	1.30	(0.32–5.34)
5–9	1	3	3.54	(0.32–38.9)
≥10	4	5	7.17	(1.69–30.4)

* Source: Gardner MJ, et al. (1990)

In viewing the findings, one is immediately struck by the apparently strong dose-response trend. However, other factors need to be considered before accepting this effect as causal. One obvious concern is the small numbers of cases in the highest exposure groups which creates statistical instability in the data. Statistical imprecision of risk estimates does not in itself invalidate observations, although the case for causation is strengthened when the data are based on large numbers. The issue of biological plausibility deserves attention. There were no prior epidemiologic data from which this effect could have been predicted. Nonetheless, a mechanism of paternal germ cell mutation is plausible.

Another important concern would be the distorting influence of confounding factors. The cases and controls in the study were matched on gender and age, and other factors of relevance, including antenatal x-rays, maternal viral infections, and Down's syndrome were also taken into account in the analysis. The conclusions that can be reached from this study are limited, and there is a need for further investigation of the issue to determine whether these essentially unexpected results appear consistently in other investigations. The findings clearly give direction to future experimental research on radiation carcinogenesis.

Example 3: Neurobehavioral Effects of Lead in Children

Lead has long been recognized as capable of causing neurological and psychological deficits in cases of poisoning. However, the effects of low-dose chronic exposures resulting from ambient sources are less well understood. Data from a follow-up study in Boston of lead exposure reveal apparent deficits of academic perfomance, reading score, and physio-logical function that correlate with lead burden, as assessed by lead concentrations in teeth (Needleman et al., 1990). The implications of these findings are profound, yet there remains the concern that the trends may have been confounded by socioeconomic factors (e.g., family education levels) that are predictors of some of the study outcomes and are related to lead exposure. Resolving issues of confounding like this can be a difficult proposition, particularly when the measurement of the suspected confounding variables is complex.

Table 8
Academic and Neurobehavioral Outcomes in Children
According to Dentin Lead Concentration*

| | Lead level (ppm) | | | |
Outcome	<5.9 (N=30)	6.0-8.2 (N=31)	8.3-22.2 (N=30)	>22.2 (N=31)
Class standing (percentile)	0.60	0.59	0.48	0.45
Reading score (correctly read words)	143.8	142.7	140.2	135.2
Reaction time (msec)				
–Preferred hand	246.6	255.5	267.3	275.1
–Non-preferred hand	241.2	238.2	258.4	261.2
Finger tapping (no./10 sec)	46.6	47.2	45.9	43.5

* Source: Needleman HL, et al. (1990)

Example 4: Respiratory Disease and Natural Gas Refinery Emissions

Residents of a community in a rural section of Alberta, Canada had long-standing concerns about what they perceived as excessive illness due to emissions from two natural gas refineries (Dales et al., 1989). Potentially toxic exposures that subsequently received attention were sulfur dioxide and hydrogen sulfide. In 1985, an epidemiologic survey was conducted to evaluate respiratory health of the citizens of the region and citizens of a non-exposed rural area not downwind from the plants. Environmental measurements permitted classification of the areas into high, low, and no exposure to sulfur-containing pollutants.

Results of respiratory symptom prevalence among children, as reported by their parents, are given in Table 9. Children's respiratory health status is often regarded as a better index of air pollution effects than adult health status because the data for children are less prone to confounding from smoking and occupational exposures. There are clear patterns of increased symptom prevalence in the highly exposed area, and some dose-reponse trends are evident. However, pulmonary function values showed no evidence of an effect of exposure (table 10). How then can these data be interpreted, and what are the implications for community air pollution remediation?

Table 9
Prevalence of Respiratory Symptoms in Children Downwind from
Natural Gas Refineries in Alberta*

Parent reported symptom (%)	Area		
	High total Sulfation (N=113)	Low total Sulfation (N=253)	Reference (N=203)
Persistent cough	21	7	3
Persistent phlegm	14	4	3
Dyspnea	4	1	1
Wheeze	20	11	12
Wheeze with dyspnea	18	15	12
Any of above	39	24	18

* Source: Dales RE, et al; (1989)

Table 10
Respiratory Function in Children Downwind from
Natural Gas Refineries in Alberta*

Respiratory function (% predicted)	Area		
	High total sulfation (N=88)	Low total sulfation (N=255)	Reference (N=160)
FEV_1	104.3 (1.4)[†]	102.6 (0.8)	100.0 (0.9)
FVC	106.9 (1.2)	103.1 (0.8)	103.5 (0.9)
MMEFR	99.8 (3.0)	100.1 (1.4)	91.8 (1.9)

† Mean (S.E.)
* Source: Dales RE, et al; (1989)

One interpretation is that air pollution caused symptoms among children in exposed areas, but not reductions in ventilatory function. It may be that symptoms are more sensitive indicators of early respiratory toxicity in children than functional changes. Alternatively, one might wonder whether there has been a systematic bias in symptom reporting by parents in the exposed areas. Perhaps knowledge and concern of ambient pollution exposures served as a strong motivator for parents in exposed areas to observe and report more thoroughly their children's symptoms than parents in the reference area. If this were indeed the case, then it could be argued that biased symptom reporting and the absence of an apparent effect on lung function indicate no respiratory hazard from exposure. A further consideration in reviewing these findings is that exposure measurement error may have resulted in misclassification and thus an inability to detect true effects if any had occurred. This study is typical of many epidemiologic studies prompted by legitimate health concerns in that the absence of internally consistent associations, and the possibility of reporting bias make conclusions tenuous.

PERSPECTIVE

From an epidemiologist's perspective, the answer to the original question posed in this review: "What do epidemiologic studies tell us?," is "quite a lot." That answer, coming from and epidemiologist, is not completely unbiased. Nevertheless, it seems reasonable to assert that epidemiology provides the only systematic summary of disease occurrence in populations, an that epidemiology can in many situations identify risk factors, and less frquently shed light on mechanisms of toxicity.

Establishing whether a particular observation is causal will always be a difficult scientific challenge, and one that usually requires multiple observations of similar circumstances before conclusions can be reached with confidence. This is especially true for relatively weak associations that may still be of public health consequence, i.e., a small increase in risk associated with a widely disseminated exposure. Moreover, as Nelson (1988) has emphasized, epidemiology and companion disciplines, such as toxicology, should be mutually reinforcing in characterizing hazards and their modes of action. Technological advances in epidemiologic and experimental methods will be of obvious future benefit. Equally important will be a climate of open-mindedness among scientists in various disciplines in evaluating results from research on the same topics that use different strategies.

REFERENCES

Austin, H., Delzell. E., Cole, P. (1988) Benzene and leukemia: a review of the literature and a risk assessment, Am. J. Epidemiol., 127:419.

Barbeau, A., Roy, M., Cloutier, T., Plasse, L., Paris, S. (1986) Environmental and genetic factors in the etiology of Parkinson's disease, Adv. Neurol., 46:299.

Creech, J.L., Johnson, M.N. (1974) Angiosarcoma of liver in the manufacture of polyvinyl chloride, J. Occup. Med., 22:25.

Dales, R.E., Spitzer, W.O., Suissa, S., et al. (1989) Respiratory health of a population living downwind from natural gas refineries, Am. Rev. Resp. Dis., 139:595.

Gardner, M.J., Snee, M.P., Hall, A.J., et al. (1990) Results of case-control study of leukaemia and lymphoma among young people near the Sellafield nuclear plant in West Cumbria, Br. Med. J., 300:423.

Hill, A.B. (1965) The environmental and disease: association or causation? Proc. R. Soc. Med., 58:295.

Kelsey, J.L., Thompson, W.D., Evans, A.S. (1986) Methods in Observational Epidemiology, New York: Oxford University Press.

Kilburn, K.H. (1982) Flame-attenuated fiberglass: another asbestos? Am. J. Ind. Med., 3:121.

Langston, J.W., Ballard, P., Tetrud, J.W., Irwin, I. (1983) Chronic Parkinsonism in humans due to a product of meperidine-analog synthesis, Science, 219:979.

MacMahon, B, Pugh, T.F. (1970) Epidemiology: Principles and Methods, Boston: Little-Brown and Co.

Monson, R.R. (1990) Occupational Epidemiology, 2nd ed, Boca Raton, FL: CRC Press.

Needleman, H.L., Schell, A., Bellinger, D., et al. (1990) The long-term effects of exposure to low doses of lead in children: an 11-year follow-up report, N. Engl. J. Med., 322:83.

Nelson, N. (1988) Mutual reinforcement between epidemiology and the laboratory in the study of environmental cancer, Ann. N.Y. Acad. Sci., 534:1021.

Qiao, Y.L., Taylor, P.R., Yao, S.X., et al. (1989) Relation of radon exposure and tobacco use to lung cancer among tin miners in Yunnan Province, China, Am. J. Ind. Med., 16:511.

Reed, D., Labarthe, D., Chen, K.M., Stallones, R. (1987) A cohort study of amyotrophic lateral sclerosis and Parkinsonism-dementia on Guam and Rota, Am. J. Epidemiol., 125:92.

Rothman, K.J. (1986) Modern Epidemiology, Boston: Little-Brown and Co.

Stern, M., Dulaney E., Gruber, S.B., et al. (1991) The epidemiology of Parkinson's disease: a case-control study of young-onset and old-onset patients, Arch. Neurol., 48:903.

Waxweiler, R.J., Stringer, W., Wagoner, J.K., Jones, J. (1976) Neoplastic risk among workers exposed to vinyl chloride, Ann. N.Y. Acad. Sci., 271:40.

EPIDEMIOLOGICAL STUDIES OF POWER-FREQUENCY ELECTRIC AND MAGNETIC FIELDS

M.L. McBride
Richard P. Gallagher

Division of Epidemiology
British Columbia Cancer Agency
Vancouver, British Columbia

Since 1979, when the first epidemiologic study linking childhood cancer and power lines was published, the question of whether extremely low frequency electric and magnetic fields, including power frequency fields (EMF), have health effects has been a scientific and public issue. Currently, there is no definitive scientific evidence of an association between power-frequency and various health outcomes. This paper summarizes the epidemiologic research to date, describes the limitations of such research, and outlines current research efforts to clarify the relationship between power-frequency fields and health effects.

Electric and magnetic fields are everywhere. Electric fields result from the strength of an electric charge; magnetic fields result from the motion of the charge. For example, the earth has an electric field, due to electric currents in its centre; molecules in our bodies and everywhere are held together by EF; messages in our nervous system involve fields.

Power-frequency (50 or 60 Hz) fields are produced wherever there is electric power. EM fields are generated by the various components of the power system, including transmission and distribution lines, building wiring, electrically-powered machinery and appliances. The strength of electric fields depends on the voltage of the source, and the strength of magnetic fields depends on the size of the current. Magnetic fields can pass through objects; electric fields cannot. Individuals are exposed to multiple sources of EMF, and level of exposure depends on duration, location relative to the source, the interaction of fields and currents from different sources, and strength of the fields.

Most of the evidence so far for possible health effects from 50 or 60 Hz EM fields has come from epidemiologic studies, while cellular, tissue, and animal studies attempt to explain these effects. Epidemiologic studies look for an association, in a human population, between exposure to a factor and a specific health outcome. While other studies are experimental in nature, epidemiologic studies are, in general, observational. Consequently, the main difficulties in the design, analysis, and interpretation of such research revolve around the question of accounting for distorting influences, or bias. Descriptive studies such as proportional incidence or mortality studies interpret routinely-collected population-based frequency data on risk factors and/or outcome, and are subject to the most undetected bias. Analytic studies use more sophisticated design and analytic methods to test specific hypotheses while adjusting for bias. Analytic studies are of two types; case-control studies, which compare the exposure experiences of groups of individuals with the disease in question and a comparison group without the disease; and cohort studies, with compare the disease outcome experience in two groups of individuals with different levels of exposure. The data is then interpreted according to certain criteria which imply a causative relationship between the factor and the health outcome (Table 1), taking into account the limitations and possible biases of the studies.

In assessing the support for causality, the first requirement is consistency of results. No one epidemiologic study can provide definitive evidence for a relationship; a consistent effect must be seen in several studies in different population groups and in different time periods. The strength of the association, or size of the relative risk, is also a factor in inferring causality. Furthermore, the relationship should be biologically coherent. Exposure to the risk factor should precede the development of the disease or health effect, and both the relationship between dose and response, and the specificity of effect, should be consistent with the proposed biological mechanism. Finally, the observed relationships should fit previous knowledge of similar agents, and be explainable in terms of a biological mechanism. These criteria need to be used to assess the evidence to date for a causal relationship.

Table 1. Epidemiologic criteria of causation

1. Consistency
2. Strength of association
3. Temporal sequence
4. Biological gradient (dose-response)
5. Specificity
6. Coherence with previous knowledge
7. Biological plausibility

There are now over 60 epidemiological studies that have examined exposures to power-frequency EM fields and cancer or other health effects. These studies can be grouped according to the health outcome of interest, the type of exposure (residential or occupational), the age group of interest (childhood or adult), and the type of study design.   The majority of the published research has cancer as its endpoint, looking at either residential or occupational exposures.   A few studies have examined reproductive effects.   Several excellent reviews have been published, including Ahlbom (1988), Aldrich and Easterley (1987), Coleman and Beral (1988), Kavet and Banks (1986), Savitz and Calle (1987), and Theriault (1990).

Nine residential studies of magnetic fields and cancer published to 1988 were reviewed by Ahlbom (1988).   There have been three studies reported since that time.   All are case-control studies.   The results of six investigations of childhood exposure are summarized in Table 2.   The most consistent findings relate to wire code, a classification of residential wiring correlated with strength of current and magnetic fields.   The Wertheimer and Leeper (1979) and Savitz et al (1988) studies both showed statistically elevated relative risks of all childhood cancers for houses with high current wire configurations.   Tomenius (1986) also found an excess risk of cancer with proximity to sources of high current in the electrical distribution system.   London et al (1991) recently reported a statistical excess of leukemia also related to high current wire configurations.   The earlier studies can be criticized for such methodologic deficiencies as a lack of information on confounders, poor choice of surrogate exposure measures, and poor choice of controls.   The two later studies (Savitz et al 1988; London et al 1991) were population-based, included data on confounders, and carried out in-house spot measurements of fields. Neither study showed any statistically significant association between measured magnetic field levels and childhood cancer risk, although in the Savitz study there was a tendency towards an increased risk under low power conditions, and in the London study the use of some appliances was positively associated with risk.   Over all the studies, there is no clear association with childhood leukemia, with most relative risks near unity.   In contrast, the risk of childhood brain tumours is consistently elevated, albeit with small numbers. An investigation of electric blanket use and cancer by Savitz et al (1990) was essentially negative; only the association with brain cancer reached significance.

Table 2. Childhood residential exposure to EMF and cancer

	All cancer	Leukemia	Brain
Wertheimer and Leeper 1979	3.2 *[1]	3.0 *[1]	2.4 *[1]
Fulton et al 1980		NS	
Myers et al 1985	NS	NS	
Tomenius et al 1986	2.1 *[3]	0.3 *[3]	3.7 *[3]
	2.1 *[4]	1.1 NS[4]	3.9 *[4]
Savitz et al 1988	1.5 *[1]	1.5 NS[1]	2.0 *[1]
	1.4 NS[3]	1.9 NS[3]	1.0 NS[3]
London et al 1991		2.1 *[1]	
		1.2 NS[3]	
Savitz et al 1990	1.3 NS[5]5	1.7 NS[5]	2.5 *[5]

*	Statistically significant relationship.
NS	No significant relationship.
1.	High current configuration according to wire code.
2.	Calculated magnetic field.
3.	Measured magnetic field.
4.	Distance to 220-kV powerline or electrical transmission facility.
5.	Electric blanket use.

The results of six studies of adult residential exposure to EMF are presented in Table 3. Five were case-control designs, and one (McDowall 1986) was a cohort study. The exposure metric varied among these studies; only one (Severson et al 1988) carried out field measurements in and out of the house. Except for the first study by Wertheimer and Leeper in 1982, there is virtually no evidence of an effect on risk of total cancers or leukemia. In particular, the Severson study was well-designed and the results were not consistent with an increased risk. A study by Preston-Martin (1988) examining electric blanket use and adult leukemia risk was unable to detect any association.

Table 3. Adult residential exposure to EMF and cancer

	All cancer	Leukemia
Wertheimer and Leeper 1982	1.4 *[1]	
Coleman et al 1989		1.3 NS[4]
McDowall 1986	1.0 NS[4]	1.4 NS[4]
Severson et al 1988		NS[3]
Preston-Martin et al 1988		0.9 NS[5]

*	Statistically significant relationship.
NS	No significant relationship.
1.	High current configuration according to wire code.
2.	Calculated magnetic field.
3.	Measured magnetic field.
4.	Distance to 220-kV powerline or electrical transmission facility.
5.	Electric blanket use.

Occupational studies of cancer and exposure to EMF have been reviewed by Savitz and Calle (1987), Coleman and Beral (1988), and Theriault (1990). Most of the early studies have been based on proportional risk ratios. Many of these studies looked at leukemia risk. Savitz and Calle (1987) combined eleven datasets and found a small excess risk of leukemia among electrical workers. Garland et al (1990) also found a barely significant excess; others (Loomis and Savitz 1989; Gallagher et al 1990) have not. As Theriault (1990) has pointed out, only one of five studies of aluminum workers demonstrated a significantly excess of acute leukemia (Milham 1979), although most standardized mortality ratios were elevated. Significantly elevated risks of all leukemia, chronic lymphocytic leukemia, and myelogenous leukemia were found in a group of underground coal miners (Gilman et al 1985), and an excess risk of leukemia was observed among electricians and welders in a naval shipyard (Stern et al 1986). In summary, the evidence for leukemia risk is weak, but consistent enough so that it cannot be dismissed on the basis of studies so far.

Of eight case-control studies of brain tumours and occupational exposure to EMF, half showed significantly increased risks; one of three studies measuring proportional risk ratios found an excess risk; and none of four cohort studies found elevated brain tumour rates among presumably exposed occupations. Two studies of telecommunications workers (Vagero 1985; DeGuire et al 1988) have observed excess risks of melanoma; other cohort studies of electrical workers have shown elevated risks of cancers of the pharynx and lung, or of all cancers, but no excess of leukemia. Six other cohort studies have been negative for any cancer risk. These data are obviously less convincing than those for leukemia.

There have been several case-control and cohort studies that have examined the relationship between occupational or environmental exposures to EMF and adverse reproductive outcomes. Two of five case-control studies reported excess risks of CNS tumours in children of workers in various EMF-exposed occupations. Increased rates of spontaneous abortions and congenital malformations were observed in two Scandinavian cohort studies. Numbers of events were very small in these studies, however.

There are several methodological weaknesses in the epidemiologic studies conducted so far that limits their interpretation. A major area of uncertainty is exposure assessment. The exposure of interest is magnetic fields; however, most published studies have relied on surrogate measures such as wire coding or job titles. The primary surrogate measure used in residential studies, wire coding, does not account for all the possible magnetic field sources, nor does it take personal exposure levels into account. It is the measure that has most consistently been related to cancer risk, however. Job titles, used in occupational studies, will tend to misclassify many workers in terms of magnetic field exposure, and again do not take personal variation in exposure into account. Actual field measurements (spot or area) have been considered to be better than these surrogates, but they also vary over time and space, as do personal measurements. Only the latest two published childhood residential studies have directly measured magnetic field levels. However, this nondifferential misclassifi- cation would tend to reduce rather than elevate estimates of risk (Copeland et al 1977).

A second concern is the lack of data on confounders, that is, agents that are correlated both to magnetic fields and/or wire codes, and to cancer. This issue is difficult to address adequately, because there are very few agents known or suspected to cause childhood cancer or leukemia, and the effect of confounding in occupational settings is difficult to assess. Savitz et al have provided a comprehensive review of this issue (Savitz et al 1989). In the two residential studies sponsored by the New York State Power Lines Project (Savitz et al 1988; Severson et al 1988) several potentially confounding demographic and lifestyle variables were assessed, but none were found to be influencing the results. This implies that confounding by these variables may not be a factor in other studies.

Other deficiencies in studies to date include small numbers of subjects and weak statistical methods. Furthermore, the excess risks reported have, in the majority of studies, been relatively small, close to the limits at which risks are detectable, and difficult to interpret as separate from the effects of unidentified factors.

Although a number of studies at the cellular, tissue, and animal levels have demonstrated effects with exposure to various components of ELF, and some epidemiologic studies have observed relationships between power-frequency ELF and cancer, virtually none of the criteria for establishment of causation have been fulfilled. Currently there are several new epidemiologic studies underway (IARC 1990) that are attempting to address the limitations and deficiencies of earlier work. The results of the latest childhood cancer study (London et al 1991) have focussed attention back to residential wiring as a potential surrogate for some other factor than ELF, possibly relating to some neighbourhood characteristic. Until new research information is available, it is impossible to conclude from the available evidence that there is a causal relationship between EMF and cancer.

References:

Ahlbom A. A review of the epidemiologic literature on magnetic fields and cancer. Scand J Work Environ Health 1988;14:337-343.

Aldrich TE, Easterley CE. Electromagnetic fields and public health. Environ Health Perspect 1987;75:159-171.

Bradford-Hill. Criteria for causation in chronic diseases.

Coleman M, Beral V. A review of epidemiological studies of the health effects of living near or working with electricity generation and transmission equipment. Int J Epidemiol 1988;17:1-13.

Coleman M, Bell CM, Taylor HL, Primic-Zakelj M. Leukaemia and residence near electricity transmission equipment: a case-control study. Br J Cancer 1989;60:793-798.

Copeland, KT, Checkoway H, McMicheal AJ, et al. Bias due to misclassification in the estimation of relative risk Am J Epidemiol 1977;105:488-495.

DeGuire L, Theriault G, Iturra H, Provencher S, et al. Increased incidence of malignant melanoma of the skin in workers in a telecommunications industry. Br J Ind Med 1988;45:824-828.

Fulton JP, Cobb S, Preble L, Leone L, et al. Electrical wiring configurations and childhood leukemia in Rhode Island. Am J Epidemiol 1980;111:292-296.

Gallagher RP, McBride ML, Band PR, et al. Occupational electromagnetic field exposure, solvent exposure, and leukemia. J Occup Med 1990;32:64-65.

Garland FC, Shaw E, Gorham ED, Garland CF, et al. Incidence of leukemia in occupations with potential electromagnetic filed exposure in United States Navy personnel. Am J Epidemiol 1990;132:293-303.

Gilman PA, Ames RG, McCawley A. Leukemia risk among US white male coal miners. J Occup Med 1985;27:669-671.

International Agency for Research on Cancer, Ad Hoc Working Group. Extremely low-frequency electric and magnetic fields and risk of human cancer. Bioelectromagnetics 1990;11:91-99.

Kavet RI, Banks RS. Emerging issues in extremely-low-frequency electric and magnetic field health research. Environ Res 1986;39:286-404.

London SJ, Thomas DC, Bowman JD, Sobel E, Cheng TC, Peters JM. Exposure to residential electric and magnetic fields and risk of childhood leukemia. Am J Epidemiol 1991;134:923-937.

Loomis and Savitz 1989.

McDowall ME. Mortality of persons resident in the vicinity of electricity transmission facilities. Br J Cancer 1986;53:271-279.

Milham S. Mortality in aluminum reduction plant workers. J Occup Med 1979;21:475-480.

Myers A, Cartwright RA, Bonnell JA, Male JC, et al. Overhead power lines and childhood cancer. In: Proceedings of the International Conference on electric and magnetic fields in Medicine and Biology. London, 1985.

Preston-Martin S et al 1988.

Savitz DA, Calle EE. Leukemia and occupational exposuse to electromagnetic fields: Rewiew of epidemiologic surveys. J Occup Med 1987;29:47-51.

Savitz DA, John EM Kleckner RC. Magnetic field exposure from electric appliances and childhood cancer. Am J Epidemiol 1990;131:763-773.

Savitz DA, Pearce NE, Poole C. Methodological issues in the epidemiology of electromagnetic fields and cancer. Epidemiol Rev 1989;11:59-78.

Savitz DA, Wachtel H, Barnes FA, John EM, Tvrdik JG. Case-control study of childhood cancer and exposure to 60-Hz magnetic fields.

Am J Epidemiol 1988;128:21-38.

Severson RK, Stevens RG, Kaune WT, Thomas DB, et al. Acute nonlymphocytic leukemia and residential exposure to power frequency magnetic fields. Am J Epidemiol 1988;128:10-20.

Stern FB, Waxweiler RA, Beaumont JJ, et al. A case-control study of leukemia at a naval nuclear shipyard. Am J Epidemiol 1986;123:980-992.

Theriault G. Cancer risks due to exposure to electromagnetic fields. Recent Results in Ca Res 1990;120:166-180.

Tomenius L. 50-Hz electromagnetic environment and the incidence of childhood tumors in Stockholm county. Bioelectromagnetics 1986;7:191-207.

Vagero D, Ahlbom A, Olin R, Sahlsten S. Cancer morbidity among workers in the telecommunications industry. Br. J Ind Med 1985;42:191-195.

Wertheimer N, Leeper E. Electrical wiring configurations and childhood cancer. Am J Epidemiol 1979;109:273-284.

Wertheimer N, Leeper E. Adult cancer related to electrical wires near the home. Int J Epidemiol 1982;11:345-355.

WORK WITH VISUAL DISPLAY UNITS
Do exposure to electromagnetic radiations or fields influence the occurrence of adverse health reactions?

Ulf Bergqvist and Bengt Knave
Dept of Neuromedicine
National Institute of Occupational Health
Solna, Sweden

The spread of VDU use and adverse health reactions

Visual display units (VDUs) - also known as visual display terminals, television monitors etc - have been in existence and use in certain types of occupational settings for several decades. It was not, however, until the latter part of the seventies that they began to spread out in "ordinary" workplaces such as newspaper production and insurance companies. Since then, the spread of these working tools have been fast and far-reaching. In Sweden, it was estimated that, in 1990, about 30-35% of all occupationally employed used a VDU daily (SCB, 1990). This spread has by no means been limited to the so-called "developed world". The increasing use of VDUs in 3rd world countries prompted the World Health Organization in 1985 to organise a WHO Working Group to discuss occupational health aspects in the use of VDUs.

In the seventies, health concerns of VDU use were primarily centered on effects of presumed or observed changes in work practices, and on the possible effects of VDU work on the eyes, primarily in terms of eye fatigue or asthenopia. The first scientific report on eye fatigue in relation to VDU work appeared in 1974 (Hultgren & Knave, 1974). From the middle of the eighties, increased attention has also been given reactions due to various psychosocial factors in VDU work situation. Another concern has arisen due to the constrained ergonomic situations at many VDU work stations, and the frequent occurrence of highly repetitive work movements in certain types of VDU jobs.

In 1987, the World Health Organization published a review of the scientific literature concerning VDUs and health. The summaries were updated in 1990 (WHO, 1990), after the second conference on Work With Display Units in Montréal, Canada. In this review, ergonomic and organizational factors of VDU work were clearly implicated in the causation of certain discomforts. Thus, there are established health concerns directed towards the VDU work situation - as described above. They will not be dealt with further, unless where they may have a direct relationship with other health concerns discussed below in relation to electromagnetic radiation and fields.

445

Concern about radiation emission from VDUs

In the later part of the seventies, concerns appeared about whether radiation emissions from VDUs might have the capacity for causing pathological reactions, "injury". These concerns originated in the similarity of VDUs with television receivers, as both are based on Cathode Ray Tubes (CRTs).

In the sixties, some television recievers based on CRTs were found to emit X-ray radiation, primarily not due to the CRT, but to another component - the high voltage shunt regulator. The replacement of these regulators with other, solid state type of circuitry, effective shielding of the CRT, and mandatory tests have virtually ended the television reciever as a source of X-ray radiation exposure (Bureau of Radiological Health, 1981).

The first explicit health concern arose due to the diagnosis of some "radiation-induced cataract" cases among VDU operators. This was followed by observations of unusually high occurrences of adverse pregnancy outcomes in some workplaces. In the beginning of the eighties, another concern arose in Norway - that of VDUs´ causing certain skin reactions among operators. A last concern has, mainly in Sweden, arisen out of the skin reaction issue, that of "electric hypersensitivity". A number of individuals have experienced various symptoms, which they have attributed to exposure to electromagnetic fields, including those emitted from VDUs.

Exposure to electromagnetic radiation and fields at VDU work stations

Most visual display units in use are based on the cathode ray tube, i e the common television receiver technology. Other, less common types of VDUs are based on other principles, e g various liquid crystal technologies. These latter types of VDUs will not be discussed here further, as the radiation concern is by and large focused on CRT-based VDUs.

The following description emission of various electromagnetic radiations and fields from CRT-based VDUs are based on "Video Display Units - Radiation Protection Guidance" (ILO, in press):

1. X-ray radiation is produced within the CRT. The glass material of the tube, however, effectively prevents any leakage of X-ray radiation outside the tube. Thus, X-ray emission from VDUs is not detectable.
2. Long wavelength ulraviolet radiation, visible light and infrared radiation are emitted from CRT-based VDUs. These emissions are small compared to the most stringent general public or occupational standards, and are also small compared to other sources such as sunlight or many

artificial light sources. There is a general consensus that emissions of optical radiation from VDUs can be disregarded as a health hazard.

3. High radiofrequency fields (generally in the MHz region) are produced within the VDUs, due to various signal traffic processes. Although they are important in terms of information secrecy, the emissions are many orders of magnitudes below current standards or concern.

4. Low frequency electromagnetic fields are produced within the VDUs, primarily by the deflection coils and associated power supply and circuitry. Thus, emission of electric and magnetic fields are basically found in two frequency bands; ELF at 50-80 Hz - corresponding to the basic power supply and to the horizontal sweep, and VLF at 15-35 kHz - corresponding to the vertical sweep. Exposures associated with these emissions are orders of magnitudes below most current occupational and general public standards.

5. Depending on the humidity of surrounding air, electrostatic fields can be found around VDUs. The total electrostatic fields at a VDU work station is, however, also contributed to by the static charge of the operator. The highest measured fields are well below current standards (USSR).

For reasons discussed below, the concern about electromagnetic radiation and field emissions from VDUs is focused on the low frequency fields. Table 1 lists typical exposure levels for these fields around CRT-based VDUs in office environments. In the VLF band (15-35 kHz), exposures to electric and magnetic fields at a VDU work station are essentially determined by the VDU´s emission levels. In the ELF band (50-80 Hz), however, a number of other sources exist in many offices, which may cause similar or higher exposures than does the VDU.

Risk assessment of exposure to electromagnetic radiations or fields at VDU work stations

The emissions of electromagnetic radiation and fields from CRT-based VDUs are well known, and do, as described above, fall considerably below those required to exceed current occupational or general public exposure standards. Thus, judging by current standards and their radiological-biological rationale, CRT-based VDUs do not constitute any radiation-related health risk.

Any contrary statement would require the existence of mechanisms producing biological effects at levels orders of magnitude below those presently known. The possibility of such effects - and mechanisms - are currently under discussion, mainly in relation to exposures to ELF electric or magnetic fields, but to some degree also in the VLF region. Regardless of this general discussion, the reported cases of cataracts, adverse pregnancy outcomes,

<u>Table 1</u>. Exposure levels of electric and magnetic fields at VDU work stations.

Measure-ment	ELF - 50-80 Hz		VLF - 15-35 kHz		
	E-field V/m	B-field μT	E-field V/m	B-field μT	
VDU:					
0.5 m from VDU	20 (0-330)	0.21 (0-1.16)	1.5 (0-15)	0.03 (0-0.14)	a/
0.3 m from VDU	1.8	0.35	3.7	0.06	b/
Facial, VDU	4 (0-30)	0.12 (0.006-0.4)	<1 (0-7)	0.004 (0-0.16)	c/
Abdominal, VDU	0.60	0.08	0.30	0.01	d/
Non-VDU:					
Background	20 (0-100)	0.07 (0-1.00)	ND	ND	e/
Abdominal, non-VDU e/	0.40	0.06	0.20	0.002	f/

<u>Notes for table 1</u>

a/ Median and range for 145 VDU work stations, 50 cm from the VDUs, (Sandström et al., 1991).

b/ Geometric means of 48 VDU work stations, 30 cm from the VDUs, with operators absent (Schnorr et al., 1991).

c/ Median and range for 216 VDU work stations, at operator's normal facial positions, but with operators absent (Bergqvist, unpublished).

d/ Geometric means of the same work stations as in b/, but at abdominal positions, with operators present (Schnorr et al., 1991).

e/ Median and range for the same offices as in a/, measured at 5 points in the office, away from VDU (Sandström et al., 1991).

f/ Geometric means of 48 work stations without VDUs, at abdominal positions with workers present (Schnorr et al., 1991).

For all measurement in a/ to d/, the exposure levels are those due to VDUs and to other sources at or close to the VDU work stations. All measured values are given in rms levels. ND=not determined skin problems and "electric hypersensitivity" in relation to VDU work necessitates a detailed discussion of whether these adverse health reactions differ between VDU and non-VDU work, and, if so, whether exposure at VDU work stations to electromagnetic phenomena can explain such differences, or whether they are explainable by other VDU work factors.

In light of the above, it is apparent that this discussion should:
- Be centered on emissions in two low frequency bands; ELF (50-80 Hz) and VLF (15-35 kHz), respectively, and at very low exposure levels compared to current standards.
- Recognize that an adverse effect due to such exposure requires the existence of hitherto unknown interaction mechanisms between such electric or magnetic fields and biological systems.
- At least at present, be conducted in ignorance of the specific requirements of such (presumed) interaction mechanisms.

Development of lens opacities and cataracts

Some case reports from 1980 (Zaret, 1984), have described "radiation induced cataracts" among 10 individuals working with CRT equipment. Critique has, however, also been directed towards these diagnoses; "of the 10 anecdotal cases, only 4 had significant lenticular opacities, and each of them had known preexisting disease or exposure to cataractogenic agents" (National Research Council, 1983). Another problem in the CRT/VDU-directed suggestion of cataract formation is that conditions around VDU work stations do not contain known cataractogenic factors to excess, including radiological ones (WHO, 1987, Marriott & Stuchly, 1986).

Some epidemiological studies were made in the first part of the 1980-ies (Smith et al., 1982, Frank, 1983, Canadian Labour Congress, 1982, Böös et al., 1985), where the presence of cataracts and/or lens opacities were compared between VDU and non-VDU workers. In only one study was an excess of lens opacities with VDU work suggested (Böös et al., 1985).

The results of that 1981 study have now been reanalysed, with adjustments carried out as to the age of the worker (Bergqvist, Böös, Calissendorff & Nyman, unpublished memorandum). For both all lens opacities and for pathological lens opacities, increased odds ratios for VDU work remained after the reanalysis. However, these opacities had been diagnosed by three different ophthalmologists, and results were strongly dependent on which ophthalmologist had made the diagnoses. In order to eliminate these uncertainties, the VDU workers were reexamined in 1987, at which time also measurements were made of various electric and magnetic fields (see Bergqvist, unpublished in table 1). The following results were obtained:

1. No differences were found between VDU and non-VDU workers as to all lens opacities without dilated pupil, nor with all and pathological lens opacities with dilated pupil. The VDU exposure were characterized as to accumulated man-years of VDU work.

2. A difference as to pathological lens opacities in non-diluted pupil was found, the odds ratio for many years vs no or little VDU work was 2.7, with a 95% confidence interval of (0.7-11). Thus, the possibility that this finding was due to random variation could not be dismissed.

3. Attempts were made to relate lens opacities to measured levels of optical radiation and electric or magnetic fields at VDU work stations, and to different VDU types, without success. Based on this, a chance explanation of the "non-significant" increase above (2.) was deemed relevant.

Recently, a large Italian study of some 30 000 individuals have been reported (Bonomi & Bellucci, 1989). In this study, there were no differences between VDU workers and non-VDU workers as to cataracts.

In summary, the suggestion that VDU workers are at a higher risk of developing lens opacities and cataracts have no current scientific backing.

VDU work and pregnancy outcomes

<u>Origin of the concern</u>: In July 1980, a Canadian newspaper report appeared, where four VDU operators who had worked with VDUs during pregnancy, all gave birth to infants with birth defects. Several such clusters - most with unusually high spontaneous abortion occurrences - have since been reported. One explanation for these clusters are that a number of groups of both increased and decreased occurrences should occur by chance alone, but only clusters of increased occurrences are likely to be (selectively) reported. The number of reported clusters does not appear to be excessive of what should be expected, based on this rationale [Bergqvist, 1984].

Other types of explanations were, however, also looked for. One concern did center on radiation emissions from VDUs. Radiation emission surveys did not, however, reveal high emission levels from the VDUs in question. In 1982, a report from a Spanish group of investigators of chick embryo malformation due to exposure to ELF magnetic fields (Delgado et al., 1982) did change the focus from radiation to low frequency magnetic fields. In more recent years, attention has also been given several non-radiation/field factors at VDU work stations, such as stress or ergonomic conditions.

<u>Experimental studies on effects of magnetic fields on reproductive processes</u>: The current scientific literature include a number of reports on experiments with various animal systems and different exposures to low fre-

quency magnetic fields. Most of these studies can be classified as to whether they have exposed pregnant rodents or avian embryos to 15-20 kHz saw-tooth, 50/60 Hz sinusoidal, or other types of magnetic fields.

Following the first report by Delgado et al. (1982), several other groups have attempted to replicate the increased occurrence of malformations in chick embryos exposed to pulsed magnetic fields. The most concerted effort was the so-called "Henhouse project", where six laboratories attempted to verify the effect by using identical equipment and standardized experimental procedures. The results of all these studies have, however, not been consistent. While significant results have been obtained in some experiments, e g by Juutilainen et al. (1986), Martin (1988) and two of the laboratories involved in the "Henhouse project" (Berman et al., 1990), others have failed to do so (Sandström et al., 1987, Maffeo et al., 1984, Maffeo et al., 1988) and four of the laboratories involved in the "Henhouse project"). It is apparent that, apart from the magnetic field exposure, other, uncontrolled factors influence the study results. For a review, see e g Juutilainen (1991). In our opinion, it is not clear whether these other factors modify the response to the magnetic field exposures, or whether they constitute an alternative explanation of the found effects.

Only one of the avian experiments (above) has been conducted with VLF fields similar to those around VDUs, that of Sandström et al. (1987). In this study, no significant effect of the fields were found. Of more importance, however, is the advocated cautionary approach to the use of non-mammalian systems when evaluating human reproduction risks. Regardless of whether this cautionary approach is accepted or not, however, the avian embryo studies have failed to produce real support for adverse reproductive human risks due to magnetic field exposures from VDUs.

Likewise, the rodent studies do show some varied results. Studies have been conducted with 50/60 Hz sinusoidal or pulsed magnetic fields - often at rather high exposure levels - without significant results (Zecca et al, 1985, Rivas et al, 1985, Rommereim et al, 1990) although some scent-marking behaviour changes has been noted in one study (McGivern et al, 1990). Other studies have used VDU-like, saw-tooth 15-20 kHz fields - see table 2.

As seen in table 2, the predominant results from these studies are an absence of adverse effects. Some results are noteworthy, however: 1/ The consistent (over several replicas) increase in resorption among exposed mice by Frölén et al. (1989), 2/ the increase in minor skeletal anomalies among exposed rats from both Huuskonen et al. (1990) and Stuchly et al. (1988) and 3/ the absence of any effects from the largest and probably most stringently controlled mice study, by Walsh et al. (1989).

<u>Table 2</u>. Results of experimental studies with rodents using VLF saw-tooth magnetic fields

| Study | Experimental conditions | | | | Results | |
	Species	Max exposure a)	Analysis b)	Malformation c)	Resorption c)	Other d)
Tribukait et al., 1987	C3H (mice)	15 µT	foetus	yes c)	no	no
Frölén et al., 1989	CBA (mice)	15 µT	foetus b)	no	yes	still-birth d)
Huuskonen et al., 1990	Wistar (rats)	15 µT	b)	yes c)	no	yes d)
Stuchly et al., 1988	SD (rats)	66 µT a)	litter (foetus)	no c)	no	no
Walsh et al., 1989	CD-1 (mice)	200 µT a)	litter	no	no	no

<u>Notes for table 2</u>
a) All used saw-tooth magnetic fields of 18 to 20 kHz, with peak-to peak measurements of field strength. The Stuchly et al. and Walsh et al. studies also used (lower) exposure levels compatible with the other studies.
b) Statistical analysis based on the foetus or the litter. The Frölén et al. study was later reported (by Juutilainen, 1991) on a "per litter" basis. The abstract currently available for the Huuskonen et al. study is not clear on this.
c) yes=significant increase among exposed animals (generally at the highest exposure level). no=failure to find such an increase. In the Tribukait et al. study, an increase was seen only in the first replicate (of two). In the Stuchly et al. and the Huuskonen et al. studies, significant increases in minor skeletal anomalities were seen, but (for the Stuchly et al. study) only when analysis was performed based on the foetus (this is unclear for the Huuskonen et al. study).
d) Includes (generally) number of implants, foetal survival, stillbirths, maternal and/or foetal body weight. The Frölén et al. study stillbirth increase was only seen in the first replicate. The Huuskonen et al. positive results were: increased mean number of implantations and (thus?) more living foetuses, and higher mean uteri weight (with content) among exposed.

The results have been discussed in terms of some methodological problems:
- All studies have been performed with different species of mice (C3H, CBA or CD-1) or rats (Sprague-Dawley or Wistar). Thus, no study is a strict replicate of any other, despite similar exposure regimes. How this species difference would influence the studied endpoints is unclear.
- The experimental conditions and study protocols are (at present) incomplete from several studies. Thus, the possibility that other factors may have influenced the results can not be currently evaluated. One suggested factor is noise: A 20 kHz field generator may generate what (to mice) is audible sound. High 18-20 kHz noice have been reported to affect e g resorption (Cook et et al, 1982).
- A common preference of teratological statistical analysis is to base statistical comparisons on the litter as the observational unit, since to use the individual foetus as the observational unit requires that the fate of individual foetuses within a single litter can be shown to be independent. Unfortunately, some of the studies in table 2 have used the foetus as the unit, despite the fact that in at least one of them, there was a distinct "litter" effect (Tribukait et al., 1987). Of the significant findings reported in table 2, only the increased resorption by Frölén et al have been verified by a "per litter" analysis (Juutilainen, 1991).

For further discussions of these rodent experiments, the readers are referred e g to Kavet & Tell (1991). Other animal species (cows, fishes, sea urchins, toads and flies) have also been studied in isolated studies, with varying results.

In a few studies, human cells have been exposed in vitro to (primarily) 50 or 60 Hz magnetic fields. In one (Nordensson et al., 1989), an increase in chromosome aberration frequency in amniotic cells was seen, while in a replicate attempt (Galt, 1990), a decrease was found. Lymphocyte cells were not affected by exposure in a third study (Livingston et al., 1991).

In conclusion, experimental studies have not been very successful in ascertaining whether low frequency magnetic fields as found around VDUs should be regarded as a teratogenic risk for humans.

<u>Epidemiological studies</u>: Table 3 lists a number of epidemiological studies that have been reporting on the possibility of a link between womens´ use of a visual display terminal during pregnancy and various pregnancy outcomes. In addition to these studies, several other studies have been conducted on this issue. These latter studies have not been able to find any difference in pregnancy outcomes attributable to VDU work, but various problems in their design or too small size precludes any real conclusion to be drawn from them. A few studies are at present (fall of 1991) ongoing, or have not yet been fully reported.

<u>Table 3</u>. Epidemiological studies on VDT use and pregnancy

Study	Description		Results a/		
	Type of study, a/	No of pregnancies	SA	CD	FG
Kurppa et al., 1985	CC	2 950	ND	no	ND
Ericson et al., 1985	Cohort	4 347	no	no	no
Ericson & Källén, 1986	CC	1 385	no c/	? c/	ND
Butler & Brix, 1986	Cohort	817	no	ND	ND
McDonald et al., 1988	CC	104 649	? c/	no c/	no
Goldhaber et al., 1988	CC	1 583	yes c/	no	ND
Nurminen & Kurppa, 1988	CC	1 475	X	X	X
Bryant & Love, 1989	CC	980	no	ND	ND
Brandt & Niel-sen, 1990 d/	Case-base	4 628	no	no	no
Windham et al., 1990	CC	1 361	no c/	ND	? c/
Tikkanen et al., 1990	CC	1 555	ND	no e/	no
Schnorr et al., 1991	Cohort	882	no	ND	ND

<u>Notes for table 3</u>

a/ SA=spontaneous abortion, CD=serious malformation, FG=endpoints related to fetal growth, e g birthweight. For each type of endpoint the outcome of the study is described as: yes=a substantial difference between VDU and non-VDU women was seen, no=no such difference or ?=results could be described both ways. ND=not determined. X=other endpoints studied, see text.

b/ CC=case-control type of study.

c/ see text for further discussion.

d/ also Nielsen & Brandt (1990) and Nielsen et al. (1989).

e/ refers to cardiovascular defects.

As seen from table 3, the majority of these studies have not been able to demonstrate an increased risk of spontaneous abortion, malformation or reduced fetal growth. There are, however, several results worthy of detailed comments.

For spontaneous abortion, a number of studies failed to show differences between VDU and non-VDU workers, or between workers with different VDU work hours (Ericson et al., 1985, Ericson & Källén, 1986, Butler & Brix, 1986, Bryant & Love, 1989, Nielsen & Brandt, 1990, Windham et al., 1990 or Schnorr et al., 1991). The study by Lindbohm et al. (1990) did not find any differences, but the analysis has not yet been completed, hitherto reported results did not include measured exposures to electric or magnetic fields. In the study by McDonald et al. (1988), a small increase in spontaneous abortion was seen, but the design of the study did tend to exaggerate somewhat the odds ratio for VDU exposed compared to non-exposed. Correcting for this small selection bias reduced or eliminated the excess odds ratios (McDonald et al., 1988, Bergqvist, 1984).

The study by Goldhaber et al. (1988) was performed at three Kaiser Permanente clinics (KPC) in Northern California. Here, increased odds ratios for spontaneous abortion were seen among clerical workers, but not among other types of workers, such as professional or managerial (where actually a decrease was seen). In the study by Windham et al. (1990), there was a small increase in spontaneous abortion (primarily early ones) with VDU work. Some pregnancies in the latter study were ascertained via two Kaiser Permanente clinics. Separating the women according to KPC ascertainment or not, it was found that the small increase in the Windham et al. study was confined to KPC members. The authors have not found any definite cause of this finding, but discussed among other influences the possible involvement of early prenatal care, and the fact that VDU workers tended to come in for such care earlier. It appears possible that some sort of selection mechanism may be in force; "Adjustment for time prenatal care began decreased the VDT associations for all SABs (spontaneous abortion) in our study to about 1."

Most studies that examined serious malformations failed to find any increase in these - treated as a group - in relation to VDU work (Kurppa et al., 1985, Ericson et al., 1985, McDonald et al., 1988, Goldhaber et al., 1988 or Brandt & Nielsen, 1990). In the study by Ericson and Källén (1986), the crude odds ratio for birth defects (including malformation and stillbirth) were increased for VDU operators compared to non-VDU working women. After controlling for smoking and stress, however, the odds did not really differ between VDU and non-VDU women.

In some studies, occurrences of specific malformations varied between VDU and non-VDU workers: Higher VDU occurrences were found for hydrocephalus (Brandt & Nielsen, 1990), renal malformations (McDonald et al., 1988) and cardiovascular defects (Kurppa et al., 1985). Lower VDU occurrences were also found; central nervous system (Kurppa et al., 1985, Brandt & Nielsen, 1990) and extremities malformation (Brandt & Nielsen, 1990). In general, the small size of the studies did not permit an efficient determination of such occurrences, though. The excess cardiovascular malformation in the Kurppa et al. study prompted the authors to perform a separate study of such malformations (Tikkanen et al, 1990); no such excess malformation could be verified.

Several studies have also investigated various endpoints related to foetal growth, such as birth weight or intrauterine growth retardation, generally without being able to detect any difference between VDU and non-VDU women (Ericson et al., 1985, McDonald et al., 1988, Nielsen et al., 1989, Tikkanen et al., 1990). In one study, somewhat elevated risks for intrauterine growth retardation was found, however (Windham et al., 1990).

Other endpoints have also been investigated, such as preterm delivery or stillbirth (Ericson et al., 1985, Nielsen et al., 1989, Tikkanen et al., 1990, McDonald et al., 1988, Nurminen & Kurppa, 1988), or threatened abortion, placental weight and maternal blood pressure (Nurminen & Kurppa, 1988), without evidence of increased risks for VDU women.

Several types of methodological problems have been discussed in relation to these studies. Selection bias have already been referred to above from the McDonald et al. and the Windham et al. studies. Recall bias was demonstrated by Bryant & Love (1989), and have also been hinted at by Windham et al. (1990). Both of these problems makes it difficult to ascertain especially early spontaneous abortions, as suggested by the Windham et al. study.

Only the study by Schnorr et al. (1991) have explicitly measured electric and magnetic fields around VDU work stations. These field levels were not associated with spontaneous abortion. As the authors pointed out, these (lack of) findings are primarily relevant for the VLF fields. The study by Lindbohm et al. (1990) has also measured electric and magnetic fields at VDU work stations, but as yet (fall of 1991), the results of the pregnancy-field analysis have not appeared.

Stress, smoking and ergonomic situations are other factors often discussed as causes of adverse pregnancy outcomes. Both stress and smoking appear to be so implicated in several of these VDU-specific studies (e g Ericson & Källén, 1985, Brandt & Nielsen, 1990, Schnorr et al., 1991). The study by

Nielsen & Brandt (1990) failed, on the other hand, to implicate ergonomic situations at VDUs in the pregnancy outcome.

<u>Concluding remarks</u>: There are no convincing data from animal experiments to suggest that electric and magnetic fields as specifically found around VDUs would constitute a teratologic risk for humans. Likewise, the epidemiological studies have failed to show a difference in pregnancy outcomes between women working and not working with VDUs during pregnancy. It is indicated but not conclusively shown, judging from the results of some of these studies, that job conditions such as stress may have some influence on certain outcomes, however.

Skin reactions among VDU operators

Early reports on skin problems and VDU work originated in Norway, but were followed by reports also from Sweden. The skin problems were often described as rashes or itching sensations. These report were followed by some epidemiological studies, that did by and large confirm an excess of subjective skin symptoms, but could - in general - not verify an excess of diagnosed skin disorders due to VDU work (Berg et al., 1990a, Knave et al., 1985a, Lidén & Wahlberg, 1985, Svensson & Svensson, 1987, Lidén & Wahlberg, 1990, Sandström et al., 1991). Neither could Berg et al. (1990b) find histological changes in facial skin biopsies from patients with supposedly VDU associated skin complaints.

Early case reports and studies (Wedberg, 1987, Knave et al., 1985b) suggested a relationship between symptoms and electrostatic phenomena in the workplace - primarily the electrostatic charge on the operator. No indications exist, on the other hand, of a link between the electrostatic field from the VDU and skin problems - despite several attempts to find such (Swanbäck & Bleeker, 1989, Knave et al., 1985b, Lidén & Wahlberg, 1990, Sandström et al., 1991). There is a similar lack of indications for VLF magnetic field from VDUs and skin problem relationships (Swanbäck & Bleeker, 1989, Sandström et al., 1989). Operators using either CRT displays or plasma displays - with the latter presumably causing lower field exposures - reported the same prevalence of symptoms (Koh et al. 1990).

Two Swedish studies are presently in the final analytical stage concerning VDU work and skin problems, both includes measurements of ELF and VLF electric and magnetic fields (Sandström et al., 1991, Bergqvist, unpublished). (See table 1 for reported levels of these fields.) Neither study has as yet (fall of 1991) reported the results of multivariate analysis, thus both are described only in terms of crude analysis results, see table 4.

<u>Table 4</u>. Crude analysis results of studies on electric and magnetic fields around VDUs and subjective skin complaints.

| Study | Odds ratios (with 95% confidence intervals | | | |
| | ELF (50-80 Hz) | | VLF (15-35 kHz) | |
	E-field	B-field	E-field	B-field
VDU-fields				
Sandström et al.	1.30	2.7	0.65	1.58
1991	(0.5-3.3)	(1.0-6.9)	(0.3-1.6)	(0.7-3.8)
Bergqvist,	0.63	0.78	1.07	0.75
unpublished	(0.3-1.2)	(0.42-1.47)	(0.57-1.99)	(0.41-1.39)
Background fields				
Sandström et al,	3.0	1.17	ND	ND
1991	(1.2-7.2)	(0.5-2.9)		

Until the full analysis of these studies are complete, it is not possible to determine to what degree these two studies are compatible or not. Based on data as presented in table 4, it appears as if VDU-specific fields (especially VLF fields) are not really potent in causing rashes. An evaluation of the impact of the "background electric 50 Hz field" as measured by Sandström et al., must also await further analysis.

In summary, there does seem to be an overrepresentation of certain subjective symptoms such as rashes among VDU operators in Sweden - but not of diagnosed skin diseases. The causes of such rashes are at present unclear, but studies hitherto performed have not been able to identify VDU-specific electric or magnetic fields as causal factors. Other factors such as humidity, temperature, stress, contact allergens, bias, secondary reactions to eye fatigue and expectation reactions are also considered. One study has preliminary suggested that 50 Hz electric fields not associated with VDUs could be responsible. It is, however, not clear how this factor would then be responsible for a supposedly VDU-specific reaction.

"Electric hypersensitivity"

During the last 5 or 6 years, attention has in Sweden been given a small group of VDU operators with more pronounced reactions, both in terms of skin problems (rashes, hot, warm and burning sensations, pain, itching etc) and often in combination with functional symptoms from the nervous system (dizziness, tingling, tiredness etc). These symptoms have appeared not only in connection with VDU work, closeness to other electric appliances and artificial lighting has also been reported as a triggering factor. Several individuals are severely afflicted and have been on extended sick-leaves. (Knave et al., 1989). The national register for work-related accidents or disorders has, in the last few years, recieved notification of about 30-40 similar cases per year.

The etiology of this phenomenon is at present unknown, suspicion has been directed to a number of widely different factors - and it may well be that several of these are jointly responsible. Based on the experiences of the affected individuals, it is currently not feasible to exclude the possibility that exposure to weak, low frequency electric or magnetic could be one of such factors. A recent provocation study have also indicated that some of these individuals may be able to perceive such fields (Wennberg et al., 1990).

Other, complementary or contributory, factors under consideration are variations in melatonin secretions, porphyrinuria (e g in relation to a pronounced "sensitivity to light" among several afflicted individuals), allergic disposition, personality and disparity between the individual and the organization. Further investigations are necessary, before any generalisable solutions can be formulated. At the same time, individual solutions are urgently needed in a number of Swedish workplaces. Programs for such individualized solutions have been formulated and applied, with some success in enabling afflicted persons to return to work.

Summary and conclusions

In this paper, several health concerns allegedly caused by exposure in VDU situations to electric or magnetic fields have been reviewed. For neither cataract formation nor for adverse reproductive outcomes has epidemiological studies verified an increased occurrence among VDU operators, nor has studies been able to show relevant electric or magnetic field involvement in such processes.

Subjective skin reactions appear, on the other hand, to be linked to VDU work in some Swedish studies - although no such link has been found for diagnosed skin disorders. The factors responsible for these skin rashes are presently unclear, but electric or magnetic fields from VDUs do not appear to be primary factors. A smaller number of more severely afflicted individuals with skin and/or nervous system symptoms of unknown but apparently complex etiology has also appeared - it is at present not possible to verify nor discard the possibility of electric or magnetic field involvement.

References

Berg, M., Lidén, S. & Axelson, O. (1990a) Facial skin complaints and work at visual display units. An epidemiological study of office employees, J. Am .Acad. Dermatol., 22: 621-625.

Berg, M., Hedblad, M.-A. & Erhardt, K. (1990b) Facial skin complaints and work at visual display units. A histopathologic study, Acta Derm. Venereol., 70: 216-220.

Bergqvist, U. (1984) Video display terminals and health, Scand. J. Work Environ. Health., 10 (Suppl 2): 1-87.

Berman, E., Chacom, L., House, D., Koch, B.A., Leal, J., Løvstrup, S., Mantiply, E., Martin, A.H., Martucci, G.I., Mild, K.H., Monahan, J.C., Sandström, M., Shamsaifar, K., Tell, R., Trillo, M.A., Ubeda, A. & Wagner, P. (1990) Development of chicken embryos in a pulsed magnetic field, Bioelectromagnetics, 11:169-187.

Bonomi, L. & Belluci, R. (1989) Consideration on the ocular pathology in 30.000 personnel of the Italian Telephone Company (SIP) using VDTs, Bolletino di Oculistica, 68 (Suppl 7):85-98.

Böös, S.R., Calissendorff, B.M., Knave, B., Nyman, K.G. & Voss, M. (1985) Work with video display terminals among office employees. III Ophthalmologic factors, Scand. J. Work. Environ. Health, 11:475-481.

Brandt, L.P.A. & Nielsen, C.V. (1990) Congenital malformations among children of women working with video display terminals, Scand. J. Work Environ. Health, 16:329-333.

Bryant, H.E. & Love, E.J. (1989) Video display terminal use and spontaneous abortion risk, Int J Epidem, 18:132-138.

Bureau of Radiological Health (1981) An evaluation of radiation emission from video display terminals. Publ No FDA 81-8153. Rockville, Md: US Dept of Health and Human Services.

Butler, W. J. & Brix, K. A. (1986) Video display terminal work and pregnancy outcome in Michigan clerical workers. In Pearce, B.G. (ed). Allegations of Reproductive Hazards from VDUs, pp. 31-52. Nottingham: Humane Technology.

Canadian Labour Congress (1982) Towards a more humanized technology; exploring the impact of Video Display Terminals on the health and working conditions of Canadian office workers. Ottawa, Que: Labour Education and Study Centre.

Cook, R.O., Nawrot, P.S. & Hamm, C.W. (1982) Effects of high-frequency noise on prenatal development and maternal plasma and uterine catecholamine concentrations in the CD-1 mouse, Toxicol. Appl. Pharmacol., 66: 338-348.

Delgado, J.M.R., Leal, J., Monteagudo, J.L. & Garcia, M.G. (1982) Embryological changes induced by weak, extremely low frequency electromagnetic fields, J. Anat., 134:533-551.

Ericson A., Källén, B. & Westerholm, P. (1986) No increased risk of foetal damage among women with VDT work (in Swedish), Läkartidningen, 82:2180-2184.

Ericson, A. & Källén, B. (1986) An epidemiology study of work with video screens and pregnancy outcomes: II. A Case-Control Study, Am. J. Ind. Med., 9:459-475.

Frank, A.L. (1983) Effects on health following occupational exposure to video display terminals. Dept. of Preventive Medicine and Environmental Health, Report 40536-0084, Lexington, KY: University of Kentucky.

Frölén, H. & Svedenstål N.-M. (1988) Influence of pulsed magnetic fields on foetal development in mice. Supplementary and extended study to an earlier reported investigation (in Swedish). Dept of Pathology. Uppsala: Swedish Agricultural University.

Galt, S. (1990) Optical fiber scattering and biological electromagnetic effects. School of Electrical and Computer Engineering, Technical Report No. 198. Gothenburg: Chalmers University of Technology.

Goldhaber, M. K., Polen, M.R. & Hiatt, R.A. (1988) The risk of miscarriage and birth defects among women who use visual display terminals during pregnancy, Am. J. Ind. Med., 13:695-706.

Huuskonen, H., Juutilainen, J., Komulainen, H. & Saali, K. (1990). Effects of low-frequency alternating magnetic fields on foetal development in rats. p E2/4 In: URSI/IEEE XVI Convention on Radio Science, Kuopio: Dept. of Physics, University of Kuopio.

Hultgren, G. & Knave, B. (1974) Discomfort glare and disturbances from light reflections in an office landscape with CRT display terminals, Appl. Ergon., 5:2-8.

ILO (in press) Video display units - radiation protection guidance, Geneva: International Labour Office.

Juutilainen, J., Harri, M., Saali, K. & Lahtinen, T. (1986) Effects of 100-Hz magnetic fields with varying waveforms on the development of chick embryos, Radiat. Environ. Biophys., 25:65-74.

Juutilainen, J. (1991), Effects of low-frequency magnetic fields on embryonic development and pregnancy, Scand. J. Work Environ. Health, 17: 149-158.

Kavet, R. & Tell, R.A. (1991), VDTs: Field levels, epidemiology, and laboratory studies, Health Physics, 61: 47-57.

Knave B., Wibom, R., Voss, M., Hedström, L. & Bergqvist, U. (1985a) Work at video display terminals. An epidemiological health investigation of office employees. I. Subjective symptoms and discomforts, Scand. J. Work Environ. Health., 11:457-466.

Knave, B., Wibom, R., Bergqvist, U., Carlsson, L.L.W., Levin, M.I.B. & Nylén, P.R. (1985b) Work at video display terminals. An epidemiological health investigation of office employees. II. Physical exposure factors, Scand. J. Work Environ. Health., 11:467-474.

Knave, B., Bergqvist, U. & Wibom, R. (1989). Symptoms and subjective discomforts at "sensitivity to electricity" (in Swedish). Undersöknings-rapport 1989:4. Solna, Sweden: National Institute of Occupational Health.

Koh, D., Goh, C.L., Jeyaratnam, J., Kee, W.C. & Ong, C.N. (1990) Dermatological symptoms among visual display unit operators using plasma displays and cathode ray tube screens, Ann. Acad. Med. Singapore, 19:617-620.

Kurppa K., Holmberg, P.C., Rantala, K., Nurminen, T. & Saxén, L. (1985) Birth defects and exposure to video display terminals during pregnancy, Scand. J. Work Environ. Health., 11:353-356.

Lidén, C. & Wahlberg, J.E. (1985) Work at video display terminals. An epidemiological health investigation of office employees. V. Dermatological examination, Scand. J. Work Environ. Health., 11:489-493.

Lidén, C. & Wahlberg, J.E. (1990) Skin diseases and VDU work. In Berlinguet, L. & Berthelette, D. (eds), Work with Display Units 89: Selected Papers from the Second International Scientific Conference on Work with Display Units, Montréal, Québec, Canada, September 1989, pp. 173-179. Amsterdam: North Holland.

Lindbohm, M.-L., Sallmén, M., Hietanen, M., Taskinen, H., Pekkarinen, M., Ylikoski, M. & Hemminki, K. (1990) Spontaneous abortion and work with video display terminals. In: 39. Nordic Occupational Health meeting. pp 131-132. Helsinki: National Institute of Occupational Health.

Livingston, G.K., Witt, K.L., Gandhi, O.P., Chatterjee, I. & Roti Roti, J.L. (1991) Reproductive integrity of mammalian cells exposed to power frequency electromagnetic fields, Environ. Mol. Mutagen, 17: 49-58.

Maffeo, S. Miller, N.W. & Carstensen, E.L. (1984) Lack of effect of weak low frequency electromagnetic fields on chick embryogenesis, J. Anat., 139:613-618.

Maffeo, S., Brayman, A.A., Miller, N.W. & Carstensen, E.L. (1988) Weak low frequency electromagnetic fields and chick embryogenesis: failure to reproduce findings, J. Anat., 157:101-104.

Marriott, I.A. & Stuchly, M.A. (1986). Health aspects of work with visual display terminals, J. Occup. Med., 28,:833-848.

Martin, A.H. (1988). Magnetic fields and time dependent effects on development, Bioelectromagnetics, 9:393-396.

McDonald, A. D., McDonald, J.C., Armstrong, B., Cherry, N., Nolin, A.D. & Robert, D. (1988). Work with visual display units in pregnancy, Brit. J. Ind. Med., 45,:509-515.

McGivern, R.F., Sokol, R.Z. & Adey. W.R. (1990) Prenatal exposure to a low-frequency electromagnetic field demasculinizes adult scent marking behavior and increases accessory sex organ weight in rats, Teratology, 41:1-8.

National Research Council (1983). Video displays, work and vision. Committee on Vision, Panel on impact of video viewing on vision of workers, Washington DC: National Academy Press.

Nielsen, C.V. & Brandt, L.P.A. (1990) Spontaneous abortion among women using video display terminals, Scand. J. Work Environ. Health, 16:323-328.

Nielsen, C.V., Brandt, L., Helsborg, L., Waldström, B. & Nielsen, L.T. (1989) The effect of VDT work on the course of pregnancy (in Danish). Department of Social Medicine, Aarhus, Denmark: University of Aarhus.

Nordensson, I., Hansson Mild, K., Sandström, M. & Mattsson, M.O. (1989) Effect of low frequency magnetic fields at a chromosomal level in human amniotic cells (in Swedish). Undersökningsrapport 1989:25. Solna, Sweden: National Institute of Occupational Health.

Nurminen, T. & Kurppa, K. (1988). Office employment, work with video display terminals, and course of pregnancy, Scand. J. Work Environ. Health, 14:293-298.

Rivas, L., Rius, C., Tello, I. & Oroza, M.A. (1985). Effects of chronic exposure to weak electromagnetic fields in mice, IRCS Med. Sci., 13:661-662.

Rommereim, D.N., Rommereim, R.L., Buschbom, R.L. & Anderson, L. (1990) Developmental toxicity study in rats exposed to 60 Hz horizontal magnetic fields. In: Bioelectromagnetics Society. Twelfth annual meeting, June 1990. p 20-1. San Antonio, Texas: Bioelectromagnetics Society.

Sandström, M., Hansson Mild, K. & Løvstrup, S. Effects of weak pulsed magnetic fields on chick embryogenesis. In Knave, B. & Widebäck, P.-G. (eds), Work with Display Units 86: Selected Papers from the International Scientific Conference on Work with Display Units, Stockholm, Sweden, May 1986, pp. 135-140. Amsterdam: North-Holland.

Sandström, M., Hansson Mild, K., Lönnberg, G., Stenberg, B. & Wall, S. (1991). The office illness project in northern Sweden. Electric and magnetic fields: a case referent study among VDT workers (in Swedish). Undersökningsrapport 1991:12. Solna, Sweden: National Institute of Occupational Health.

Sandström, M., Stenberg, B. & Hansson Mild, K. (1989) Provocation tests with ELF and VLF electromagnetic fields on patients with skin problems associated with VDT work. Presented at the Second International Scientific Conference on Work with Display Units, Montréal, 11-14 September.

SCB (1990). Use of Computers (in Swedish). Report DF 1:1990, Stockholm: Statistics Sweden.

Schnorr, T.M., Grajewski, B.A., Hornung, R.W., Thun, M.J., Egeland, G.M., Murray, W.E., Conover, D.L. & Halperin, W.E. (1991) Video display terminals and the risk of spontaneous abortion, N. Eng. J. Med., 324:727-733.

Smith, A.B., Tanaka, S., Halperin, W. & Richards, R.D. (1982) Report of a cross-sectional survey of video display terminal (VDT) users at the Baltimore Sun. Cincinnatti, OH: National Institute for Occupational Safety and Health.

Stuchly, M.A., Ruddick, J., Villeneuve, D., Robinson, K., Reed, B., Lecuyer, D.W., Tan, K. & Wong, J. (1988) Teratological assessment of exposure to time-varying magnetic fields, Teratology, 38:461-466.

Swanbäck, G. & Bleeker, T. (1989) Skin problems from visual display units, Acta. Derm. Venereol., 69:46-51.

Svensson, E. & Svensson, J. (1987) Prolonged VDT work resulted in mild skin discomforts - a tendency for increase in wintertime (in Swedish), Läkartidningen, 84:1843-1845.

Tikkanen, J., Heinonen, O.P., Kurppa, K. & Rantala, K. (1990) Cardiovascular malformations and maternal exposure to video display terminals during pregnancy, Eur. J.Epidemiol., 6:61-66.

Tribukait, B., Cekan, E. & Paulsson, L.E. (1987). Effects of pulsed magnetic fields on embryonic development in mice. In Knave, B. & Widebäck, P.-G. (eds), Work with Display Units 86: Selected Papers from the International Scientific Conference on Work with Display Units, Stockholm, Sweden, May 1986, pp. 129-134. Amsterdam: North-Holland.

Walsh, M.L., Agnew, D., Harvey, S., Wiley, M., Corey, P., Charry, J. & Kavet, R. (1989) Magnetic field rodent reproductive study (MFRRS). Presented at the Annual Review of Research on Biological Effects of 50 & 60 Hz Electric and Magnetic Fields (Contractors Review), Portland OR, November 1989. Also reported by Kavet & Tell (1991).

Wedberg, W.C. (1987) Facial particle exposure in the VDU environment: The role of static electricity. In Knave, B. & Widebäck, P.-G. (eds), Work with Display Units 86: Selected Papers from the International Scientific Conference on Work with Display Units, Stockholm, Sweden, May 1986. pp. 151-159. Amsterdam: North-Holland.

Wennberg, A., Franzén, O. & Paulsson, L.E. (1990). Detection of electric and magnetic fields. An investigation of individuals with reported "electric hypersensitivity". Undersökningsrapport 1990:20. Solna, Sweden: National Institute of Occupational Health.

Windham, G.C., Fenster, L., Swan, S.H. & Neutra, R.R. (1990) Use of video display terminals during pregnancy and the risk of spontaneous abortion, low birthweight, or intrauterine growth retardation., Am. J. Ind. Med., 18:675-688.

WHO (1987). Visual display terminals and workers health. Offset publication No. 99. Geneva: World Health Organization.

WHO (1990). Update on Visual Display Terminals and Workers' Health. WHO/OCH/ 90.3. Geneva: World Health Organization.

Zaret, M.M. (1984) Cataracts and visual display units. In: Pearce, B. (ed), Health Hazards of VDTs?, pp. 47-54. Chichester: Wiley

Zecca, L., Ferrario, P. & Dal Conte, G. (1985). Toxicological and teratological studies in rats after exposure to pulsed magnetic fields, Bioelectrochem. Bioenerg., 14:63-69.

GUIDELINES AND STANDARDS

M H Repacholi

Visiting Scientist
Australian Radiation Laboratory
Victoria, Australia

Abstract

A standard is a general term, incorporating both regulations and guidelines and is defined as a set of specifications for equipment or rules laid down to promote the safety of an individual or group of people. A regulation is normally promulgated under a legal statute and is referred to as a mandatory standard. A guideline does not generally have any legal force and is issued for guidance only - a voluntary standard. Statements of policy on the safe use of NIR are considered to be guidelines.

Standards can specify limits of exposure (exposure standards) and other safety rules for personal protection and/or specify details on the performance, construction, design or functioning of a device, or methods of testing its performance (device performance or emission standards).

Standards development should preferably be preceded by the preparation of, or reference to a document that identifies the hazards of human exposure to the particular NIR. Such a document would summarize the experimental data on exposure of various biological systems, the known mechanisms of interactions of NIR with biological systems, a health-hazard analysis and an assessment of the various national and international standards. Such an assessment forms an important basis for recommendations. This paper describes the types of NIR standards and guidelines which currently exist and a brief summary of the IRPA and some national standards in each part of the NIR spectrum.

Introduction

At sufficiently high intensities or field strengths, all of the NIR are capable of producing biological effects in cells or tissues which could result in adverse health consequences. Standards are developed because of need. When persons are exposed to sufficient amounts of NIR to cause adverse health effects or devices are fabricated such that injury can result from their use, the need for standards generally becomes self-evident. The potential severity of the hazard should influence whether a standard is made either as a guideline or regulation.

The development of standards requires a realistic assessment of reported biological effects from exposure to NIR. However, to define clearly a dose-response relationship in biological systems exposed to many of the NIR presents a very complex problem, not only because of the wide range of frequencies or wavelengths and intensities that are used, but also because of the large number of physical and biological variables that must be considered and their interrelationships.

Before developing standards for NIR devices or personnel exposure to the various NIR, the following points must be considered:

1. Does the NIR really present a problem to humans? Are there confirmed biological effects from exposing human beings or other mammalian systems to the NIR? If so, do these confirmed effects constitute a health hazard?

2. What sources of NIR could be potentially hazardous; how rapidly are they proliferating and how can they best be controlled?

3. Will the existence and implementation of regulations or guidelines alleviate the problem?

4. What is the acceptable risk to health of exposure to NIR, considering the benefits that can accrue from its use?

5. What would be the economic impact of such controls on device manufacturers, retailers and users?

6. What national and/or international standards exist; what is their scientific basis; what is the philosophy of acceptable risk associated with their adoption?

Thus, before proceeding with the development of standards, one must review thoroughly the available scientific literature on the NIR biological effects, study the various sources of NIR and evaluate the current national and international standards on the subject. This procedure will generally provide answers to the above questions with varying degrees of satisfaction. It is then up to the individual regulatory agencies or standard-setting organizations to determine whether there is a need for standards and, if so, whether compulsory regulations or voluntary guidelines should be developed.

Development and implementation of regulations to limit the exposure to NIR can provide beneficial effects to the health of exposed persons and industry in the following ways:

1.   The existence of regulations would serve as a signal to industry and the general public that there is a concern about NIR exposure and that they should become aware of the potential hazards.

2.   Regulations provide goals to be achieved at the planning stages by manufacturers of devices and by organizations involved in the installation and construction of NIR facilities.

3.   Devices or facilities producing the NIR in excess of the levels permitted by the regulations can then be identified and appropriate remedial action taken.

4.   Regulations form the basis for safe working practices to ensure that employees are not exposed to excessive levels of the NIR.

Safe-use guidelines have a number of advantages over regulations:

1.   They can be introduced more rapidly.

2.   Modifications to the guidelines can be made quickly if necessary.

3.   They can be made with more flexibility to adjust to changes in technology.

On the other hand, safe-use guidelines have limitations:

1.   They are voluntary and thus need not be followed.

2.   They are less useful to manufacturers, since regulations provide detailed requirements on design, construction, or functioning.

3.   Regulations, since they are mandatory, protect good manufacturers from "fly-by-night" operations.

Standards

To protect the general population, patients and persons occupationally exposed to NIR, two types of standards are generally promulgated:

1.   Exposure standards apply to personnel protection and generally refer to maximum levels to which whole or partial body exposure is permitted. An example is the IRPA radiofrequency field standard (IRPA 1988) limiting the RF exposure of workers to 4W/kg.

2.   Emission standards refer to equipment or devices and specify maximum NIR emission or leakage levels, sometimes at a specified distance. Detailed specifications on the design, construction, functioning and performance of the device are usually given to ensure that the maximum exposure levels are not exceeded. An example of this standard is that for microwave ovens which specifies that the leakage levels should not exceed $5mW/cm^2$ at 5 cm from the surface of the oven.

Exposure standards differ from emission standards in that they are usually derived from biological effects data and normally represent an exposure level below which adverse health effects have not been detected. Maximum levels of NIR emission for a device would be set below that provided in an exposure standard, since personnel could conceivably be exposed to fields of radiation from more than one device at a time, or because the device is or will be used in an uncontrolled environment.

Standards development should preferably be preceded by a document that summarizes the experimental data on exposure of various biological systems to the NIR, the known mechanisms of interactions of the NIR with biological systems and an assessment of the various national and international standards. This criteria document then forms an important scientific basis for regulations from which the maximum exposure levels incorporated into standards can be determined. There are many standards today that have been developed without such a rationale or criteria document and it is therefore often difficult to determine the scientific basis on which they are founded.

To develop exposure limits, certain criteria must be met from the scientific data before they can be considered. These have already been discussed by Repacholi and Stolwijk (1991). Once thresholds of exposure are identified (above which, adverse effects are likely, but below which there is little likelihood of any adverse health impact) safety factors may be incorporated, the magnitude of which will depend on the completeness of our biological effects data and the population to be protected. Safety factors in health protection standards do not guarantee safety, but represent an attempt to compensate for unknowns and uncertainties.

Ultrasound

AIRBORNE

Ultrasound is commonly used in many industrial processes, including cleaning, drilling, mixing and emulsification. Most of these invariably emit airborne ultrasound, not only at the ultrasonic operating frequency, but also at sub-harmonics in the audible acoustic range. Many industrial applications use high ultrasonic intensities that produce cavitation, observed as a type of boiling action in the liquid, which produces high audible noise levels.

In industry, many workers have complained of subjective symptoms (nausea, tinnitus, headaches, fatigue, etc.) when operating devices such as ultrasound cleaning tanks. Hearing loss may occur from exposure to very high intensities of airborne ultrasound, but no well-defined threshold for this effect has been determined.

Effects on the general public appear to be mediated by nervous reaction. Many people are unable to enter commercial establishments having an intrusion alarm (where the alarm is turned off, but the airborne ultrasound is still radiating) because they immediately suffer headaches or feel nauseated.

Airborne ultrasound is usually quantified in terms of sound pressure level (SPL) in decibels (dB), such that:

$$\mathrm{SPL(dB)} = 20 \log_{10}(p/p_r)$$

where p is the root mean square acoustic pressure and p_r is equivalent to approximately the lowest level of audible sound perceived by humans at the most sensitive frequency (approx. 1 kHz), and is normally taken as equal to 20 micropascals (μPa). Twenty (20) μPa is equivalent to an acoustic intensity $I_r = 10^{-12}$ W/m^2 in the air. An octave band contains a range of frequencies, the upper frequency limit being twice the lower frequency limit. The centre or mid frequency used to designate each octave is twice the centre frequency of the preceding octave band. One-third octave bands are used to geometrically split an octave band into three parts and the mid frequency is used to designate each band.

The IRPA/INIRC, in conjunction with the World Health Organization, Geneva, drafted a document entitled "Environmental Health Criteria for Ultrasound (UNEP/WHO/IRPA 1982). This document forms the primary scientific data base for the development of the IRPA/INIRC Airborne Ultrasound Standard as shown in Tables 1-3 (from IRPA 1984).

LIQUID-BORNE ULTRASOUND

The IRPA/INIRC worked in conjunction with the World Health Organization to review the scientific literature on ultrasound (UNEP/WHO/IRPA 1982).
This document presented an in-depth review of the biological effects literature and a health risk assessment of human exposure to ultrasound. One of the key recommendations of this review was: "The

Table 1:   **Limits for continuous (approx. 8 h/day) occupational exposure to airborne ultrasound**

Mid frequency of one-third octave band (kHz)	Sound pressure level (dB re: 20 μPa)
20	75
25	110
31.5	110
40	110
50	110
63	110
80	110
100	110

Table 2: Modification to occupational exposure limits given in Table 1 for exposure durations not exceeding 4 hours per day.

Total exposure duration (h)	Correction to SPL (dB)
2-4	+3
1-2	+6
0-1	+9

Table 3: Limits for continuous (24 h/day) exposure of the general public to airborne ultrasound

Mid frequency of one third-octave band (kHz)	Sound pressure level (dB re: 20 μPa)
20	70
25	100
31.5	100
40	100
50	100
63	100
80	100
100	100

establishment of guidelines on the performance of diagnostic ultrasound equipment is recommended -including requirements concerning image quality and stability, and quality assurance measures. However, at present there does not appear to be a need to limit the output exposure levels of diagnostic ultrasound equipment, other than to recommend strongly that the lowest output levels be used commensurate with image quality, adequate to obtain the necessary diagnostic information".

Even today there does not appear to be a need to limit human exposure to ultrasound and so the IRPA/INIRC has not done so. However, guidelines on equipment specifications, performance and measurement methods have been developed by the International Electrotechnical Commission (IEC) and organisations such as the Americal Institute for Ultrasound in Medicine (AIUM). Probably the most widely accepted guidance on exposure to liquid-borne ultrasound comes from the AIUM which issued the statement given in Table 4 (from Dunn 1991).

Ultraviolet Radiation

Ultraviolet radiation occupies that portion of the electromagnetic spectrum from 100 to 400 nm and is used in a wide a variety of medical, industrial and cosmetic purposes. The most significant adverse health effects of exposure to UVR have been reported at wavelengths below 315 nm, known collectively as actinic ultraviolet. The IRPA/INIRC guideline (IRPA 1985, 1989) has been limited to wavelengths greater than 180 nm where UVR is transmitted through air. The most restrictive limits are for exposure to radiation having those wavelengths less than 315 nm. For ultraviolet in the spectral region (315-400 nm) the total radiant exposure incident on the unprotected eye should not exceed 1.0 J cm^{-2}(10 kJ m^{-2}) within an 8-hour period.

For the actinic UV spectral region from 180 to 315 nm, the limits for radiant exposure incident upon the unprotected skin or eye within an 8-hour period are quite complicated and readers are referred to the original IRPA (1989) guidelines for more detail

These guidelines for exposure limits were designed to protect against such effects as erythema and delayed effects on the skin, as well as photokeratoconjunctivitis, cataract and retinal effects on the eye. A detailed rational is given in IRPA (1989).

TABLE 4 : AMERICAN INSTITUTE OF ULTRASOUND IN MEDICINE STATEMENT ON MAMMALIAN "IN VIVO" BIOLOGICAL EFFECTS. APPROVED OCTOBER 1987

A review of bioeffects data supports the following statement:

In the low megahertz frequency range there have been (as of this date) no independently confirmed significant biological effects in mammalian tissues exposed "in vivo" to unfocused ultrasound with intensities[a] below 100mW/cm^2 or to focused[b] ultrasound with intensities below 1 W/cm^2. Furthermore, for exposure times[c] greater than one second and less than 500 seconds for unfocused ultrasound, or 50 seconds for focused ultrasound such effects have not been demonstrated even at higher intensities, when the product of intensity and exposure time is less than 50 joules/cm^2.

[a] Free-field spatial peak, temporal average (SPTA) for continuous wave exposures and for pulsed-mode exposures with pulses repeated at a frequency greater than 100 Hz.

[b] Quarter-power (-6 dB) beam width smaller than four wavelengths or 4 mm, whichever is less at the exposure frequency.

[c] Total time including off-time as well as on-time for repeated pulse exposures.

Laser

Adverse health effects of exposure to laser radiation are of particular concern in the visible and near infrared (400 nm-1400 nm) where retinal injury can occur, although adverse biological effects are theoretically possible across the entire optical spectrum from 180 nm to 1 mm. Within this region, exposure limits vary enormously because of variations in biological effects and the different critical structures of the eye potentially at risk. IRPA/INIRC has drafted exposure limits for this entire wavelength region for exposure durations between 1 ns and 8 hours. The biological effects induced by optical radiation are essentially the same for coherent and incoherent sources for any given wavelength, exposure area and duration, however there is a necessity to treat lasers as a special case because few conventional optical sources can approach the radiant intensities and irradiances of lasers.

The exposure limits developed by IRPA/INIRC (IRPA 1988) are quite complicated, having a number of modifying factors which account for the particular exposure situation. Readers are referred to the IRPA (1988) standard for complete details.

Radiofrequency Fields

Radiofrequency electromagnetic energy is used in a variety of applications in industry, commerce, medicine, research and the home. At sufficiently high intensities, exposure to radiofrequency electromagnetic fields can produce a variety of adverse health effects (UNEP/WHO/IRPA 1991). Such effects include, overloading of the thermoregulatory response, thermal injury, altered behavioral patterns, convulsions and decreased endurance. Exposure limits are needed to protect against these adverse health effects of radiofrequency field exposure.

The IRPA/INIRC guidelines on limits (IRPA 1988a) were developed on the basis that an SAR of 4 W/kg was considered to be an RF energy absorption rate above which there was an increasing likelihood that adverse health consequences could occur. Below this level, no adverse health effects have been established from acute exposures. Incorporating a safety factor of 10 to allow for possible consequences of long-term exposure, 0.4 W/kg was used as the basic limit for deriving exposure limits for occupational exposure. A further safety factor of 5 was incorporated to derive general public limits.

According to the IRPA (1988) guidelines, occupational exposure to radiofrequency fields at frequencies below and up to 10 MHz should not exceed the levels of unperturbed RMS electric and magnetic field strengths given in Table 5, when the squares of the electric and magnetic field strengths are averaged over any 6-minute period during the working day, provided that the body-to-ground current does not exceed 200 mA and that any hazards of radiofrequency burns are eliminated. For pulsed fields, a conservative approach consists of limiting pulsed electric and magnetic field strengths as averaged over the pulse width to 32 times the values given in Table 5.

Occupational exposure to frequencies above 10 MHz should not exceed a SAR of 0.4 W/kg when averaged over any 6-minute period and over the whole body, provided that in the extremities (hands, wrists, feet and ankles) 2 W per 0.1 kg shall not be exceeded and that 1 W per 0.1 kg shall not be exceeded in any other part of the body.

The limits of occupational exposure given in Table 5 for the frequencies between 10 MHz and 300 GHz are the working limits derived from the SAR value of 0.4 W/kg. They represent a practical approximation of the incident plane wave power density needed to produce the whole body average specific absorption rate of 0.4 W/kg. These limits apply to whole body exposure from either continuous or modulated electromagnetic fields from one or more sources, averaged over any 6-minute period during the working day.

Exposures of the general public to frequencies above 10 MHz should not exceed a SAR of 0.08 W/kg when averaged over the whole body and over any 6-minute period.

The limits of radiofrequency exposure to the general public given in Table 6 for the frequencies between 10 MHz and 300 GHz are derived from the SAR value of 0.08 W/kg. They represent a practical approximation of the incident plane wave power density needed to produce the whole body average SAR of 0.08 W/kg.

The limits given in Table 6 apply to whole body exposure from either continuous or modulated electromagnetic fields from one or more sources, averaged over any 6-minute period during the 24-hour day.

The full rationale for the IRPA/INIRC exposure limits is given in IRPA (1988a).

A recent summary of the RF standards existing in different countries of the world has been given by Repacholi (1990).

TABLE 5: Occupational exposure limits to radiofrequency (RF) electromagnetic fields (From IRPA 1988a).

Frequency	Unperturbed RMS field strength		Equivalent plane wave power density	
f(MHz)	Electric E(V/m)	Magnetic H(A/m)	P_{eq}(W/m^2)	P_{eq}(mW/cm^2)
0.1-1	614	1.6/f	-	-
>1-10	614/f	1.6/f	-	-
>10-400	61	0.16	19	1
>400-2000	3f$^{\frac{1}{2}}$	0.008f$^{\frac{1}{2}}$	f/40	f/400
>2000-300000	137	0.36	50	5

TABLE 6: General public exposure limits to radiofrequency electromagnetic fields (from IRPA 1988a).

Frequency	Unperturbed RMS field strength		Equivalent plane wave power density	
f(MHz)	Electric E(V/m)	Magnetic H(A/m)	P_{eq}(W/m^2)	P_{eq}(mW/cm^2)
0.1-1	87	0.23/f$^{\frac{1}{2}}$	-	-
>1-10	87/f$^{\frac{1}{2}}$	0.23/f$^{\frac{1}{2}}$	-	-
>10-400	27.5	0.073	2	0.2
>400-2000	1.37f$^{\frac{1}{2}}$	0.0037f$^{\frac{1}{2}}$	f/200	f/2000
>2000-300000	61	0.16	10	1

50/60 Hz Fields

The IRPA/INIRC worked in conjunction with the WHO to review the literature on biological effects of 50/60 Hz electric and magnetic fields on biological systems. This resulted in two publications, one on ELF electric fields (UNEP/WHO/IRPA 1984) and the other on magnetic fields (UNEP/WHO/IRPA 1987). It was concluded that the only established mechanism of action of 50/60 Hz electric and magnetic fields is the introduction of internal electric fields and currents in the exposed biological system. When the scientific literature was reviewed to determine the current densities that produced biological effects, it was found that:

(a) Between 1 and 10 mA/m^2 (induced by magnetic flux densities above 0.5 and up to 5 mT at 50/60 Hz), minor biological effects have been reported.

(b) Between 10 and 100 mA/m^2 (above 5 and up to 50 mT at 50/60 Hz) there are well established effects, including visual and nervous system effects.

(c) Between 100 and 1000 mA/m^2 (above 50 and up to 500 mT at 50/60 Hz) stimulation of excitable tissue is observed and there are possible health hazards.

(d) Above 1000 mA/m^2 (greater than 500 mT at 50/60 Hz) extra systoles and ventricular fibrillation can occur (acute health hazards).

Some epidemiological reports, present data indicative of an increase in the incidence of cancer among children, adults and occupational groups. The studies suggest an association with exposure to weak 50 or 60 Hz magnetic fields. These associations cannot be satisfactorily explained by the available theoretical basis for the interaction of 50/60 Hz electromagnetic fields with living systems. The magnetic flux densities in some epidemiological studies suggesting an increased cancer incidence are at values near $0.25\mu T$. This magnetic flux density would induce a current density that is well below those levels normally occurring in the body. The epidemiological studies are not conclusive. Although these epidemiological data cannot be dismissed, there must be additional studies before they can serve as a basis for health hazard assessment. Furthermore, scant laboratory evidence is available to support the hypothesis that there is an association between 50/60 Hz fields and increased cancer risk.

With all available evidence, the IRPA/INIRC drafted the limits to 50/60 Hz fields as shown in Table 6. These limits are derived using the criterion that the induced current density should not exceed 10 mA/m^2, the upper level of endogenous currents normally found in humans. The IRPA (1990) limits are very similar to those recommended by other national bodies as shown in Table 7. Details other standards are shown in Table 8 from Repacholi (1991).

Static Magnetic Fields

Static magnetic fields exert Lorentz forces on moving ionic charge carriers and thus give rise to induced electric fields and currents. This is the basic interaction causing magnetically-induced blood flow potentials. Static magnetic fields also give rise to magneto-orientation effects and magnetomechanical translation on paramagnetic and ferromagnetic materials.

The IRPA/INIRC worked in conjunction with WHO to review the literature on static magnetic fields (UNEP/WHO/IRPA 1987) and concluded that human exposure up to 2000 mT does not appear to have any adverse health impact. Using this review and more recent literature, the IRPA/INIRC is in the process of drafting guidelines on limits of exposure to static magnetic fields. The IRPA/INIRC draft limits for continuous whole body exposure to static fields are shown in Table 9. Other national recommendations on static field exposure limits are also shown in Table 9.

The Australian National Health and Medical Research Council guidelines for magnetic resonance facilities have limits of exposure to static magnetic fields for patients, workers and the general public. Details of these are shown in Table 10.

TABLE 7:   IRPA (1990) limits of exposure to 50/60 Hz electric and magnetic fields.

Exposure characteristics	Electric field strength kV m^{-1}(rms)	Magnetic flux density mT (rms)
Occupational		
Whole working day	10	0.5
Short term	30[a]	5[b]
For limbs	-	25
General public		
Up to 24 h d^{-1} [c]	5	0.1
Few hours per day[d]	10	1

[a] The duration of exposure to fields between 10 and 30 kV m^{-1} may be calculated from the formula t $\leq$80/E, where t is the duration in hours per work day and E is the electric field strength in kV m^{-1}.

[b] Maximum exposure duration is 2 h per work day.

[c] This restriction applies to open spaces in which members of the general public might reasonably be expected to spend a substantial part of the day, such as recreational areas, meeting grounds, and the like.

[d] These values can be exceeded for a few minutes per day provided precautions are taken to prevent indirect coupling effects.

TABLE 8:   50/60 Hz Standards and Guidelines Limits for Continuous General Public and Occupational Exposure

Standard	Electric Field (kV/m)		Magnetic Field (mT)	
	Public	Occupat.	Public	Occupat.
IRPA (1990) Australia	5	10	0.1	0.5
NHMRC (1989)	5	10	0.1	0.5
Germany (1989	20.6	20.6	5.024	5.024
UK NRPB (1989)	12.28	12.28	2.0	2.0

USSR (1975)	-	5	-	-
USSR (1985)	-	-	-	1.76
USA ACGIH (1991)	-	25	-	1.0(60 Hz)
				1.2(50 Hz)
Poland (1980)	-	15	-	-

TABLE 9:   Standards/Guidelines on Static Magnetic Field Limits for Continuous Whole Body Exposure

	Magnetic Flux Density (mT)	
	Public	Occupational
Australia NH&MRC (1991)	10	200
UK, NRPB (1989)	2	2
USA, ACGIH (1991)	-	60
IRPA/INIRC Proposed (1991)	10	200

TABLE 10:   Magnetic Field Exposure Limits NH&MRC 1991 MR Guidelines

Exposure characteristics	Magnetic Flux Density
Patients	
Whole Body (duration of examination)	2T
Extremities	5T
Staff	
Whole Body	
- average over working day	200mT
- peak value	2T
Extremities (few mins per day)	5T
Visitors	
Continuous exposure	10mT

REFERENCES

ACGIH (1990)   American Conference of Governmental and Industrial Hygienists. Notice of intended change - sub-radiofrequency (30 kHz and below) and static electric fields. Appl. Occup. Environ. Hyg. 5(10):734-737.

DUNN, F (1991) Ultrasound, IEEE Trans. Ed. 34(3): 266-268.

GERMANY (1989) Safety from electromagnetic fields: Limits of field strengths for protection of persons in the frequency range from 0 to 30 kHz. Standard DIN VDE 0848, part 4, Deutsche Elektrotechnische Kommission im DIN and VDE (DKE), Berlin.

IRPA (1984) International Radiation Protection Association, International Non-Ionizing Radiation Committee. Interim guidelines on limits of human exposure to airborne ultrasound. Health Physics 46:969-974.

IRPA (1988) International Radiation Protection Association, International Non-Ionizing Radiation Committee. Guidelines on limits of exposure to laser radiation of wavelengths between 180 nm and 1 mm. Health Physics 54:573-574.

IRPA (1988a) International Radiation Protection Association, International Non-Ionizing Radiation Committee.Guidelines on limits of exposure to radiofrequency electromagnetic fields in the frequency range from 100 kHz to 300 GHz. Health Physics 54:115-123.

IRPA (1989) International Radiation Protection Association, International Non-Ionizing Radiation Committee. Guidelines on limits of exposure to ultraviolet radiation of wavelengths between 180 nm and 400 nm (incoherent optical radiation). Health Physics 56:971-972.

IRPA (1990) International Radiation Protection Association, International Non-Ionizing Radiation Committee. Interim guidelines on limits of exposure to 50/60 Hz electric and magnetic fields. Health Physics 58(1): 113-122.

NH&MRC (1989) Interim Guidelines on Limits of Exposure to 50/60 Hz Electric and Magnetic Fields. Radiation Health Series No. 30. National Health and Medical Research Council, Canberra, Dec.

NRPB (1989)   National Radiological Protection Board.   Guidance on standards: Guidance as to restrictions on exposure to time varying electromagnetic fields and the 1988 recommendations of the International Non-Ionizing Radiation Committee. NRPB-GS11, NRPB, Chilton, UK.

REPACHOLI, M.H. (1990)   Radiofrequency field exposure standards: Current limits and relevant bioeffects data. In: Biological effects and medical applications of electromagnetic energy. O. P. Gandhi ed. Prentice Hall, New Jersey pp 9-27.

REPACHOLI, M.H. (1991) 50/60 Hz standards and guidelines. Radiation Protection in Australia 9(4): 116-123.

REPACHOLI, M.H. and STOLWIJK, J.A.J. (1991) Criteria for evaluating scientific literature and developing exposure limits. Radiation Protection in Australia 9(3): 79-84.

UNEP/WHO/IRPA (1982)   United Nations Environment Programme, World Health Organisation, International Radiation Protection Association, Ultrasound, Environmental Health Criteria 22. WHO, Geneva.

UNEP/WHO/IRPA (1984)   United Nations Environment Programme/World Health Organization/International Radiation Protection Association, Extremely Low Frequency (ELF) Fields, Environmental Health Criteria 35, WHO, Geneva.

UNEP/WHO/IRPA (1987), United Nations Environment Programme/World Health Organisation/International Radiation Protection Association, Environmental Health Criteria 69. Magnetic Fields. WHO, Geneva

UNEP/WHO/IRPA (1991), United Nations Environment Programme/World Health Organization/International Radiation Proteciton Association, Electromagnetic fields 300 Hz to 300 GHz, Environmental Health Criteria, WHO, Geneva (in press).

USSR (1975)   "Occupational Safety Standards System. Electrical Fields of Current Industrial Frequency of 400 kV and above. General Safety Requirements". Standard no. 12.1.002-75, National Standards Committee, Moscow (in Russian).

USSR (1985)   "Maximum Permissible Levels of Magnetic Fields with the Frequency 50 Hz". Document 3206-85, USSR Ministry of Health, Moscow (in Russian).